CATEGORIES, ALLEGORIES

North-Holland Mathematical Library

VOLUME 39

NORTH-HOLLAND
AMSTERDAM · NEW YORK · OXFORD · TOKYO

CATEGORIES, ALLEGORIES

Peter J. FREYD
Andre SCEDROV
University of Pennsylvania
Philadelphia, PA 19104-6395, USA

1990

NORTH-HOLLAND
AMSTERDAM · NEW YORK · OXFORD · TOKYO

ISBN 0 444 70368 3 (hardbound)

Published by:

ELSEVIER SCIENCE PUBLISHERS B.V.
P.O. Box 211
1000 AE Amsterdam
The Netherlands

Sole distributors for the USA and Canada:

ELSEVIER SCIENCE PUBLISHING COMPANY, INC.
655 Avenue of The Americas
New York, NY 10010
USA

Library of Congress Cataloging in Publication Data

Freyd, Peter J.
Categories, allegories / Peter J. Freyd, Andre Scedrov.
p. cm.--(North-Holland mathematical library; v. 39)
Includes index.
ISBN 0-444-70368-3: -- ISBN 0-444-70367-5 (pbk.):
1. Categories (Mathematics) 2. Allegories (Mathematics)
I. Ščedrov, Andrej. 1955- . II. Title. III. Series.
QA169.F73 1989
511.3--dc20 89-8823
CIP

Transferred to digital printing 2006

To Pam and Bonnie

PREFACE

This book is about general concepts and methods that occur throughout mathematics and now increasingly in theoretical computer science. Our purpose is to give a thorough introduction to categories that emphasizes the geometric nature of the subject and explains its connections to mathematical logic. The book is intended for that curious person who has seen some basic topology and algebra and feels drawn to learning and exploring more mathematics. We also include, in small print, extensive additional material, some of which is an enticement for further study, and some of which will be of interest even to the experts.

Part one contains a detailed treatment of the fundamentals of Geometric Logic, which combines four central ideas: natural transformations, sheaves, adjoint functors, and topoi. Natural transformations, together with functors and categories, were introduced by Eilenberg and Mac Lane in the 1940's in their work in algebraic topology and homological algebra. The notion of sheaf, a key notion in algebraic topology, was formulated in the late 1940's by Lazard and developed by Cartan, and in another form by Leray. Adjoint functors were introduced by Kan in the late 1950's and developed by the first author and by Lawvere in the late 1950's and early 1960's. Lawvere made a crucial discovery that logical rules of inference are instances of adjoint functors. The notion of topos arose in the work of Grothendieck and his school in algebraic geometry in the 1950's and early 1960's. It was refined by Lawvere and Tierney in the late 1960's and in that form it presents a categorical framework for higher-order logic and set theory.

A special feature of our approach is a general calculus of relations presented in part two. This calculus offers another, often more amenable framework for concepts and methods discussed in part one. Some aspects of our calculus of relations find their origin in the relational calculi of Peirce and Schroeder from the last century, and in the 1940's in the work of Tarski and others on relational algebras. We believe that the representation theorems we discuss are an original feature of our approach.

We thank Saunders Mac Lane and Andreas Blass for their comments on a draft of the manuscript. Any omissions or errors are, of course, ours.

Philadelphia, 1972–1990

CONTENTS (and introduced notions)

In the list of notions, alternate words appear in brackets.

Chapter One: CATEGORIES

Chapter Two: ALLEGORIES

APPENDICES

A NOTE TO THE READER

Parenthesized section numbers indicate sections that may be skipped in the first reading.

CHAPTER ONE

CATEGORIES

1.1. BASIC DEFINITIONS

The theory of CATEGORIES is given by two unary operations and a binary partial operation. In most contexts lower-case variables are used for the 'individuals' which are called *morphisms* or *maps*. The values of the operations are denoted and pronounced as:

$\Box x$ *the source of* x ,

$x\Box$ *the target of* x ,

xy *the composition of* x *and* y .

The axioms:

$$xy \text{ is defined} \quad \text{iff} \quad x\Box = \Box y ,$$

$$(\Box x)\Box = \Box x \quad \text{and} \quad \Box(x\Box) = x\Box ,$$

$$(\Box x)x = x \quad \text{and} \quad x(x\Box) = x ,$$

$$\Box(xy) = \Box(x(\Box y)) \quad \text{and} \quad (xy)\Box = ((x\Box)y)\Box ,$$

$$x(yz) = (xy)z .$$

1.11. The ordinary equality sign $=$ will be used only in the symmetric sense, to wit: if either side is defined then so is the other and they are equal. A theory, such as this, built on an ordered list of partial operations, the domain of definition of each given by equations in the previous, and with all other axioms equational, is called an ESSENTIALLY ALGEBRAIC THEORY.

1.12. We shall use a venturi-tube $\succeq$ for *directed equality* which means: if the left side is defined then so is the right and they are equal. The axiom that $\Box(xy) = \Box(x(\Box y))$ is equivalent, in the presence of the earlier axioms, with $\Box(xy) \succeq \Box x$ as can be seen below.

1.13. $\Box(\Box x) = \Box x$ because $\Box(\Box x) = \Box((\Box x)\Box) = (\Box x)\Box = \Box x$. Similarly $(x\Box)\Box = x\Box$.

The following are equivalent properties on a morphism e

$$\textit{there exists } x \textit{ such that } e = \Box x\,,$$

$$\textit{there exists } x \textit{ such that } e = x\Box\,,$$

$$e = \Box e\,,$$

$$e = e\Box\,,$$

$$\textit{for all } x, \quad ex \succeq x\,,$$

$$\textit{for all } x, \quad xe \succeq x\,.$$

Such an e is called an IDENTITY MORPHISM.

1.14. The theory of MONOIDS is usually given by a constant 1 and a binary operation xy with equations

$$1x = x\,,$$

$$x1 = x\,,$$

$$x(yz) = (xy)z\,.$$

Given such we obtain a category by defining $\Box x = x\Box = 1$. In the other direction, given a non-empty category satisfying any one of the following further equations:

$$\Box x = \Box y\,,$$

$$\Box x = y\Box\,,$$

$$x\Box = y\Box\,,$$

we define 1 as the unique identity morphism and obtain a monoid. Monoids, therefore, may be viewed as special cases of categories.

1.15. Given any set we may define $\Box x = x\Box = xx = x$ and obtain a category. Conversely, any category which satisfies either of the further equations $\Box x = x$ or $x\Box = x$, necessarily so arises. Such are called DISCRETE CATEGORIES.

1.16. A category that satisfies the further equation $\Box x = x\Box$ may be viewed as a disjoint union of monoids (one for each identity morphism). Note that a discrete category is, in this context, a disjoint union of trivial monoids.

1.17. x is LEFT-INVERTIBLE if there exists y such that yx is an identity morphism (necessarily $x\square$) and is RIGHT-INVERTIBLE if there exists z such that xz is an identity morphism (necessarily $\square x$). x is an ISOMORPHISM if it is both left and right invertible. From

$$y = y(xz) = (yx)z = z$$

we see that an isomorphism has a unique left-inverse, a unique right-inverse and they are equal (because any left-inverse is equal to any right-inverse). We introduce a unary partial operation x^{-1} called the INVERSE of x, defined iff x is an isomorphism, with directed equations:

$$\square x = (\square x)^{-1},$$

$$x^{-1}x \succeq x\square,$$

$$xx^{-1} \succeq \square x,$$

$$(x^{-1})^{-1} \succeq x,$$

$$x^{-1}y^{-1} \succeq (yx)^{-1}.$$

The isomorphisms of a category form a subcategory in which x^{-1} is everywhere defined. A category in which all morphisms are isomorphisms is called a GROUPOID. If further, it is a monoid, it is, of course, called a GROUP.

1.18. Given categories **A** and **B**, a function $F\colon \mathbf{A} \to \mathbf{B}$ is a FUNCTOR if

$$\square x = y \quad \textit{implies} \quad \square(Fx) = Fy,$$

$$x\square = y \quad \textit{implies} \quad (Fx)\square = Fy,$$

$$xy = z \quad \textit{implies} \quad (Fx)(Fy) = Fz.$$

Equivalently, F is a functor iff

$$F(\square x) = \square(Fx),$$

$$F(x\square) = (Fx)\square,$$

$$F(xy) \succeq (Fx)(Fy).$$

The directed equality is necessary. Note that any function between discrete categories is a functor, but that $F(xy) = (Fx)(Fy)$ holds only for *separating functions* between discrete categories, i.e., those which preserve inequality.

1.181. Any functor preserves left-invertibility and right-invertibility. Moreover, $F(x^{-1}) \succeq (Fx)^{-1}$.

1.182. A function $F\colon \mathbf{A} \to \mathbf{B}$ is a CONTRAVARIANT FUNCTOR if

$$F(\Box x) = (Fx)\Box\,,$$

$$F(x\Box) = \Box(Fx)\,,$$

$$F(xy) \succeq (Fy)(Fx)\,.$$

For any $\mathbf{A}$ we define its OPPOSITE CATEGORY, $\mathbf{A}^\circ$, by keeping the same elements but twisting the structure. The source-operation on $\mathbf{A}^\circ$, denoted here as $\boxdot$, is defined by $\boxdot x = x\Box$, the target operation by $x\boxdot = \Box x$, the composition by $x \circ y = yx$. The identity function $\mathbf{A} \to \mathbf{A}^\circ$ is a contravariant isomorphism. Its inverse is the identity function $\mathbf{A}^\circ \to \mathbf{A}$. (Which is the same as $\mathbf{A}^\circ \to \mathbf{A}^{\circ\circ}$ since $\mathbf{A}^{\circ\circ} = \mathbf{A}$.)

Any contravariant functor $F\colon \mathbf{A} \to \mathbf{B}$ uniquely factors as $\mathbf{A} \to \mathbf{A}^\circ \to \mathbf{B}$ where $\mathbf{A}^\circ \to \mathbf{B}$ is COVARIANT (i.e., not contravariant). Thus the contravariant functors from $\mathbf{A}$ to $\mathbf{B}$ may be construed as the covariant functors from $\mathbf{A}^\circ$ to $\mathbf{B}$.

1.19. We shall denote the identity morphisms in $\mathbf{A}$ by $|\mathbf{A}|$. A functor $F\colon \mathbf{A} \to \mathbf{B}$ clearly induces a function $|\mathbf{A}| \to |\mathbf{B}|$.

1.1(10). The general notion of isomorphism [1.17] when specialized to the "category of categories and functors" yields the notion of ISOMORPHISM OF CATEGORIES, more immediately defined as a one-to-one onto functor. Note that the directed equality $F(xy) \succeq F(x)F(y)$ that appears in the definition of functor [1.18] may be replaced with a true equality: that is, for an isomorphism, $F(xy)$ is defined iff $F(x)F(y)$ is defined. The inverse of F is the unique functor G such that $Gy = x$ iff $Fx = y$.

1.2. BASIC EXAMPLES AND CONSTRUCTIONS

Most categories are presented in terms of an auxiliary two-sorted theory, the two sorts called *objects* and *proto-morphisms* (upper case variables are conventionally reserved for objects and lower case for proto-morphisms); a unary operation defined on objects with proto-morphisms as values, denoted by 1_A; a ternary predicate called the SOURCE-TARGET PREDICATE (or the ARROW PREDICATE), denoted $A \xrightarrow{x} B$, pronounced:

x may be construed as going from A to B ;

a ternary predicate denoted $[xy = z]$, pronounced:

z is a composition of x and y .

We use

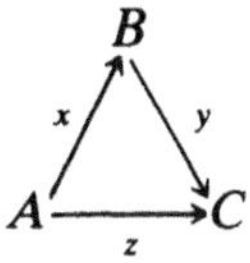

to denote the conjunction:

$$A \xrightarrow{x} B, \quad B \xrightarrow{y} C, \quad A \xrightarrow{z} C \quad \text{and} \quad [xy = z] .$$

The axioms:

For all A, $A \xrightarrow{1_A} A$.

For all $A \xrightarrow{x} B$,

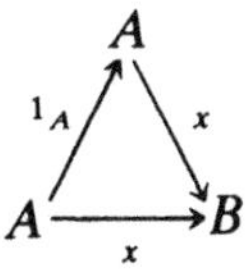

and

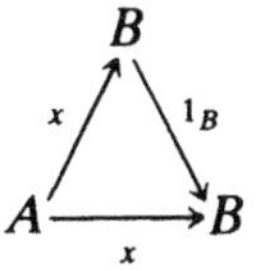

.

For all $A \xrightarrow{x} B$ and $B \xrightarrow{y} C$ there exists z such that

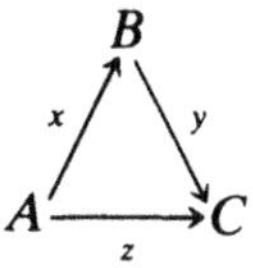

.

For all x, y, z, s, t, u, v, A, B, C, D

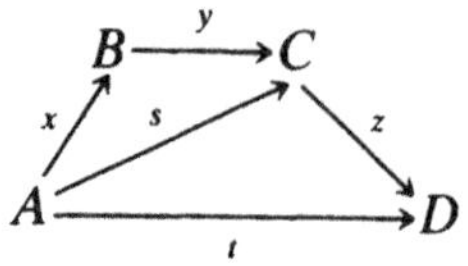

and

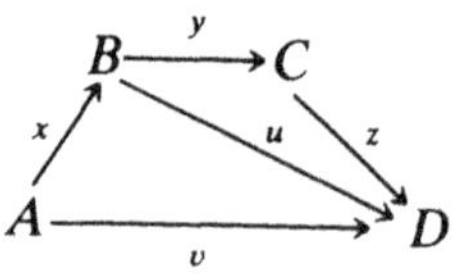

imply $t = v$.

1.21. With a few important exceptions, $[xy = z]$ is a partial operation on x, y. We may recast the axioms in this case:

For all A, $A \xrightarrow{1_A} A$.

$A \xrightarrow{x} B$ *implies* $1_A x = x$ *and* $x 1_B = x$.

$A \xrightarrow{x} B$ *and* $B \xrightarrow{y} C$ *imply that* xy *is defined and that* $A \xrightarrow{xy} C$.

$A \xrightarrow{x} B$, $B \xrightarrow{y} C$, $C \xrightarrow{z} D$ *imply that* $(xy)z = x(yz)$.

(The most important exception appears in [B.31]. Also see [1.383].)

1.22. One then considers the category whose morphisms are the given *instances* of the source-target predicate:

$$\Box(A \to B) = A \xrightarrow{1_A} A\,.$$

$$(A \to B)\Box = B \xrightarrow{1_B} B\,.$$

$$(A \xrightarrow{x} B)(B \xrightarrow{y} C) = A \xrightarrow{z} C \quad \text{for } [xy = z]\,.$$

Categories thus obtained are sometimes named after the objects and sometimes after the morphisms. We shall use the phrase 'category of . . .' to indicate the former and the phrase 'category composed of . . .' to indicate the latter.

1.23. Typically one describes a functor between two categories (presented according to 1.2–1.22) by describing a function F between the objects and then a rule that assigns to each $A \xrightarrow{x} B$ a proto-morphism x^* such that $FA \xrightarrow{x^*} FB$. It seems never to cause confusion to label x^* as Fx.

1.24. Different presentations in terms of objects and protomorphisms [1.2–1.22] can give rise to isomorphic categories and the choice of the most convenient presentation is often a matter of style.

1.241. To define the CATEGORY OF SETS, $\mathscr{S}$, we take the objects to be sets (in whatever universe of sets is chosen). There are several choices for the proto-morphisms. We may, for example, take sets of ordered pairs as proto-morphisms and define $A \xrightarrow{f} B$ by

$$\langle a, b\rangle \in f \quad \Rightarrow \quad a \in A \text{ and } b \in B\,;$$

$$\textit{for all } a \in A \textit{ there exists unique } b \textit{ such that } \langle a, b\rangle \in f\,;$$

$$\langle a, c\rangle \in (fg) \textit{ iff there exists } b \textit{ such that } \langle a, b\rangle \in f \textit{ and } \langle b, c\rangle \in g\,.$$

In this case composition is everywhere defined. Note that each proto-morphism that ever appears in the source-target predicate appears repeatedly. (For example, $\emptyset \xrightarrow{\emptyset} A$ for all A.)

Without changing the resulting category, we can eliminate all proto-morphisms that never appear by keeping only those f such that $\langle a, b\rangle \in f$ and $\langle a, b'\rangle \in f$ imply $b = b'$.

1.242. For the CATEGORY OF GROUPS, $\mathscr{G}$, we take the objects to be groups: that is, an object is a set together with a group structure. The proto-morphisms may be taken as the morphisms of $\mathscr{S}$. The source-target predicate is given by the definition of group-homomorphism. (If groups are viewed as special cases of categories [1.17] then group-homomorphism, of course, coincides with functor.)

Again, each proto-morphism that ever appears in the source-target predicate appears repeatedly. It is inconvenient here to remove the proto-morphisms that never appear. (f appears iff it is non-empty and $\{a \mid fa = b\}$ and $\{a \mid fa = b'\}$, whenever non-empty, are equinumerous. Cf. [1.381–1.383] for more examples.)

1.243. Typically a new category **A** is constructed from an old category **B** by specifying a new class of objects $\mathscr{O}$, whose elements will be denoted in upper case; a function $\mathscr{O} \to |\mathbf{B}|$, the values of which are denoted $|A|$; taking **B** as the class of proto-morphisms; specifying a source-target predicate such that $A \xrightarrow{x} B$ implies $\Box x = |A|$ and $x\Box = |B|$; verifying that $A \xrightarrow{|A|} A$ and that $A \xrightarrow{x} B$, $B \xrightarrow{y} C$ imply $A \xrightarrow{xy} C$.

It goes without saying that the composition on proto-morphisms is given by the existing composition on **B**.

A category **A** constructed in this manner from **B** is said to be FOUNDED on **B**.

The functor $U: \mathbf{A} \to \mathbf{B}$ that forgets the additional information given in the construction of **A** is called the FORGETFUL FUNCTOR.

A category **A** founded on $\mathscr{S}$ is called a CONCRETE CATEGORY, and the forgetful functor $U: \mathbf{A} \to \mathscr{S}$ is called the UNDERLYING SET FUNCTOR in this case.

1.244. The objects of a concrete category are usually described as sets with some sort of structure: $|A|$ is then understood to be the *underlying set*, $A \xrightarrow{f} B$ is usually defined in terms of 'preserving' the structure. The restrictions are then that identity functions preserve structure and that a composition of structure-preserving functions is again structure-preserving.

1.245. Suppose that **B** is the one-morphism category and that **A** is founded on **B**. The entire structure of **A** is contained in the source-target predicate: if $*$ denotes the unique morphism in **B** then necessarily $1_A = A \xrightarrow{*} A$ and $[\, * \; * \; = \; * \,]$. $A \xrightarrow{*} B$ viewed as a binary predicate on objects is reflexive and transitive. Conversely, given a class $\mathscr{O}$ with a PRE-ORDERING, that is, a reflexive and transitive relation $\leqslant$ we may define a category using just one protomorphism [1.2–1.22] where $A \xrightarrow{*} B$ iff $A \leqslant B$. Functors between such categories coincide with order-preserving functions.

1.246. Groups may be viewed as categories in which every morphism is an isomorphism and which have one identity morphism.

The fact that both groups and POSETS (pre-ordered sets in which $A \leqslant B$ and $B \leqslant A$ imply $A = B$) can be viewed as special cases of categories is more than a curiosity: it has allowed a transfer between techniques, most clearly to be seen in the theory of cohomology for categories that specializes correctly in these two extremal cases. (The posets of historical interest were the lattices of open subsets of a topological space. Čech cohomology is the resulting cohomology theory.)

1.25. Given a category **A**, take $|\mathbf{A}|$ as the objects, **A** as the protomorphisms, and define $A \xrightarrow{x} B$ iff $A = \Box x$ and $B = x\Box$. **A** is isomorphic to the category composed of instances [1.22] of this arrow predicate. The circle is not necessarily closed in the other direction, when **A** itself is already given as in [1.2–1.22] by the arrow predicate [1.24, 1.241, 1.242].

1.251. The ARROW NOTATION is usual even when a pure category is being discussed. A diagram

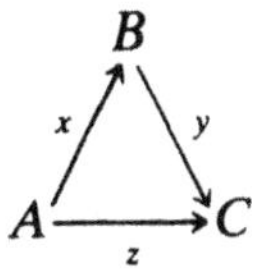

means $xy = z$.

We shall not bother to say that such a diagram 'commutes': rather we shall assume that all diagrams are statements of equations unless we say otherwise. We shall say otherwise by inserting a *puncture mark*:

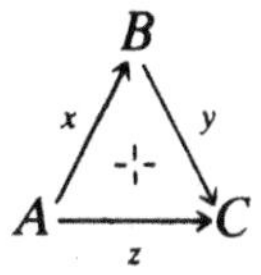

means $A \xrightarrow{x} B$, $B \xrightarrow{y} C$, $A \xrightarrow{z} C$

but does not imply $xy = z$. Nor does it imply $xy \neq z$.

Each puncture mark in a complex diagram removes just one equation. For example,

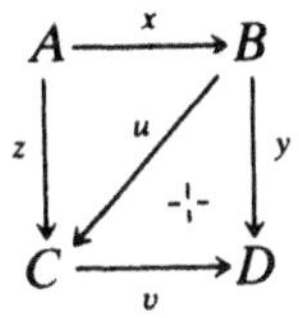

implies $xy = zv$ and $xu = z$;

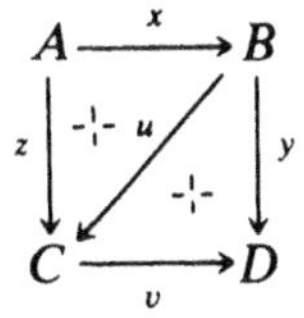

still implies an equation, namely $xy = zv$.

1.26. Given $B \in \mathbf{A}$ we define the SLICE CATEGORY $\mathbf{A}/B$ as the category founded on $\mathbf{A}$ with $\{x \mid x\Box = B\}$ as objects; $|x| = \Box x$ as underlying function; $x \xrightarrow{y} z$ iff $x = yz$. In the arrow notation, the objects of $\mathbf{A}/B$ are morphisms $C \xrightarrow{x} B$ in $\mathbf{A}$ with target B, and morphisms in $\mathbf{A}/B$ are triangles in $\mathbf{A}$:

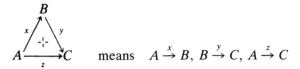

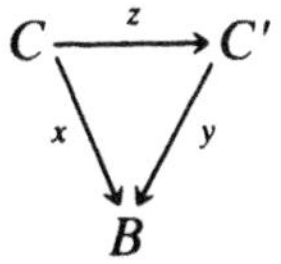

(Remember, all triangles are commutative triangles unless said not to be.)

The forgetful functor is denoted $\Sigma: \mathbf{A}/B \rightarrow \mathbf{A}$.

1.261. For B in the category of sets $\mathscr{S}$, we may interpret the objects of $\mathscr{S}/B$ as B-indexed families of pairwise-disjoint sets, $\{A_i\}_{i\in B}$. (For $f: A \to B$, $A_i = \{a \mid fa = i\}$.) The morphisms of $\mathscr{S}/B$ are B-indexed families of functions $\{f_i\}_{i\in B}$. The obvious functor $\mathscr{S}/B \to \mathscr{S}^B$ (the elements of $\mathscr{S}^B$ are functions from B to $\mathscr{S}$) is an equivalence of categories as defined below. This example is misleading. Let Ring be the category of rings (unless stated otherwise, all rings have units) and ring homomorphisms, and $\mathbb{Z}$ the ring of integers. Then Ring/$\mathbb{Z}$ is called the category of augmented rings: rings equipped with a ring homomorphism to $\mathbb{Z}$ (called augmentation), and whose morphisms are ring homomorphisms preserving the augmentation. This category is isomorphic to the category of rings with or without units (consider augmentation ideals) [1.384]: an example where adding more structure results in less structure.

1.262. Let $\mathscr{L}\mathscr{H}$ be the category whose objects are topological spaces and whose morphisms are continuous maps that are LOCAL HOMEOMORPHISMS: $f: X \to Y$ if for every $x \in X$ there is an open subset $U \subset X$ such that $x \in U$ and that f restricted to U is a homeomorphism onto an open subset of Y. (That is, if there exists open $V \subset Y$ and an isomorphism $\theta: U \to V$, as defined in the category composed of continuous maps, such that $U \to X \xrightarrow{f} Y = U \xrightarrow{\theta} V \to Y$.) For any $x \in X$, $\{y \in Y \mid fx = y\}$ is a discrete subset of Y.

$\mathscr{L}\mathscr{H}/Y$ is the category of LAZARD SHEAVES over Y, denoted $\mathscr{Sh}(Y)$. Its objects may be viewed as *continuously* Y-indexed families of pairwise-disjoint sets, its maps are Y-indexed families of functions which, in concert, are continuous. Note that if $X_1 \xrightarrow{f} X_2 \to Y$ is a local homeomorphism and if $X_2 \to Y$ is a local homeomorphism then f is a local homeomorphism iff it is continuous. *Thus the maps is $\mathscr{Sh}(Y)$ from $X_1 \to Y$ to $X_2 \to Y$ may be defined as the continuous maps $X_1 \to X_2$ such that*

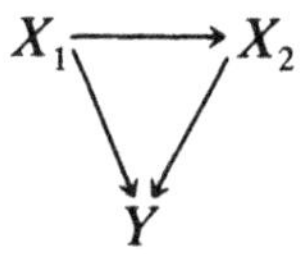

1.263. The dual construction has one well-known specialization. Let 1 denote a one-element object in $\mathscr{S}$. The *counter-slice category* $1\backslash\mathscr{S}$ (whose objects are functions from 1) is usually called the *category of pointed sets*. If $\mathscr{S}$ is replaced with the category of topological spaces and continuous maps the resulting construction yields the *category of pointed spaces*.

1.27. Let $\mathbf{A}$ be a SMALL CATEGORY, that is, a category whose underlying class of morphisms is an object in $\mathscr{S}$. The FUNCTOR CATEGORY $\mathscr{S}^{\mathbf{A}}$ is founded on $\mathscr{S}^{|\mathbf{A}|}$: the objects of $\mathscr{S}^{\mathbf{A}}$ are functors from $\mathbf{A}$ to $\mathscr{S}$; given $F \in \mathscr{S}^{\mathbf{A}}$ we define $|F| \in |\mathscr{S}^{|\mathbf{A}|}|$ as the function that sends A in $|\mathbf{A}|$ to FA in $\mathscr{S}$; an $|\mathbf{A}|$-indexed family of functions $\{\eta_{\mathbf{A}}: FA \to GA\}$ is a morphism in $\mathscr{S}^{\mathbf{A}}$ if it is a NATURAL TRANSFORMATION of functors, that is, if for every $A \xrightarrow{f} B$ in $\mathbf{A}$ it is the case that

$$\begin{array}{ccc} FA & \xrightarrow{\eta_A} & GA \\ {\scriptstyle Ff}\downarrow & & \downarrow{\scriptstyle Gf} \\ FB & \xrightarrow[\eta_B]{} & GB \end{array}$$

In particular, a natural transformation η is an isomorphism in the functor category iff each η_A is an isomorphism. We then write $G = F^{\eta}$, and call it a CONJUGATE of F.

1.271. Lazard sheaves and functor categories provide the two most important families of examples in *Geometric Logic*. If $\mathbf{A}$ is a one-object category, that is, if we view $\mathbf{A}$ as a monoid M, then $\mathscr{S}^{\mathbf{A}}$ is usually called the CATEGORY of M-SETS: its objects may be viewed as sets, each with a specific M-action; its morphisms as functions that respect the action. If M is a group, this coincides with the ordinary notion of a group actions and 'equivariant' maps.

We may generalize: for any small $\mathbf{A}$ define a RIGHT $\mathbf{A}$-SET as a set X together with a unary operation from X to $|\mathbf{A}|$ denoted $x\Box$, and a binary partial operation from X and $\mathbf{A}$ to X, denoted xa defined iff $x\Box = \Box a$. The axioms:

$$x(x\Box) = x\,,$$

$$(xa)\Box \succeq a\Box\,,$$

$$x(ab) = (xa)b\,.$$

(The directed equality may be replaced with $(xa)\Box = ((xa)\Box)(a\Box)$.)

A map between right $\mathbf{A}$-sets is a function f such that

$$f(x\Box) = (fx)\Box\,,$$

$$f(xa) \succeq (fx)a\,.$$

A right **A**-set X gives rise to a functor $F: \mathbf{A} \to \mathscr{S}$ as follows: for $A \in |\mathbf{A}|$, FA is the set $\{x \mid x\square = A\}$; for $A \xrightarrow{a} B \in \mathbf{A}$, Fa is the function that sends $x \in FA$ to $xa \in FB$. If Y is another right **A**-set and G its corresponding functor then the maps from X to Y are is one-to-one correspondence with the natural transformations from F to G. We obtain a functor

$$(\textit{right } \mathbf{A}\textit{-sets}) \to \mathscr{S}^{\mathbf{A}}$$

which is not, in general, an isomorphism of categories but is, as defined in the next section, an equivalence of categories. (A functor $F: \mathbf{A} \to \mathscr{S}$ arises from a right **A**-set iff its values are pairwise-disjoint: $FA \cap FB = \emptyset$ for $A \neq B$. If **A** is a monoid, then right **A**-sets coincide with functors $\mathbf{A} \to \mathscr{S}$.)

1.272. A small category **A** is automatically a right **A**-set. The resulting functor $C: \mathbf{A} \to \mathscr{S}$ (called the CAYLEY REPRESENTATION) sends $A \in |\mathbf{A}|$ to the set of morphisms with A as target. $C(A \xrightarrow{x} B)$ carries $y \in C(A)$ to $yx \in C(B)$. Since C is one-to-one it establishes an isomorphism between **A** and a subcategory of $\mathscr{S}$. (In particular, every small category is isomorphic to a concrete category.) If **A** is a group and if the target of C is cut down to the category of permutations of the unique object in $C(A)$, this is the classical Cayley representation.

Any universally quantified elementary sentence in the predicates of category theory true for $\mathscr{S}$ is clearly true for any subcategory of $\mathscr{S}$, hence true for every small category. If **A** were not small but contained a counter example for such a sentence we could take the subcategory generated by the elements appearing in the counterexample, thus obtaining a counterexample in a small category, and then use the Cayley representation to move the counterexample into $\mathscr{S}$. Hence:

Every universally quantified elementary sentence in the predicates of category theory true for the category of sets is true for all categories.

This is the first of our examples of a representation theorem yielding a completeness theorem.

1.273. The category of contravariant functors from **A** to $\mathscr{S}$ is usually denoted as $\mathscr{S}^{\mathbf{A}^\circ}$. LEFT **A**-SETS are to contravariant functors as right **A**-sets are to covariant functors. A left **A**-set is a set X together with a unary operation from X to $|\mathbf{A}|$, denoted $\square x$, and a binary partial operation from **A** and X to X, denoted ax, defined iff $a\Box = \Box x$. The axioms:

$$(\Box x)x = x ,$$

$$\Box(ax) = \Box a ,$$

$$(ab)x = a(bx) .$$

A left **A**-set may be regarded as a right $\mathbf{A}^\circ$-set.

1.274. If F and G are isomorphic as objects in $\mathscr{S}^{\mathbf{A}}$, then there exists an isomorphism-valued function $\theta: |\mathbf{A}| \to \mathscr{S}$ such that $F^\theta = G$.

A NATURAL EQUIVALENCE of functors is usually defined as an isomorphism in $\mathscr{S}^{\mathbf{A}}$. Naturally equivalent functors are exactly the same as conjugate functors.

1.28. A morphism e is an IDEMPOTENT if $ee = e$ (necessarily $\Box e = e\Box$). Given a class $\mathscr{E}$ of idempotents in a category **A**, $\mathscr{S}\!plit(\mathscr{E})$ is defined with **A** as proto-morphisms [1.2–1.22], with $\mathscr{E}$ as objects, and with $A \xrightarrow{x} B$ iff $Ax = x = xB$. (We are using upper-case for the idempotents that happen to be in $\mathscr{E}$.) Note that the identity morphism on A is $A \xrightarrow{A} A$. The forgetful operation $\mathscr{S}\!plit(\mathscr{E}) \to \mathbf{A}$ is not, in general, a functor (if $\mathscr{E}$ is not contained in $|\mathbf{A}|$, then $\mathscr{S}\!plit(\mathscr{E}) \to \mathbf{A}$ does not preserve identity morphisms).

If $|\mathbf{A}| \subset \mathscr{E}$ we obtain a functor $A \to \mathscr{S}\!plit(\mathscr{E})$ that sends x to $(\Box x) \xrightarrow{x} (x\Box)$.

1.281. We say that $\langle A \xrightarrow{x} B,\ B \xrightarrow{y} A\rangle$ SPLITS an idempotent $A \xrightarrow{e} A$ if $xy = e$ and $yx = 1_B$. Every e in $\mathscr{E}$ acquires a canonical splitting in $\mathscr{S}\!plit(\mathscr{E})$: $\langle(\Box e) \xrightarrow{e} e,\ e \xrightarrow{e} (e\Box)\rangle$. If $|\mathbf{A}| \subset \mathscr{E}$, then $\mathbf{A} \to \mathscr{S}\!plit(\mathscr{E})$ is universal in the following sense: given $T: \mathbf{A} \to \mathbf{B}$ and a choice for each $e \in \mathscr{E}$ of a splitting $\langle x, y\rangle$ in **B** of Te there exists unique $\mathscr{S}\!plit(\mathscr{E}) \to \mathbf{B}$ that carries the canonical splitting of e to the chosen splitting.

Every $A \xrightarrow{e} A$ in $\mathscr{S}\!plit(\mathscr{E})$ such that $e \in \mathscr{E}$ also has a canonical splitting: $\langle A \xrightarrow{e} e,\ e \xrightarrow{e} A\rangle$. If $\mathscr{I}$ is the class of all idempotents then all idempotents split in $\mathscr{S}\!plit(\mathscr{I})$.

1.282. For a fixed idempotent $A \xrightarrow{e} A$ there can be many splittings. There is, however, a certain kind of uniqueness familiar in category theory. If $A \xrightarrow{x} B$, $B \xrightarrow{y} A$ and $A \xrightarrow{x'} B'$, $B' \xrightarrow{y'} A$ split the same idempotent, then there is a canonical isomorphism from B to B'. Indeed there is a unique map $B \to B'$ such that

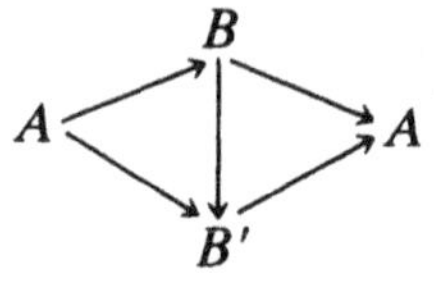

and it is an isomorphism. Its inverse is the unique map such that

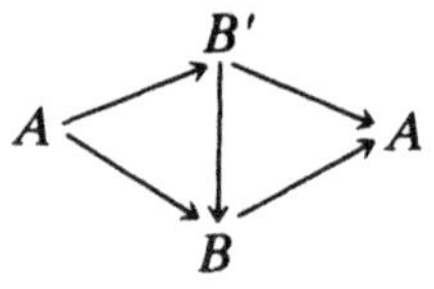

($B \to B'$ is easily constructed as $B \to A \to B'$.)

1.283. For a semi-group S and set of idempotents $\mathscr{E} \subset S$ the category $\mathit{Split}(\mathscr{E})$ is STRONGLY CONNECTED, that is, for every ordered pair of objects (A, B) there exists $A \to B$. Indeed, $\mathit{Split}(\mathscr{E})$ is equipped with a canonical choice of such maps: take $A \xrightarrow{AB} B$.

Let $\mathbf{A}$ be a strongly connected category. For every function $[\ ,\ ]: |\mathbf{A}| \times |\mathbf{A}| \to \mathbf{A}$ such that $A \xrightarrow{[A,B]} B$ and $[A, A] = 1_A$ we may construct a semi-group S with a class of idempotents $\mathscr{E}$ such that $\mathit{Split}(\mathscr{E})$ is isomorphic to $\mathbf{A}$ where the isomorphism carries the canonical choice to the given choice: let S be the semigroup whose elements are the morphisms of $\mathbf{A}$ with multiplication defined by $x \cdot y = x[x\Box, \Box y]y$; let $\mathscr{E} = |\mathbf{A}|$. One may adjoin a unit 1 to S in order to obtain a monoid. (Do not adjoin 1 to $\mathscr{E}$.)

1.284. Say that a function $F: \mathbf{A} \to \mathbf{B}$ between categories is a PRE-FUNCTOR if $F(xy) \succeq F(x)F(y)$. (A pre-functor need not preserve identity morphisms.) The forgetful operation $\mathit{Split}(\mathscr{E}) \to \mathbf{A}$ is a pre-functor and is universal in the following sense: Let $F: \mathbf{B} \to \mathbf{A}$ be a pre-functor. Note that $F(1_B)$ is idempotent. If $F(|\mathbf{B}|) \subset \mathscr{E}$, then there exists a unique *functor* $\mathbf{B} \to \mathit{Split}(\mathscr{E})$ such that

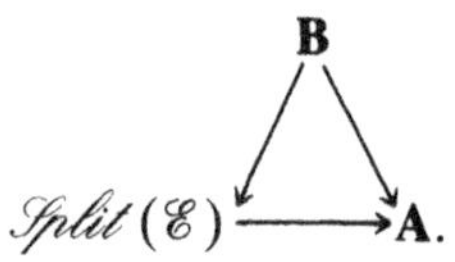

If $\mathscr{E}$ is the set of all idempotents in $\mathbf{A}$, then $\mathit{Split}(\mathscr{E}) \to \mathbf{A}$ is the universal pre-functor.

1.3. EQUIVALENCE OF CATEGORIES

1.31. Given objects A, B in a category $\mathbf{A}$ we denote the set of maps from A to B by $\mathbf{A}(A, B)$. In most cases the prefix $\mathbf{A}$ in the notation is redundant and is omitted.

A functor $F: \mathbf{A} \to \mathbf{B}$ induces for each pair $A, B \in \mathbf{A}$ a function $(A, B) \to (FA, FB)$.

F is an EMBEDDING if $(A, B) \to (FA, FB)$ is one-to-one for all A, B.

F is FULL if $(A, B) \to (FA, FB)$ is onto for all A, B.

A FULL SUBCATEGORY is one whose inclusion functor is full.

F has a REPRESENTATIVE IMAGE if for all $B \in |\mathbf{B}|$ there exists $A \in |\mathbf{A}|$ and an isomorphism $FA \to B$.

F is an EQUIVALENCE FUNCTOR if it is a full embedding with a representative image.

1.32. Each of the properties on F above is preserved under conjugation of functors. Each is preserved under composition of functors. Moreover each has a cancellation principle: if $\mathbf{A} \xrightarrow{F} \mathbf{B} \xrightarrow{G} \mathbf{C}$ is an embedding then so is F; if $\mathbf{A} \xrightarrow{F} \mathbf{B} \xrightarrow{G} \mathbf{C}$ is full then so is G; if $\mathbf{A} \xrightarrow{F} \mathbf{B} \xrightarrow{G} \mathbf{C}$ has a representative image then so does G.

A STRONG EQUIVALENCE between $\mathbf{A}$ and $\mathbf{B}$ is a pair of functors $F: \mathbf{A} \to \mathbf{B}$, $G: \mathbf{B} \to \mathbf{A}$ such that FG and GF are each conjugate to the appropriate identity functor. From the above remarks FG and GF are both equivalence functors hence F and G are both equivalence functors. As will be seen, the axiom of choice implies (indeed, is equivalent to) the statement that every equivalence functor is half of a strong equivalence. See [1.362].

1.33. A property is preserved by a functor F if whenever it holds in the domain it continues to hold after application by F. It is REFLECTED by F if whenever it holds after application it already holds in the domain.

All functors preserve isomorphisms; if F is an embedding and *reflects* isomorphisms it is called FAITHFUL. Full embeddings, note, are faithful. It seems to be a general principle that almost any property of interest is reflected by faithful functors that preserve it.

1.331. *If F reflects left-invertibility it reflects isomorphisms*: if Fx is an isomorphism it is left-invertible, hence x is left-invertible: if $yx = 1$ then $Fy = (Fx)^{-1}$ and Fy is left-invertible, hence y is left-invertible; if $zy = 1$ then $z = x$ and $y = x^{-1}$.

Because it is one-to-one and reflects left-invertibility, *the Cayley representation is faithful.*

We may thus extend the metatheorem of [1.272] to include sentences in which the invertibility of a morphism appears as a primitive predicate.

1.332. The *contravariant Cayley representation* for a small category $\mathbf{A}$ is obtained by taking $\mathbf{A}$ as a left $\mathbf{A}$-set. The resulting functor, $C^\circ: \mathbf{A} \to \mathscr{S}$, sends $A \in \mathbf{A}$ to the set of morphisms with A as source. $C^\circ(A \xrightarrow{x} B)$ carries $y \in C(B)$ to $xy \in C(A)$.

Let $\mathscr{P} : \mathscr{S} \to \mathscr{S}$ be the *contravariant power-set functor*: $\mathscr{P}(S)$ is the set of subsets of S; $\mathscr{P}(S_1 \xrightarrow{f} S_2)$ carries $S' \in \mathscr{P}(S_2)$ to its inverse image in $\mathscr{P}(S_1)$; that is, $y \in (\mathscr{P}f)(S')$ iff $f(y) \in S'$.

$\mathbf{A} \xrightarrow{C^\circ} \mathscr{S} \xrightarrow{\mathscr{P}} \mathscr{S}$ is covariant and reflects right-invertibility: if $C^\circ(A \xrightarrow{x} B)$ is right-invertible, then it is one-to-one and $\{1_A\} \in C^\circ(A)$ cannot be carried to $\emptyset \in \mathscr{P}C^\circ(B)$ (because $\mathscr{P}C^\circ(x)\emptyset = \emptyset$); hence there exists $y \in C^\circ(B)$ such that $C^\circ(x)$ carries y to 1_A; that is, x is right-invertible.

Define $F: \mathbf{A} \to \mathscr{S}$ by $F(A) = C(A) \times \mathscr{P}C^\circ(A)$. F reflects both left and right invertibility and we may adjoin two more predicates to the metatheorem of [1.272].

1.333. A functor between posets is always an embedding. It is faithful iff it is one-to-one on objects. It is full iff the ordering on the domain is induced by the ordering on the range. It is an equivalence functor iff it is an isomorphism of posets.

1.34. Objects A and B are said to be ISOMORPHIC if there exists an isomorphism $A \xrightarrow{\sim} B$. The decorated arrow $\xrightarrow{\sim}$ is used to denote an isomorphism. $A \simeq B$ denotes the existence of an isomorphism. $\simeq$ is an equivalence relation on objects preserved by any functor, reflected by full embeddings (but contrary to the general principle above, not reflected by all faithful functors). An equivalence functor induces a one-to-one correspondence between the isomorphism-types of the two categories.

1.341. If $F: \mathbf{A} \to \mathbf{B}$ is an equivalence functor and if for every A in $\mathbf{A}$ the isomorphism class of A in $\mathbf{A}$ is equinumerous with the isomorphism class of FA in $\mathbf{B}$, then one may use the axiom of choice to prove that $\mathbf{A}$ and $\mathbf{B}$ are isomorphic categories, indeed that F is conjugate to an isomorphism: to wit, there is a one-to-one and onto map $|\mathbf{A}| \xrightarrow{F'} |\mathbf{B}|$ with an isomorphism $FA \xrightarrow{\theta_A} FA'$, for each $A \in |\mathbf{A}|$. Then $F' = F^\theta$.

1.35. If $\mathbf{A}$ is founded on $\mathbf{B}$ [1.243] we obtain a FORGETFUL FUNCTOR $U: \mathbf{A} \to \mathbf{B}$ that sends $A \in \mathbf{A}$ to $|A| \in \mathbf{B}$ and $A \xrightarrow{x} B$ in $\mathbf{A}$ to x in $\mathbf{B}$. U is always an embedding. Forgetful functors are sometimes called *groundings*, sometimes *foundation functors*. (U stands for *underlying*. We avoid F because it is reserved in this context for a functor $\mathbf{B} \to \mathbf{A}$ called the *free functor*.)

1.36. Given a category $\mathbf{B}$, a class $\mathscr{C}$ and an *onto* function $T: \mathscr{C} \to |\mathbf{B}|$ we define a category $[T]$ founded on $\mathbf{B}$ with $\mathscr{C}$ as objects, T as underlying function and with $A \xrightarrow{x} B$ defined in the most inclusive sense: $A \xrightarrow{x} B$ iff $TA = \Box x$ and $TB = x\Box$. $[T]$ is called an INFLATION of $\mathbf{B}$. The resulting forgetful functor $[T] \to \mathbf{B}$ is a full embedding and literally onto, hence an equivalence functor. ($[T]$ may be viewed as the result of artificially replicating the objects of $\mathbf{B}$.)

If $S: |\mathbf{B}| \to \mathscr{C}$ is such that $|\mathbf{B}| \xrightarrow{S} \mathscr{C} \xrightarrow{T} |\mathbf{B}|$ is the identity function we obtain a functor $\mathbf{B} \to [T]$ that sends x to $(\Box x) \xrightarrow{x} (x\Box)$, called an INFLATION CROSS-SECTION. $\mathbf{B} \to [T] \to \mathbf{B}$ is the identity functor. $[T] \to \mathbf{B} \to [T]$ is canonically conjugate to the identity functor via the isomorphism-valued function $\theta: \mathscr{C} \to [T]$ that sends A to $A \xrightarrow{1} S(TA)$.

If $T: \mathscr{C} \to |B|$ is a left-invertible, then $[T] \to \mathbf{B}$ is a strong equivalence. The axiom of choice implies that every inflation of $\mathbf{B}$ is strongly equivalent with $\mathbf{B}$.

1.361. *Any equivalence functor is a composition of an inflation cross-section, followed by an isomorphism, followed by an inflation forgetful functor.*

Given an equivalence functor $F: \mathbf{A} \to \mathbf{B}$ define $\mathscr{C}$ to be the set of triples $\langle A, \theta, B \rangle$ where $A \in \mathbf{A}$, $B \in \mathbf{B}$ and $\theta: FA \to B$ is an isomorphism. Let $T: \mathscr{C} \to |\mathbf{A}|$ send $\langle A, \theta, B \rangle$ to A and let $S: |\mathbf{A}| \to \mathscr{C}$ send A to $\langle A, 1, FA \rangle$. T is onto and S is a cross section. Let $T': \mathscr{C} \to |\mathbf{B}|$ send $\langle A, \theta, B \rangle$ to B. Since F has a representative image, T' is onto.

$[T]$ and $[T']$ are isomorphic categories: send $\langle A, \theta, B \rangle \xrightarrow{x} \langle A', \theta', B' \rangle$ in $[T]$ to $\langle A, \theta, B \rangle \xrightarrow{\theta^{-1} x \theta'} \langle A', \theta', B' \rangle$ in $[T']$.

1.362. The axiom of choice implies that T', above, is left-invertible. Hence the existence of an equivalence functor $\mathbf{A} \to \mathbf{B}$ would imply the existence of an equivalence functor $\mathbf{B} \to \mathbf{A}$. Moreover the resulting pair of functors yields a strong equivalence between $\mathbf{A}$ and $\mathbf{B}$.

1.363. We say that two categories $\mathbf{A}, \mathbf{B}$ are EQUIVALENT CATEGORIES if they have isomorphic inflations. [1.361] says that the existence of an equivalence functor between two categories implies that they are equivalent categories.

The relation on categories of being equivalent categories is an equivalence relation: clearly reflective and symmetric; not so clearly transitive. It suffices to show that any two inflations of a category, $\mathbf{B}$, have, themselves, isomorphic inflations. Given onto functions $T_1: \mathcal{O}_1 \to |\mathbf{B}|$ and $T_2: \mathcal{O}_2 \to |\mathbf{B}|$ define $\mathcal{O}_3 = \{\langle A, B\rangle \mid T_1A = T_2B\}$ ($\mathcal{O}_3$ is the 'pullback', as will be defined in [1.431], of T_1 and T_2.) The obvious projections from $\mathcal{O}_3$ to $\mathcal{O}_1$ and $\mathcal{O}_2$ are onto and $\mathcal{O}_3 \to \mathcal{O}_1 \to |\mathbf{B}| = \mathcal{O}_3 \to \mathcal{O}_2 \to |\mathbf{B}|$. The induced inflation of B with $\mathcal{O}_3$ as objects may be regarded as an inflation of $[T_1]$ and of $[T_2]$.

1.364. A category is SKELETAL if isomorphic objects are equal. Any equivalence functor between skeletal categories is necessarily an isomorphism. Given a category $\mathbf{A}$ and a skeletal category $\mathbf{A}'$, we say that $\mathbf{A}'$ is a SKELETON of $\mathbf{A}$ if there exists an equivalence functor $\mathbf{A}' \to \mathbf{A}$ and we say that $\mathbf{A}'$ is a CO-SKELETON if there exists an equivalence functor $\mathbf{A} \to \mathbf{A}'$.

As will be clear from the proof, every occurence of the word 'category' below may be replaced with the word 'groupoid'.

Each of the following is equivalent to the axiom of choice:

(a) *Equivalent categories are strongly equivalent.*

(b) *Every category has a skeleton.*

(c) *Every category has a co-skeleton.*

(d) *Any two skeletons of a category are isomorphic.*

(e) *Any two co-skeletons of a category are isomorphic.*

For convenience we add:

(f) *Given a non-empty family* $\{S_\iota\}_I$ *of equinumerous sets there exists* $0 \in I$ *and a family of isomorphisms of the permutation groups* $\{\theta_\iota: \mathrm{Aut}(S_0) \to \mathrm{Aut}(S_\iota)\}_I$.

(g) *Given a non-empty family* $\{S_\iota\}_I$ *of equinumerous infinite sets, there exists a family* $\{x_\iota\}_I$ *such that* $x_i \in S_i$ *for all* $i \in I$.

BECAUSE: The axiom of choice implies (a) by [1.362] and easily implies (b). (a) and (b) together yield (c). (a) and (b) each imply (e) and (c) implies (d). We will show that (e) implies (d) implies (f) implies (g) and, finally, that (g) implies the axiom of choice.

(e) implies (d) as follows: If $\mathbf{A}_0$ and $\mathbf{A}_1$ are skeletons of a common category, then they are equivalent categories, that is, there is another category which appears as an inflation of both of them and, hence, has both $\mathbf{A}_0$ and $\mathbf{A}_1$ as co-skeletons. (e) implies that $\mathbf{A}_0$ and $\mathbf{A}_1$ are isomorphic.

(d) implies (f) as follows: Given a non-empty family $\{S_\iota\}_I$ of equinumerous sets, choose $0 \in I$ and let $\mathbf{A}$ be the category founded on $\mathcal{S}$ whose set of objects is $I \times \{0, 1\}$ with forgetful functor defined by $|\langle i, 0\rangle| = S_0$, $|\langle i, 1\rangle| = S_\iota$ and source-target predicate defined by $\langle i, k\rangle \xrightarrow{f} \langle i', h'\rangle$ iff $i = i'$ and f is an isomorphism of sets. For $k = 0, 1$ define $\mathbf{A}_k$ as the full subcategory of objects of the form $\langle i, k\rangle$. Clearly both $\mathbf{A}_0$ and $\mathbf{A}_1$ are skeletons of $\mathbf{A}$ and an isomorphism $\mathbf{A}_0 \to \mathbf{A}_1$ clearly yields a family of isomorphisms as demanded by (f).

Some group theory is needed to show that (f) implies (g). Recall that the *support* of a permutation is the set of its non-fixed points and that a *transposition* is a permutation with exactly two elements in its support. If α is a non-trivial permutation on an infinite set then it is a transposition iff $(\alpha\beta\alpha^{-1}\beta^{-1})^6 = 1$ for all permutations β. Hence any isomorphism $\mathrm{Aut}(S) \to \mathrm{Aut}(S')$ carries transpositions to transpositions. (This statement is also true for finite S and S' unless S has exactly six elements.)

Two transpositions fail to commute iff there is a single element common to their supports. Given a non-empty family of equinumerous infinite sets $\{S_\iota\}_I$ let $\{\theta_i\colon \mathrm{Aut}(S_0) \to \mathrm{Aut}(S_\iota)\}_I$ be a family of isomorphisms as insured by (f). Choose a pair α, β of non-commuting transpositions of S_0 and define $x_\iota \in S_\iota$ as the unique element common to the supports of $\theta_\iota(\alpha)$ and $\theta_\iota(\beta)$.

(g) implies the axiom of choice as follows: Given a family $\{S_\iota\}_I$ of non-empty sets, let $A = \bigcup_I S_i$ and let A^* be the set of finite words on A. Clearly A^* may be embedded in $S_i \times A^*$ and clearly $S_\iota \times A^*$ may be embedded in $A \times A^*$. Concatenation yields an embedding of $A \times A^*$ into A^*. The Cantor–Bernstein theorem does not use the axiom of choice, hence $\{S_\iota \times A^*\}_I$ is a family of equinumerous infinite sets. A choice of elements for $\{S_\iota \times A^*\}_I$ clearly yields a choice of elements for $\{S_\iota\}_I$.

1.365. There is a missing statement in the list of equivalent statements above: every category is equivalent to a skeletal category. This appears to be weaker than the axiom of choice. If this statement is interpreted in the category of G-sets, then it is true iff G is a free group.

1.366. Define the kernel of an equivalence functor $T\colon \mathbf{A} \to \mathbf{B}$ to be the subcategory $\mathscr{K} \subset \mathbf{A}$ composed of all maps in $\mathbf{A}$ sent by T to identity maps in $\mathbf{B}$. One may easily verify:

1) $|\mathbf{A}| \subset |\mathscr{K}|$, that is, all identity maps in $\mathbf{A}$ are in $\mathscr{K}$.

2) $\mathscr{K}$ is a groupoid, that is, every map in $\mathscr{K}$ is an isomorphism whose inverse is also in $\mathscr{K}$.

3) $\mathscr{K}$ is a pre-order, that is, there is at most one $\mathscr{K}$-map from A to B, any $A, B \in \mathbf{A}$.

Any subcategory satisfying these three conditions will be called an EQUIVALENCE KERNEL. The name is justified by:

For any equivalence kernel $\mathscr{K} \subset \mathbf{A}$ there is an onto equivalence functor $\mathbf{A} \to \mathbf{A}/\mathscr{K}$ whose kernel is $\mathscr{K}$.

$\mathbf{A}/\mathscr{K}$ may be most easily constructed as a pure category by taking its morphisms to be double cosets, that is, sets of the form $\mathscr{K}x\mathscr{K} = \{kxk' \mid k, k' \in \mathscr{K}\}_{x\in\mathbf{A}}$. The composition of cosets C, C' is defined to be $CC' = \{cc' \mid c \in C, c' \in C'\}$ if non-empty (if empty then the composition is not defined). $\Box C = \mathscr{K}\{\Box c \mid c \in C\}$, $C\Box = \{c\Box \mid c \in C\}\mathscr{K}$. The canonical functor $\mathbf{A} \to \mathbf{A}/\mathscr{K}$ sends x to $\mathscr{K}x\mathscr{K}$.

If F: $\mathbf{A} \to \mathbf{B}$ *is any functor (equivalence or not) such that* $F(\mathscr{K}) \subset |\mathbf{B}|$, *that is, such that all* $\mathscr{K}$*-maps are sent by F to identity maps, then there exists a unique functor* $\mathbf{A}/\mathscr{K} \to \mathbf{B}$ such that $F = \mathbf{A} \to \mathbf{A}/\mathscr{K} \to \mathbf{B}$.

$\mathbf{A}/\mathscr{K} \to \mathbf{B}$ sends a coset C to the unique element in $F(C) = \{Fc \mid c \in C\}$.

1.367. [1.361] and [1.366] readily yield:

Any equivalence functor is the composition of an inflation cross-section, followed by a canonical functor of the form $\mathbf{A} \to \mathbf{A}/\mathscr{K}$ *followed by an isomorphism.*

(1.37.) We consider a number of important functors relating Lazard sheaves, functor categories, and **A**-sets.

(1.371.) For a set B every object in $\mathscr{S}^B$ is isomorphic to a pairwise disjoint B-indexed family $\{A_i\}_{i \in B}$ (that is, $A_i \cap A_j = \emptyset$ for $i \neq j$). The full embedding $\mathscr{S}/B \to \mathscr{S}^B$ described in [1.261] is thus an equivalence functor. Even without the axiom of choice one may define $\mathscr{S}^B \to \mathscr{S}/B$ to complete the equivalence functor to a strong equivalence.

Similarly any functor F: $\mathbf{A} \to \mathscr{S}$ is isomorphic (conjugate) to a functor with pairwise-disjoint values (that is, $FA \cap FB = \emptyset$ for $A \neq B$), hence the full embedding (*Right* **A**-*sets*) $\to \mathscr{S}^{\mathbf{A}}$ described in [1.271] is an equivalence functor and may be completed to a strong equivalence.

(The first paragraph becomes a special case of the second paragraph if B is viewed as a discrete category.)

(1.372.) We have seen that a poset **P** may be viewed as a category. It may also be viewed as a topological space: take its *ideals* as its open subsets. (All ideals in posets will be *downdeals*: $\mathscr{A} \subset \mathbf{P}$ is a downdeal if $x \in \mathscr{A}$ and $y \leq x$ imply that $y \in \mathscr{A}$. Upward ideals will be called *updeals*.) Such spaces are distinguished by the fact that they are T_0 and that arbitrary intersections of opens are open.

The category of Lazard sheaves over the space **P** *is isomorphic to the category of left* **P**-*sets*. Our two main examples $\mathscr{LH}(\mathbf{P})$ and $\mathscr{S}^{\mathbf{P}^\circ}$, in this case, coincide.

A left **P**-set X may be topologized by taking its open subsets to be the sub **P**-sets. Define f: $X \to \mathbf{P}$ by sending x to $\Box x$ and verify that f is a local homeomorphism.

A map of left **P**-sets is easily seen to be continuous. We obtain a functor (*Left* **P**-*sets*) $\to \mathscr{LH}(\mathbf{P})$.

Given a local homeomorphism f: $X \to \mathbf{P}$ note first that the open sets in X are preserved, as are those in **P**, under arbitrary intersection. For $x \in X$ there is, therefore, a minimal open neighborhood U_x of x. X

becomes a left **P**-set as follows: given $x \in X$ define $\Box x = fx$; given $y \leqslant \Box x$ define $(y \to \Box x)x$ as the unique point in U_x over y.

Given a morphism

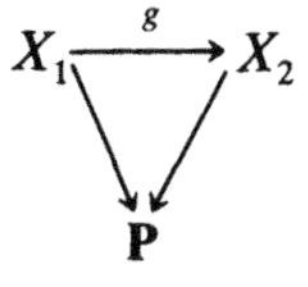

in $\mathcal{S}h(\mathbf{P})$ verify that g carries minimal open sets to minimal open sets, hence becomes a map of left **P**-sets. We obtain a functor $\mathcal{S}h(\mathbf{P}) \to$ (*Left* **P**-sets) which is readily verified to be the inverse of (*Left* **P**-*sets*) $\to \mathcal{S}h(\mathbf{P})$.

(1.373.) *An alternate description of* $\mathcal{S}h(Y)$

For a space Y let $\mathcal{O}(Y)$ be the lattice of open subsets of Y. We regard $\mathcal{O}(Y)$ as a category: the maps of $\mathcal{O}(Y)$ need not be considered abstractions; they may be taken to be the inclusion maps between open subsets of Y. For any local homeomorphism $X \to Y$ consider the left $\mathcal{O}(Y)$-set $\Gamma_*(X)$ whose elements are SECTIONS, that is, continuous functions $f: U \to X$ such that $U \xrightarrow{f} X \to Y$ is the inclusion function of an open subset U in Y. Define $\Box f = U$. Given a morphism $V \subset U$ in $\mathcal{O}(Y)$ define $(V \subset U)f$ as $V \to U \xrightarrow{f} X$. To save space we will write $(V \subset U)f$ as Vf.

Given a morphism

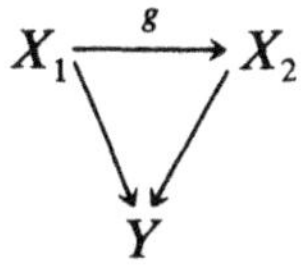

in $\mathcal{S}h(Y)$ we easily obtain a map of left $\mathcal{O}(Y)$-sets $\Gamma_*g: \Gamma_*X_1 \to \Gamma_*X_2$ (it sends $U \xrightarrow{f} X_1$ to $U \to X_1 \to X_2$) and obtain a functor $\Gamma_*: \mathcal{S}h(Y) \to \mathcal{S}^{\mathcal{O}(Y)^\circ}$.

Γ_* is an embedding. If $X_1 \xrightarrow{h} X_2$ is different from g, there must exist $x \in X_1$ such that $g(x) \neq h(x)$. By the definition of local homeomorphism there must exist a section $f: U \to X_1$ such that x is in the image of f. Hence $fg \neq fh$ and $\Gamma_*g \neq \Gamma_*h$.

Γ_* is full. Suppose $\bar{g}: \Gamma_*X_1 \to \Gamma_*X_2$ is a map of left $\mathcal{O}(Y)$-sets. We seek $g: X_1 \to X_2$. Given $x \in X$, let $f: U \to X_1$ be a section such that x is in the image of f. Let $y \in Y$ be such that $f(y) = x$. $\bar{g}(f)$ is a section $U \to X_2$. Define $g(x) = (\bar{g}(f))(y)$. Verify that this definition of $g(x)$ is independent of the choice of f. To show that $g: X_1 \to X_2$ is a local homeomorphism it sufficies to show for any section $f: U \to X_1$ that $u \xrightarrow{f} X_1 \xrightarrow{g} X_2$ is a section and that is routine.

$\Gamma_* : \mathscr{H}(Y) \to \mathscr{S}^{\mathscr{O}(Y)^\circ}$ is a full embedding but never an equivalence functor. (Given a local homeomorphism $X \to Y$ there is exactly one section f such that $\Box f = \emptyset$; clearly there are left $\mathscr{O}(Y)$-sets in which such is not the case.)

The objects of $\mathscr{S}^{\mathscr{O}(Y)^\circ}$ are often called PRE-SHEAVES over Y.

There is a very important functor $S: \mathscr{S}^{\mathscr{O}(Y)^\circ} \to \mathscr{H}(Y)$. For a left $\mathscr{O}(Y)$-set Z we define $S(Z)$ as follows: for each $y \in Y$ consider the subset $Z_y \subset Z$ of elements z such that $y \in \Box z$; consider the equivalence relation on Z_y defined by $z_1 \equiv z_2$ if there exists an open neighborhood V of y, contained in both $\Box z_1$ and $\Box z_2$, such that $Vz_1 = Vz_2$; define S_y as the set of equivalence classes $(Z_y)/\equiv$; define $S(Z)$ to be the union of the S_y's; let $S(Z) \to Y$ be the obvious function; for each $z \in Z$ consider the section $\hat{z}: \Box z \to S(Z)$ which sends $y \in \Box z$ to the equivalence class of z in S_y; topologize $S(Z)$ by taking the images of all such sections as a basis; verify that each $\hat{z}$ is a homeomorphism of $\Box z$ onto its image.

The elements of S_y are called GERMS at y. The set S_y is the STALK at y. (A stalk is made of germs, a sheaf of stalks.)

Given a map of left $\mathscr{O}(Y)$-sets $g: Z_1 \to Z_2$ we easily obtain

$$\begin{array}{ccc} S(Z_1) & \xrightarrow{S(g)} & S(Z_2) \\ & \searrow \quad \swarrow & \\ & Y. & \end{array}$$

To verify that it is a local homeomorphism it suffices to verify for each $z \in Z_1$ that

$$\begin{array}{ccc} & \Box z & \\ {\scriptstyle \hat{z}} \swarrow & & \searrow {\scriptstyle \widehat{g(z)}} \\ S(Z_1) & \xrightarrow[S(g)]{} & S(Z_2). \end{array}$$

$\mathscr{H}(Y) \to \mathscr{S}^{\mathscr{O}(Y)^\circ} \to \mathscr{H}(Y)$ is conjugate to the identity functor, that is, there is a natural homeomorphism

$$\begin{array}{ccc} S(\Gamma_* X) & \longrightarrow & X \\ & \searrow \quad \swarrow & \\ & Y. & \end{array}$$

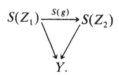

There is also a natural *transformation* $Z \to \Gamma_*(SZ)$. These transformations may be used to obtain a natural equivalence between the set-valued bi-functors $\mathscr{H}(Y)(SZ, X) \simeq \mathscr{S}^{\mathscr{O}(Y)^\circ}(Z, \Gamma_* X)$. Such a pair of functors is called an ADJOINT PAIR. S is a LEFT ADJOINT of Γ_* and Γ_* is a RIGHT ADJOINT of S. S is called THE ASSOCIATED SHEAF FUNCTOR.

(1.374.) *Sheaves as left* $\mathscr{O}(Y)$*-sets*

Which left $\mathscr{O}(Y)$-sets are isomorphic to objects of the form $\Gamma_*(X)$? An answer to this question will allow us immediately to describe $\mathscr{H}(Y)$ (up to equivalence of categories) in terms of left $\mathscr{O}(Y)$-sets, and eventually to generalize the notion of sheaf in important ways.

Given a left $\mathscr{O}(Y)$-set Z we will say that a subset $F \subset Z$ is *consistent* if for every pair $x, y \in F$ it is the case that $(\Box x \cap \Box y)x = (\Box x \cap \Box y)y$. Define $\Box F$ as $\bigcup_{x \in F} \Box x$. Given any $z \in Z$ and covering $\{U_i\}$ of $\Box z$ we obtain a consistent subset $\{U_i z\}$. Such consistent subsets will be called *realizable*. Z is called *complete* if every consistent subset is uniquely realizable, that is, if for all consistent $F \subset Z$ there exists unique $z \in Z$ such that $\Box z = \Box F$ and $(\Box x)z = x$ for all $x \in F$.

For a local homeomorphism $X \to Y$, $\Gamma_* X$ is complete: a consistent family of sections F may be 'pieced together' to obtain a unique section with domain $\Box F$ that restricts correctly to each of the given sections.

If Z is a complete left $\mathscr{O}(\Gamma)$-set, then $Z \to S(\Gamma_* Z)$ is an isomorphism: it is onto because each consistent subset is realizable; it separates because each consistent subset is uniquely realizable.

Hence a left $\mathscr{O}(Y)$-set is isomorphic to an object of the form $\Gamma_*(X)$ iff it is complete.

The definition of complete left $\mathscr{O}(Y)$-sets depends only on the lattice $\mathscr{O}(Y)$, not on Y. Two spaces with isomorphic lattices of open sets therefore have strongly equivalent categories of Lazard sheaves.

1.375. Given a space Y let $\hat{Y}$ be the corresponding T_0 space obtained by identifying points in Y that cannot be distinguished by open sets. $\mathscr{O}(Y)$ and $\mathscr{O}(\hat{Y})$ are isomorphic lattices and hence by the last section $\mathscr{H}(Y)$ and $\mathscr{H}(\hat{Y})$ are strongly equivalent categories. One may directly describe $\mathscr{H}(Y) \to \mathscr{H}(\hat{Y})$ as the functor that sends $X \to Y$ to $\hat{X} \to \hat{Y}$, where $\hat{X}$ is to X just as $\hat{Y}$ is to Y.

If Y is T_0, then for every local homeomorphism $X \to Y$, X is also T_0. It usually goes without saying, when discussing Lazard sheaves, that the spaces are assumed to be T_0.

If Y is T_1, then for every local homeomorphism $X \to Y$, X is again T_1. Not so for T_2. Given open $U \subset Y$ consider $\{1, 2\} \times Y$ and identify $\langle 1, y\rangle \equiv \langle 2, y\rangle$ iff $y \in U$. The resulting quotient space X has an obvious local homeomorphism to Y. X is T_2 iff U is both open and closed. If every space in $\mathscr{H}(Y)$ is T_2, then X is discrete (in which case $\mathscr{H}(Y) \simeq \mathscr{S}/Y \simeq \mathscr{S}^Y$.)

1.38. We consider several important examples of equivalences of categories, the first three of which do not come equipped with canonical strong equivalences [1.32], and the others do. In particular, we consider several important DUALITIES, i.e. contravariant strong equivalences.

1.381. Consideration of computer hardware leads to the *category composed of finite lists*. The proto-morphisms [1.2–1.22] are finite sequences of natural numbers; the objects are natural numbers; $n \xrightarrow{A} m$ iff the length of A, ρA, is n and $A_i < m$ for $i = 0, 1, \ldots, n-1$; AB, if defined, is the sequence such that $(AB)_i = B_{A_i}$ for $i = 0, 1, \ldots, n-1$. (In programming language APL, AB is $B[A]$.) This category is equivalent to the category of finite sets. The existence of a strong equivalence is the same as the existence of a simultaneous choice of orderings on each finite set. Note that $(AB)C$ can be defined without $A(BC)$ being defined (but $A(BC) \succeq (AB)C$).

1.382. Let K be a field. The category composed of finite K-matrices has K-valued functions as proto-morphisms [1.2–1.22]; finite sets as objects; $I \xrightarrow{f} J$ iff the domain of f is $I \times J$; and the usual rules for matrix multiplication. The category of finite dimensional K-spaces is a concrete category [1.243] with finite dimensional K-spaces as objects and with the arrow predicate given by the definition of linear transformation. The canonical equivalence functor from the former to the latter category sends a finite set I to the set of K-valued functions with domain I.

The existence of a strong equivalence between these categories is the same as the existence of a simultaneous choice of an unordered basis for all finite dimensional K-spaces.

1.383. Computer considerations lead to a different category composed of K-matrices: proto-morphisms [1.2–1.22] are finite sequences of elements of K; objects are natural numbers; $n \xrightarrow{A} m$ if $\rho A = nm$; $[AB = C]$ if there exist n, m, p such that

$$\rho A = nm, \qquad \rho B = mp, \qquad \rho C = np,$$

$$C_{in+k} = \sum_{j=0}^{m-1} A_{im+j} \cdot B_{jm+k}\,.$$

(This is, perhaps, the most familiar example is which composition of proto-morphisms is not a partial operation.)

This category allows an equivalence functor to the previous category composed of K-matrices [1.382]. The existence of a strong equivalence functor is the same as the existence of a simultaneous choice of orderings for all unordered finite bases.

1.384. Let Ring be the category of rings (unless stated otherwise, all rings have units) and ring homomorphisms. The category of augmented rings Ring/$\mathbb{Z}$ [1.261] is strongly equivalent to the category of rings with or without units. The functor in one direction carries an augmented ring to the kernel of its augmentation. In the other direction, if S is a ring with or without units, let $R = \mathbb{Z} \times S$, with componentwise addition, and with first projection as the augmentation. When defining multiplication in R, it is helpful to think of $\langle k, s\rangle$ as $k + s$ in R. Let

$$\langle k, s\rangle \cdot \langle l, r\rangle = \langle kl, \pm(s + \cdots + s) \pm (r + \cdots + r) + sr\rangle ,$$

where there are l copies of s and k copies of r. $\langle 1, 0\rangle$ is the unit in R.

This is a striking example of a situation where adding more structure results in less structure.

1.385. The category of pointed sets [1.263] is strongly equivalent to the category of sets and partial maps. Delete the distinguished point from a pointed set, and restrict a map between pointed sets to a partial map with domain $\{x \in \Box f \mid xf \neq \cdot\}$. In the other direction, add a distinguished point and let

$$k^+(x) = \begin{cases} k(x), & \text{if defined}, \\ \cdot, & \text{if } k(x) \text{ undefined or } x = \cdot . \end{cases}$$

Similarly, the category of pointed compact Hausdorff spaces is strongly equivalent to the category of locally compact Hausdorff spaces and partial maps with an open domain, such that the inverse image of a compact subset is compact (i.e. proper maps). In one direction, remove the base point. In the other direction, take the one-point compactification.

1.386. Every equivalence $F\colon \mathbf{A} \to \mathbf{B}$ induces a canonical strong equivalence between the functor categories $\mathscr{S}^{\mathbf{A}}$ and $\mathscr{S}^{\mathbf{B}}$. For a functor $S\colon \mathbf{B} \to \mathscr{S}$, let $S^{\cdot}\colon \mathbf{A} \to \mathscr{S}$ be obtained by composing with F. For a functor $T\colon \mathbf{A} \to \mathscr{S}$, let $T_{.}\colon \mathbf{B} \to \mathscr{S}$ be given by $T.(B) = ((B, F(\_)), T)$. A natural isomorphism $T.(F(\_)) \to T(\_)$ is given as follows. Because $T.(F(A)) = ((F(A), F(\_)), T)$ for $A \in |\mathbf{A}|$, let _ be A, and let $F(A) \to F(A)$ be $1_{\mathbf{A}}$ [1.272, 1.442]. We also need an isomorphism $S(B) \to (FS).(B)$, where $(FS).(B) = ((B, F(\_)), S(F(\_)))$, natural in B. Given a morphism $B \to F(A)$ in $\mathbf{B}$, apply S to get a map. Evaluate at $x \in S(B)$. The reader may check this when in need of a crossword puzzle.

1.387. Both categories in [1.382] are self-dual. For the category of matrices, the duality is given by transposition of matrices. For the category of finite-dimensional vector spaces, the duality is given by considering the vector space $(\_, K)$ of linear transformations to K. Each of these dualities induces the other by the weak equivalence of [1.382.]

1.388. The duality between the category of finite posets and order-preserving maps, and the category of finite distributive lattices and lattice homomorphisms is yet another of several examples with an underlying theme. The two-element set $\mathbf{2}$ has both structures. $(\_, \mathbf{2})$ gives the underlying set of the dual (in each direction). The structure is given pointwise, i.e. it is induced by the power of the corresponding structure on $\mathbf{2}$. (That is, given a poset P, $(P, \mathbf{2})$ is the distributive lattice of its updeals. Given a distributive lattice L, $(L, \mathbf{2})$ is the poset (ordered by inclusion) of its prime filters, i.e. updeals $\mathscr{F}$ closed under meets, such that $x \vee y \in \mathscr{F}$ implies $x \in \mathscr{F}$ or $y \in \mathscr{F}$. On both sides of this duality, the morphisms $(\_) \to ((\_, \mathbf{2}), \mathbf{2})$ (given by evaluation) are isomorphisms.

1.389. We briefly describe the STONE DUALITY between the category of boolean algebras and homomorphisms, and the category of compact Hausdorff spaces with a basis of clopen sets, and continuous maps. The two-element set **2** can be considered as an object in either category, as a two-element boolean algebra or as a discrete space with two points. Both directions of the strong equivalence are given by $(\_, \mathbf{2})$.

The dual of a boolean algebra is the set of its ultrafilters [1.634, 1.635]. It is topologized as a power of the two-point discrete space. This is the STONE SPACE of the given boolean algebra. It is a compact, Hausdorff space with a basis of clopen sets.

For any topological space X, $(X, \mathbf{2})$ is the set of its clopen subsets. It is a boolean algebra as a subalgebra of the power set of **2**.

Boolean homomorphisms $(\_) \to ((\_, \mathbf{2}), \mathbf{2})$ (given by evaluation) are isomorphisms. For a nonempty compact Hausdorff space with a basis of clopen sets, an evaluation at a point defines a homeomorphism with its double dual $((\_, \mathbf{2}), \mathbf{2})$.

1.39. The language of diagrams

We shall follow the tradition, heretofore restricted to blackboards, of diagrammatic definitions. A left-invertible map, for example, is typically described by drawing $A \to B$ then drawing

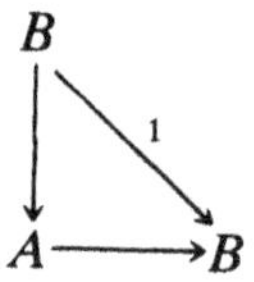

while saying 'there exists'. We shall denote this sequence as:

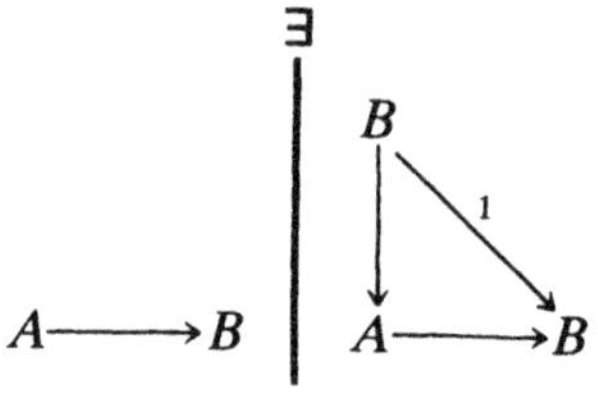

The statement that all morphisms in a category have left-inverses would be:

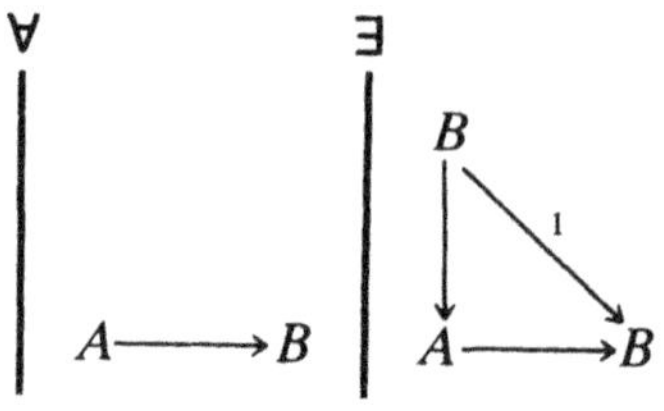

The statement that all idempotents split would be:

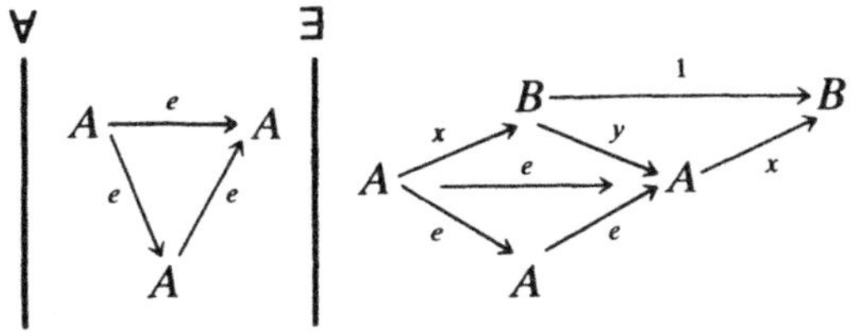

Sometimes a branch is needed. The statement that a category is *linearly ordered* would be:

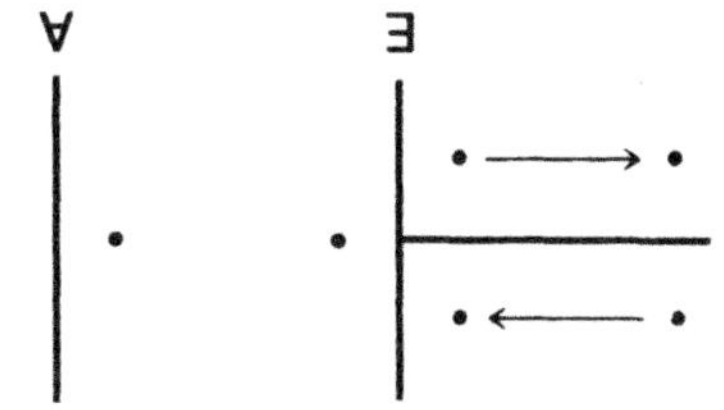

(That is, for any pair of objects A, B either $A \to B$ or $B \to A$ exists.)

1.391. Note that in the sequence of diagrams one draws on a blackboard objects are never collapsed. One may impose commutativity conditions, that is, require two paths to be 'equal' when before they were not, but one does not do so for objects. There is a good reason for this tradition. Properties on diagrams that can be encoded in the language of diagrams – with this restriction against identifying objects – are precisely the elementary properties preserved and reflected by equivalence functors.

This assertion requires a formalization of the diagrammatic language which will occupy the rest of section 1.3.

1.392. A FINITE PRESENTATION for a category is a finite set of 'morphisms' $x_1, x_2, \ldots, x_n$ and a finite set of equations thereon, involving the source, target and composition operations. The source-target equations may be expressed graphically. For example $\Box\bar{x} = \Box\bar{y}$, $\bar{x}\Box = \Box y$, $\bar{y}\Box = \Box x$ and $x\Box = y\Box$, may be expressed by the graph

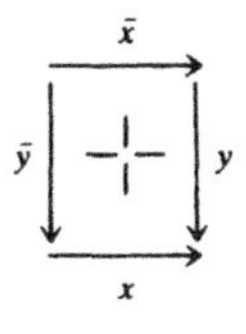

The puncture mark is present in order not to impose the equation $\bar{x}y = \bar{y}x$.

Given such a presentation we obtain a category whose identity morphisms are named by $\square x_i$ and $x_i\square$, $i = 1, 2, \ldots, n$, with all other morphisms named by strings of the form $x_{i_1}, x_{i_2} \cdots x_{i_m}$ where $x_{i_j}\square = \square x_{i_{j+1}}$ is a consequence of the given equations for $j = 1, 2, \ldots, m-1$. We define $\square(x_{i_1}x_{i_2} \cdots x_{i_m}) = \square x_{i_1}$ and $(x_{i_1}x_{i_2} \cdots x_{i_m})\square = x_{i_m}\square$ and define composition by concatenation. Two names of morphisms are identified if the given finite set of equations forces them to be identified.

1.393. The convention that a diagram commutes except where a puncture mark says that it need not does not allow us – unless it is modified – to diagrammatically describe the equations $x\square = \square y$ and $xy = \square(xy)$ (because $\cdot \underset{y}{\overset{x}{\rightleftarrows}} \cdot$ forces both xy and yx to be identity morphisms). Hence we allow labels on the vertices and arrows subject to the rules:

(1) *For any arrow labeled* 1 *the vertex labels of its beginning and end must be the same.*

(2) *For any two appearances of the same arrow label the vertex labels appearing at the beginnings of the arrows must be the same.*

(3) *For any two appearances of an arrow with the same label the vertex labels appearing at the ends of the arrows must be the same.*

The diagrams appearing in 1.39 are examples.

Caution: reversing an arrow labeled 1 can change meaning.

1.394. Corresponding, then, to the blackboard sequence of diagrams we obtain a sequence of finitely presented categories each with a functor to the next. The restriction against collapsing objects corresponds to the condition that the functors between the finitely presented categories separate objects. Note that the conventions of the last section guarantee that no objects become identified.

1.395. In an arbitrary category (e.g. the category of small categories) define a Q-SEQUENCE as a finite sequence of objects $A_0, A_1, \ldots, A_n$ (e.g. finitely presented categories), morphisms $A_{i-1} \to A_i$ (e.g. functors) and a sequence of quantifiers $Q_0, Q_1, \ldots, Q_n$ where each Q_i is either $\forall$ of $\exists$. We say that a morphism $A_0 \to B$ (e.g. functor) SATISFIES the Q-sequence if:

either $Q_0 = \forall$ *and for every*

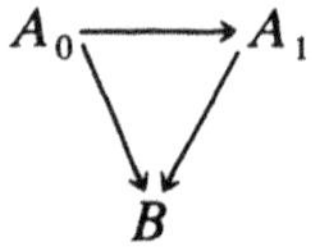

it is the case that $A_1 \to B$ *satisfies the* Q*-sequence beginning with* A_1;

or $Q_0 = \exists$ *and there exists* 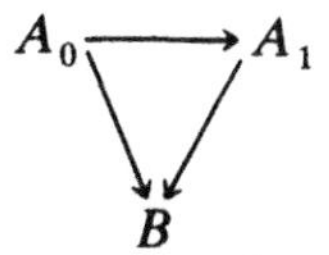

such that $A_1 \to B$ *satisfies the Q-sequence beginning with* A_1.

Note that if $n = 0$ then by the above definition $A_0 \to B$ satisfies iff $Q_0 = \forall$.

In the diagrammatic notation the quantifiers over the vertical bars officially belong on the left side of the bar. When the end quantifier is $\forall$ it is customarily omitted.

The COMPLEMENTARY Q-SEQUENCE is defined by transposing the $\forall$'s and $\exists$'s. $A_0 \to B$ satisfies a Q-sequence iff it does *not* satisfy the complementary Q-sequence.

Clearly, if $B \to B'$ is an isomorphism, then $A_0 \to B$ satisfies a given Q-sequence iff $A_0 \to B \to B'$ does. That is, isomorphisms preserve and reflect satisfaction.

1.396. Now suppose that we have two classes of morphisms $\mathscr{C}$ and $\mathscr{R}$ such that for every square

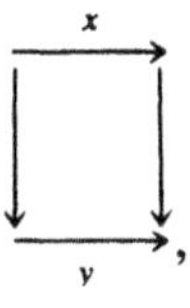 $x \in C$, $y \in \mathscr{R}$ there exists 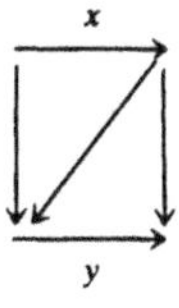

In the case at point $\mathscr{C}$ is the class of functors that separate objects, $\mathscr{R}$ the inflations.

If $A_0 \to A_1 \to \cdots \to A_n$ *is a Q-sequence in* $\mathscr{C}$, *then morphisms in* $\mathscr{R}$ *preserve and reflect satisfaction. That is, if* $B \to B' \in \mathscr{R}$, *then* $A_0 \to B$ *satisfies iff* $A_0 \to B \to B'$ *does so.*

BECAUSE: If suffices to show preservation of satisfaction (since reflecting satisfaction is equivalent to preservation of satisfaction for the complementary Q-sequence.) We do so by induction on length. The case is immediate for $n = 0$.

If $Q_0 = \exists$ and if $A_0 \to B$ satisfies, then there exists

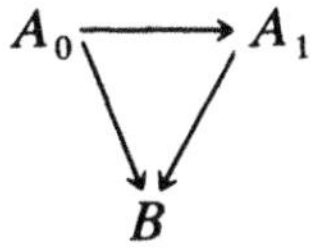

such that $A_1 \to B$ satisfies the Q-sequence beginning with A_1. By induction, $A_1 \to B \to B'$ also satisfies hence $A_0 \to B \to B'$ does so.

If $Q_0 = \forall$ and if $A_0 \to B$ satisfies we must show that for every

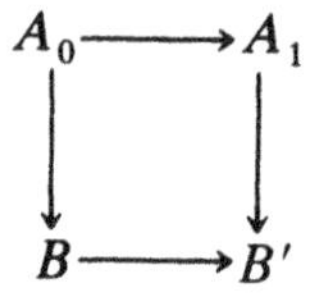

it is the case that $A_1 \to B'$ satisfies. By hypothesis on $\mathscr{C}$ and $\mathscr{R}$ there is a back-diagonal

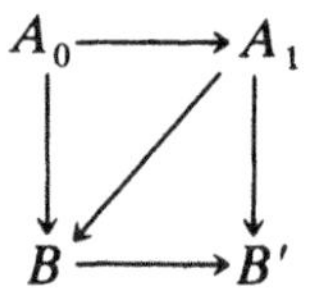

Since $A_0 \to B$ satisfies it is the case that $A_1 \to B$ satisfies. By induction on length, therefore, $A_1 \to B \to B'$ satisfies.

1.397. Suppose $B_2 \to B_3 \in \mathscr{R}$ and $(B_1 \to B_2 \to B_3) \in \mathscr{R}$. Trivially for any Q-sequence in $\mathscr{C}$, satisfaction is preserved and reflected by $B_1 \to B_2$. In the case at point, therefore, inflation cross-sections preserve and reflect Q-sequences in $\mathscr{C}$. Clearly the composition of morphisms that preserve and reflect continue to do so. By [1.361], therefore, any equivalence functor preserves and reflects Q-sequences all of whose functors separate objects.

1.398. A Q-sequence of categories beginning with the empty category describes a class of categories, namely those for which $\emptyset \to \mathbf{A}$ satisfies. The class in question is closed under cartesian products. The class of linearly ordered categories is not so closed, hence no Q-sequence can describe it. Branching is necessary.

Given a poset T and an element $i \in T$ we let T/i denote the principal downdeal $\{j \mid j \leqslant i\}$ and $i\backslash T$ the principal updeal $\{j \mid j \geqslant i\}$. T is a *tree* if T/i is finite and linearly ordered for all $i \in T$. It is a *rooted tree* if, further, there is an element $0 \in T$, called its *root*, such that $0\backslash T = T$. We denote the immediate predecessor of $i \in T$ (the penult in T/i) as i^*. The *length* of a tree is the maximal size of the T/i's.

A *Q-tree* in a given category is a rooted tree T of finite length, a collection of objects $\{A_i\}_{\iota \in T}$ and of morphisms $\{A_{i^*} \to A_i\}_{\iota \in T-\{0\}}$, together with a collection of quantifiers $\{Q_\iota\}_{i \in T}$, $Q_i = \forall$ or $\exists$, each i.

Given $i \in T$ we refer to the Q-tree *sprouting from* A_ι as that which is based on $i \backslash T$.

We easily extend the notion of satisfaction. $A_0 \to B$ satisfies a given Q-tree if:

either $Q_0 = \forall$ *and for all* $i^* = 0$ *and*

$$\begin{array}{ccc} A_0 & \longrightarrow & A_i \\ & \searrow \quad \swarrow & \\ & B & \end{array}$$

it is the case that $A_\iota \to B$ *satisfies the* Q*-tree sprouting from* A_i;

or $Q_0 = \exists$ *and there exists* $i^* = 0$ *and*

$$\begin{array}{ccc} A_0 & \longrightarrow & A_\iota \\ & \searrow \quad \swarrow & \\ & B & \end{array}$$

such that $A_\iota \to B$ *satisfies the* Q*-tree sprouting from* A_i.

Given classes $\mathscr{C}$ and $\mathscr{R}$, as above, the reasoning may be easily modified to show that $\mathscr{R}$-morphisms preserve and reflect satisfaction of Q-trees in $\mathscr{C}$. In particular, equivalence functors between categories preserve and reflect satisfaction of those Q-trees all of whose functors separate objects.

1.399. *Properties on diagrams preserved and reflected by equivalence functors are invariant under conjugation. That is, if* $F_1 \colon \mathbf{A} \to \mathbf{B}$ *and* $F_2 \colon \mathbf{A} \to \mathbf{B}$ *are conjugate then* F_1 *satisfies the property iff* F_2 *does.*

BECAUSE: We may first construct the *mapping-cylinder* of F_1, to wit, the inflation $\mathbf{B}' \to \mathbf{B}$ with the disjoint union $|\mathbf{A}| + |\mathbf{B}|$ as objects and $|\mathbf{A}| + |\mathbf{B}| \to |\mathbf{B}|$ the function that sents $A \in |\mathbf{A}|$ to $F_1 A$ and $B \in |\mathbf{B}|$ to B. $\mathbf{A} \xrightarrow{F_1} \mathbf{B} = \mathbf{A} \xrightarrow{F_1'} \mathbf{B}' \to \mathbf{B}$ where F_1' sends A to A and $A_1 \xrightarrow{x} A_2$ to $F_1 A_1 \xrightarrow{F_1 x} F_1 A_2$. Note that F_1' separates objects. Moreover $\mathbf{A} \xrightarrow{F_2} \mathbf{B} = \mathbf{A} \xrightarrow{F_2'} \mathbf{B}' \to \mathbf{B}$ where $F_2' = \mathbf{A} \xrightarrow{F_2} \mathbf{B} \to \mathbf{B}'$, $\mathbf{B} \to \mathbf{B}'$ the obvious cross-section. Hence by replacing $\mathbf{B}$ with $\mathbf{B}'$ we may specialize to the case that F_1 separates objects.

Let $\theta \colon |\mathbf{A}| \to \mathbf{B}$ be such that $F_1^\theta = F_2$. Define $\theta' \colon |\mathbf{B}| \to \mathbf{B}$ by $\theta'_{F_1 A} = \theta_A$ for all $A \in |\mathbf{A}|$ and $\theta'_B = 1_B$ for all $B \in |\mathbf{B}|$ that do not appear as $F_1 A$, any A. Let T be the θ'-conjugate of the identity functor. Then $\mathbf{A} \xrightarrow{F_1} \mathbf{B} \xrightarrow{T} \mathbf{B} = F_2$.

1.3(10). In this section we show that for any elementary property on diagrams preserved and reflected by equivalence functors there is a finitely presented Q-tree all of whose functors separate objects. (A Q-tree is finitely presented if it is finite and each category therein is finitely presented.)

1.3(10)1. *Any elementary predictate in category theory is given by a finitely presented Q-tree with a free category as root.*

A finitely generated free category corresponds to a diagram of the form:

$$\cdot \xrightarrow{x_1} \cdot \quad \cdot \xrightarrow{x_2} \cdot \quad \cdots \quad \cdot \xrightarrow{x_n} \cdot$$

The basic predicates correspond to Q-sequences as follows:

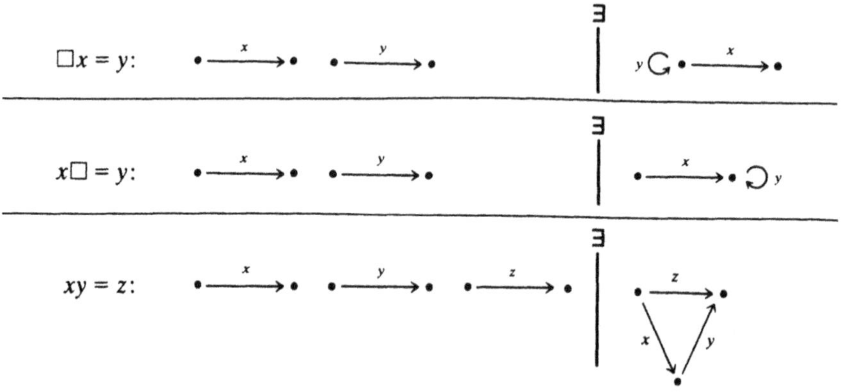

We may easily extend any Q-tree to include another free variable, say w, by disjointly adding $\cdot \xrightarrow{w} \cdot$ to each of the categories therein. Given two Q-trees on the same free root **F**, their conjunction corresponds to the Q-tree, still with **F** as root, with two branches each an identity functor and the given Q-trees sprouting therefrom. Label the root ∀. For disjunction, label the root ∃.

As already observed, negation is obtainable by transposing the ∀'s and ∃'s [1.395]. Recall that we omit the ∀ when it is at the end.

Given a Q-tree with free root **F** generated by $x_1, x_2, \ldots, x_n$ we may universally quantify with respect to x_1, by taking the free category on $x_2, \ldots, x_n$ as root with one branch, namely the inclusion into **F** from which sprouts the given Q-tree. Label the root ∀. For existential quantification, label the root ∃.

1.3(10)2. Virtually no elementary predicate on non-empty free diagrams is reflected by equivalence functors. The pathology is already present for the basic predicates. ($T: \mathbf{A} \to \mathbf{B}$ can fail to reflect $xy = z$ because it can be the case that $x\Box \neq \Box y$ in **A** but that $(Tx)\Box = \Box(Ty)$ in **B**.)

When we say that a predicate on a non-free diagram, say

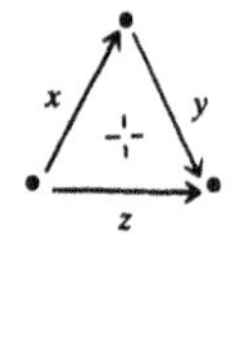
,

is preserved and reflected, we refer to a *functor* from a given finitely presented category. It may well be that a *function* $D: \mathbf{R} \to \mathbf{A}$ is not a functor but that $\mathbf{R} \xrightarrow{D} \mathbf{A} \xrightarrow{T} \mathbf{B}$ is. Such is not to be construed as a failure of T to reflect some property. Only functors are to be considered.

In the last section we found a Q-tree with free root for an arbitrary elementary predicate. If we begin with an elementary predicate on a non-free diagram we must take the equations therefrom and impose them throughout the Q-tree.

1.3(10)3. *Suppose* $\mathbf{R}$ *is the root of a given* Q*-tree and that for all equivalence functors* $\mathbf{A} \to \mathbf{B}$ *it is the case that* $\mathbf{R} \to \mathbf{A}$ *satisfies the given* Q*-tree iff* $\mathbf{R} \to \mathbf{A} \to \mathbf{B}$ *does. Then there is a new* Q*-tree still with* $\mathbf{R}$ *as root, in which all functors separate objects, and such that* $\mathbf{R} \to \mathbf{A}$ *satisfies the given* Q*-tree iff it satisfies the new* Q*-tree, all* $\mathbf{A}$.

If, further, the given Q*-tree is finitely presented then so is the new.*

The argument is indicated below. The preservation of finite presentability is left as a matter of inspection to the reader.

We shall use the axiom of choice. Its removability is a consequence of the Gödel completeness theorem.

1.3(10)4. We will say that a Q-tree is *good* if all functors therein separate objects. It is *nearly-good* if all but the initial branches from the root are good. It is *stable* if satisfaction is preserved and reflected by equivalence functors. Two Q-trees are *coextensive* if they have the same root $\mathbf{R}$ and if $\mathbf{R} \to \mathbf{B}$ satisfies the first iff it satisfies the second, for all $\mathbf{R} \to \mathbf{B}$. We wish to show that every stable Q-tree is coextensive with a good Q-tree.

We will say that two Q-trees are *S-coextensive* if they have the same root $\mathbf{R}$ and if for all *skeletal* $\mathbf{B}$ it is the case that $\mathbf{R} \to \mathbf{B}$ satisfies the first iff it satisfies the second.

LEMMA. *Every* Q*-tree is S-coextensive with a nearly-good* Q*-tree.*

The argument is by induction of the length of Q-trees. The assertion is vacuously true for very short trees. It rests on the following, non-inductive, sublemma:

Every nearly-good Q*-tree is S-coextensive with a nearly-good* Q*-tree whose root quantifier is* $\exists$.

BECAUSE: Clearly, we may assume that the given Q-tree has $\forall$ as its root quantifier.

For an equivalence relation E on the objects of $\mathbf{R}$ let $\mathbf{R}/E$ denote the category obtained by so identifying the objects. We construct a good Q-tree with $\mathbf{R}/E$ as root with the following property:

$\mathbf{R}/E \to \boldsymbol{B}$, *B skeletal, is satisfied iff* $\mathbf{R}/E \to \mathbf{B}$ *separates objects and* $\mathbf{R} \to \mathbf{R}/E \to \mathbf{B}$ *satisfies the original* Q*-tree.*

Given such, we then take the Q-tree with root $\mathbf{R}$, root quantifier $\exists$, initial branches $\mathbf{R} \to \mathbf{R}/E$ for all possible equivalence relations E on $|\mathbf{R}|$, with the Q-tree sprouting from $\mathbf{R}/E$ as described. Such is easily seen to be S-coextensive with the given Q-tree.

The $\mathbf{R}/E$-rooted Q-tree is obtained, first, by removing all initial branches (and the Q-trees sprouting therefrom) from the original Q-tree that disagree with E: that is, $\mathbf{R} \xrightarrow{T_i} \mathbf{A}_i$ is retained iff the equivalence relation induced by T_i on $|\mathbf{R}|$ is contained in E. We then reduce each category in the tree by E obtaining a good Q-tree with $\mathbf{R}/E$ as root. Next, for every distinct pair of objects in $\mathbf{R}/E$ we graft on a good Q-tree that says that the objects are *not* isomorphic. The root, recall, is quantified $\forall$.

Back to the lemma. Given a Q-tree with root quantifier $\exists$ let $\{\mathbf{R} \to \mathbf{A}_i\}$ be the set of initial branches. By induction we may assume that the Q-tree sprouting from $\mathbf{A}_i$, each i, is nearly-good. By the sublemma we may assume that each has $\exists$ as root-quantifier. The non-alternation of quantifiers of the initial branches allows us to compose the functors $\mathbf{R} \to \mathbf{A}_i \to \mathbf{A}_j$ $(j^* = i)$ and eliminate the $\mathbf{A}_i$'s from the tree. The result is a nearly-good Q-tree.

If the given Q-tree has $\forall$ as root-quantifier we may apply the above argument to the complementary Q-tree.

1.3(10)5. For any category $\mathbf{B}$ we may choose an object from each isomorphism class of objects and define $\mathbf{B}'$ as the full subcategory thereof. $\mathbf{B}'$ is skeletal and the inclusion $\mathbf{B}' \to \mathbf{B}$ is an equivalence functor.

Given a stable empty-rooted Q-tree we may, by the last section, construct a new Q-tree that is good and that is S-coextensive with the given Q-tree. (Nearly-good implies good if the root has at most one object.) We know that good Q-trees are stable [1.398]. $\emptyset \to \mathbf{B}$ satisfies the given Q-tree iff $\emptyset \to \mathbf{B}'$ does (since the given Q-tree is assumed to be stable). $\emptyset \to \mathbf{B}'$ satisfies the given Q-tree iff it satisfies the new Q-tree (by the last section). $\emptyset \to \mathbf{B}'$ satisfies the new Q-tree iff $\emptyset \to \mathbf{B}$ does (because the new Q-tree is stable). Hence the given Q-tree and the new Q-tree are coextensive.

1.3(10)6. For larger roots we must do more. Recall that $F: \mathbf{R} \to \mathbf{A}$ satisfies a stable Q-tree (with root $\mathbf{R}$) iff $F^\theta: \mathbf{R} \to \mathbf{A}$ does [1.399]. We will call this property on Q-trees *C-stability*.

Every C-stable Q-tree is S-coextensive with a good Q-tree.

BECAUSE: We may, by [1.3(10)4], assume that the given Q-tree is nearly-good. For $i \in T$ let $\mathbf{R} \to \mathbf{A}_i$ be the composition of the branches from $\mathbf{R}$ to $\mathbf{A}_i$. Let $\mathbf{R} \to \hat{\mathbf{A}}_i$ be the mapping-cylinder described in [1.399] and verify that the resulting Q-tree with the $\mathbf{A}_i$'s replaced by the $\hat{\mathbf{A}}_i$'s is coextensive with the given Q-tree.

The last paragraph of the last section can now be repeated to show that every stable Q-tree is coextensive with a good Q-tree.

1.4. CARTESIAN CATEGORIES

1.41. A MONIC PAIR of morphisms is defined by

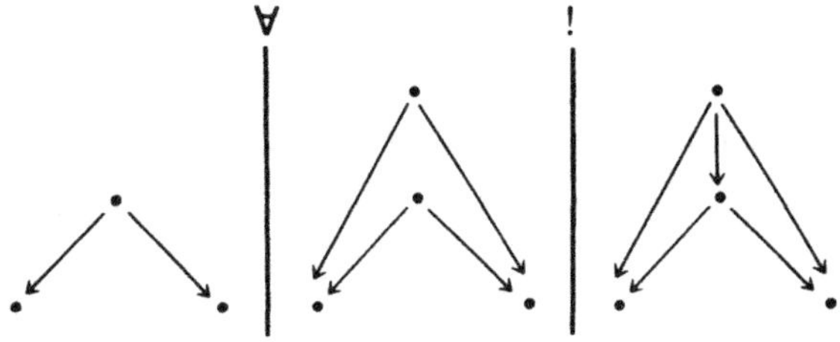

The exclamation mark is to be read 'there is at most one extension to'. It may, of course, be avoided:

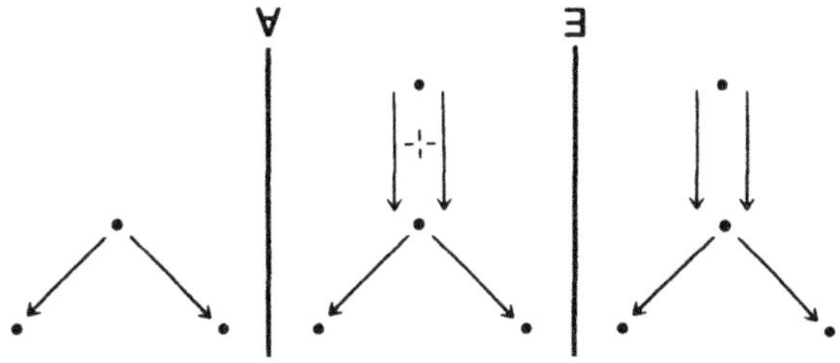

(Recall that the puncture-mark removes only one equation. A translation into the non-diagrammatic notation: x, y is a monic if

$$\Box x = \Box y \wedge \forall_{u,v}[(ux = vx) \wedge (uy = vy)] \Rightarrow u = v.)$$

An exclamation mark may be safely used as a quantifier in the diagrammatic language *provided no new vertex appears immediately thereafter.*

We will, when convenient, denote a monic pair as

A single morphism is monic if

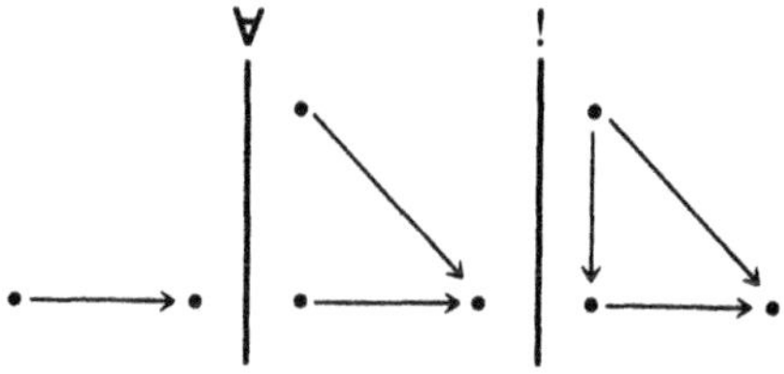

Such have been variously called *monomorphisms*, *monos*, *injections*, *inclusions*. We shall call them, simply, *monics* or *monic morphisms*. When convenient we will indicate that a morphism is monic by denoting it $\rightarrowtail$.

1.411. In most concrete categories that arise in nature, a morphism f is monic iff it is one-to-one: $(f(x) = f(y)) \Rightarrow x = y$. Such insures monic (all embeddings reflect monics), but not conversely. Consider the category of ordered fields (we do not require $0 \neq 1$); the map from the reals to the one-element field is monic.

In any functor category, $\mathscr{S}^{\mathbf{A}}$, a transformation $\alpha\colon T_1 \to T_2$ is monic iff $\alpha_A\colon T_1A \rightarrowtail T_2A$ all A. That such suffices for monic is obvious. That it is required will be shown in [1.462].

1.412. We may generalize the notion of monic pairs to monic n-tuples. Indeed, given any family $\mathscr{F}$ of morphisms with common source, we say that is a *monic family* if whenever $\forall_{x \in \mathscr{F}}\ ux = vx$ it is the case that $u = v$. A TABLE is an object T together with a monic finite sequence of morphisms $x_1, \ldots, x_n$ with T as a common source. T is called the TOP. The targets of the x_i's are called the FEET, and the x_i's themselves are called the COLUMNS.

If $\langle T'; x'_1, \ldots, x'_n \rangle$ is another table, we say that the tables are isomorphic if there is an isomorphism $\theta\colon T \xrightarrow{\sim} T'$ such that $\theta x'_i = x_i$, $i = 1, 2, \ldots, n$. (Isomorphic tables necessarily have the same sequence of feet.) An isomorphism class of tables is called a RELATION. The family of relations on a sequence of feet $A_1, A_2, \ldots, A_n$ is denoted $\mathscr{R}el(A_1, A_2, \ldots, A_n)$.

In the case $n = 1$ an isomorphism class of tables with an object A as foot is usually called a SUBOBJECT of A. The family of all such is denoted $\mathscr{S}ub(A)$.

In the case $n = 0$ an isomorphism class of tables is called a VALUE.

The family of such is denoted $\mathscr{V}al$. An object is called a SUBTERMINATOR if it is the top of a table with no columns:

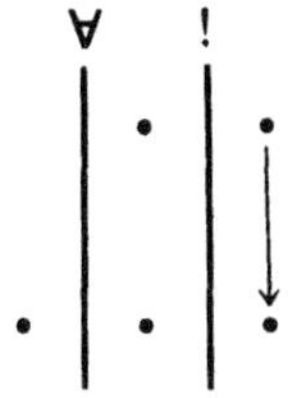

1.413. Given tables $\langle T; x_1, x_2, \ldots, x_n\rangle$ and $\langle T'; x'_1, x'_2, \ldots, x'_n\rangle$ we say that the first is CONTAINED in the second if there exists a morphism $z: T \to T'$ such that $zx'_i = x_i$, $i = 1, 2, \ldots, n$. Note that $T \to T'$ is unique and necessarily monic. Containment is a pre-ordering on tables and a partial ordering on relations; that is, if two tables are each contained in the other, they are necessarily isomorphic. $\mathscr{R}el$, $\mathscr{S}ub$ and $\mathscr{V}al$ may be viewed as posets.

1.414. Let B be an object in a category $\mathbf{A}$. $A \to B$ is monic in $\mathbf{A}$ iff when viewed as an object in $\mathbf{A}/B$ it is a subterminator. We obtain a canonical one-to-one correspondence between the posets $\mathscr{S}ub_{\mathbf{A}}(B)$ and $\mathscr{V}al_{\mathbf{A}/B}$.

In this manner the poset of values of $\mathscr{S}h(Y)$ is canonically isomorphic with the poset $\mathscr{O}(Y)$ of open subsets of Y.

1.415. In the category of sets, a table on A, B may be viewed as a listing ('tabulation'), without repetition, of the instances of a relation from A to B. Indeed, the usual extensional notion of relations on sets coincides with the categorical notion as applied to this case.

The definition of relations in terms of sets of ordered pairs is an unnecessary complication. A computer can tabulate a relation whether sets of ordered pairs exist or not.

1.42. Finite products and equalizers

1.421. An object is called a TERMINATOR (elsewhere: *final* or *terminal object*) if

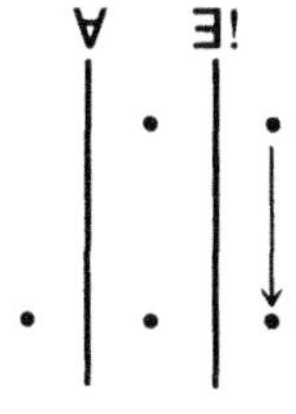

The unique-existential quantifier may, of course, be avoided:

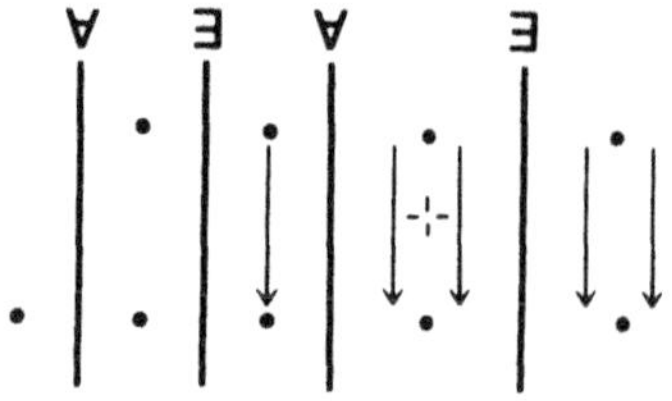

(∃! may be safely used as a quantifier in the diagrammatic language, *provided no new vertices appear immediately thereafter.*)

The property that a category has a terminator is thus stated

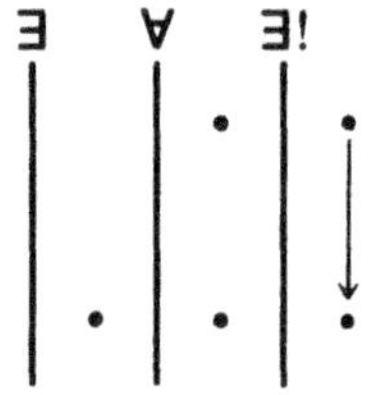

A terminator is, of course, a subterminator. It represents, if it exists, the maximum value. Hence if B, and B_2 are both terminators they are isomorphic; indeed, there is a unique map $B_1 \to B_2$ and it is an isomorphism.

We shall indicate that an object is a terminator, when convenient, by using the symbol 1 as its name. The unique morphism from a given object A to 1 will be denoted, when necessary, as p_A.

1.422. An object is a terminator in the category of sets iff it has a unique element. A functor T: $\mathbf{A} \to \mathscr{S}$ is a terminator in $\mathscr{S}^{\mathbf{A}}$ iff TA has a unique element for every $A \in |\mathbf{A}|$. There is no terminator in $\mathscr{LH}$. On the other hand, $\mathscr{Sh}(Y)$ does have a terminator; indeed, for any category $\mathbf{A}$ and object $B \in |\mathbf{A}|$ there is a terminator in $\mathbf{A}/B$, namely the identity morphism on B.

1.423. A binary PRODUCT diagram is defined by

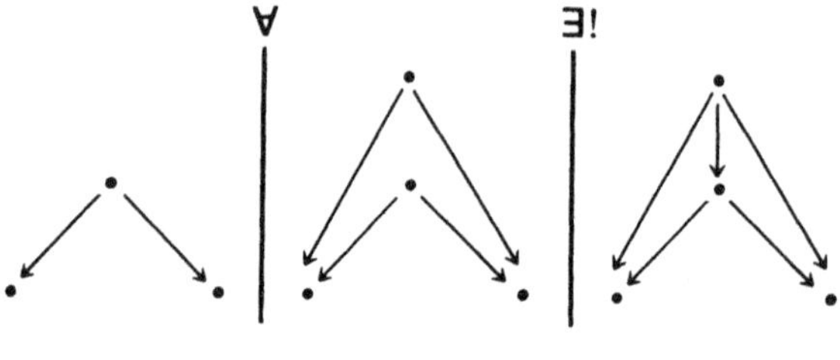

We shall indicate that an object is a product of a given pair of objects A, B by denoting it as $A \times B$. The morphisms will be denoted, when necessary, as $\ell: A \times B \to A$ and $r: A \times B \to B$. $\langle f, g\rangle: X \to A \times B$ denotes the unique morphism such that $\langle f, g\rangle \ell = f$, $\langle f, g\rangle r = g$. $(x \times y): A' \times B' \to A \times B$ denotes $\langle \ell' x, r' y\rangle$. Diagramatically we can indicate products thusly:

A category is said to *have binary products* if

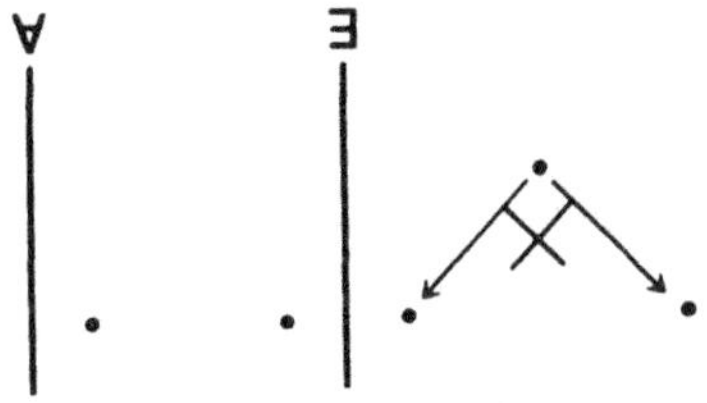

Note that if $A \times B$ exists it is a table on A, B and represents the maximum element in $\mathscr{R}el(A, B)$. Hence products, just as terminators, are unique up to isomorphism.

1.424. Sets of ordered pairs, of course, yield products in the category of sets. Given functors $T_1, T_2: \mathbf{A} \to \mathscr{S}$ we may construct $T_1 \times T_2$ in $\mathscr{S}^{\mathbf{A}}$ 'object-wise': define $(T_1 \times T_2)(A)$ as $T_1(A) \times T_2(A)$. In most examples of concrete categories one first thinks of, the products are given as sets of pairs, endowed with the additional structure. This is not always the case. In particular, in the category whose morphisms are inclusions of sets, products coincide with intersection of sets.

In a poset, viewed as a category, products coincide with greatest lower bounds.

In the category of sets and partial maps, the product of A and B is given as $AB + A + B$, where AB is the set of ordered pairs (product in $\mathscr{S}$), and $+$ is the disjoint sum operation. (However, cf. [1.385].) In the category of sets and relations, the products are given as disjoint sums.

1.425. Given an indexed set $\{A_i\}_I$ of objects, a *product* is an object P and an indexed family $\{p_i: P \to A_i\}_I$ such that for any X and family $\{x_i: X \to A_i\}_I$ there exists unique $z: X \to P$ such that $zp_i = x_i$ all $i \in I$. The usual notation for P is $\Pi_I A_i$. A product of the empty family is a terminator. Products of non-empty finite families are constructible by repeated use of binary products. Hence the existence of finite products is equivalent with the existence of a terminator and of binary products.

Note that a poset when viewed as a category has all finite products iff it is a semi-lattice.

1.426. $\begin{array}{ccc} & T & \\ {}^{x}\swarrow & & \searrow^{y} \\ A & & B \end{array}$ is monic iff $\langle x, y\rangle : T \to A \times B$ is monic and we obtain a natural correspondence between the posets $\mathscr{R}el(A, B)$ and $\mathscr{S}ub(A \times B)$. More generally, $\mathscr{R}el(A_1, A_2, \ldots, A_n)$ and $\mathscr{S}ub(A_1 \times A_2 \times \cdots \times A_n)$ are isomorphic posets, any $A_1, A_2, \ldots, A_n$. In the case $n = 0$ we obtain $\mathscr{V}al = \mathscr{S}ub(1)$.

(1.427.) Using the fact that $(F_1 \to F_2) \in \mathscr{S}^{\mathbf{A}}$ is monic if $F_1A \to F_2A$ is monic, all $A \in \mathbf{A}$ [1.462], we may analyse the poset of values of $\mathscr{S}^{\mathbf{A}}$. T is a terminator iff TA has precisely one element for all $A \in \mathbf{A}$. $F \to T$ is monic, therefore, iff FA has *at most* one element, each $A \in \mathbf{A}$. The isomorphism type represented by $F \subset T$ is distinguished by its *support*, that is, the set $\mathscr{S}pt\,F = \{A \mid FA \neq \emptyset\}$. Define $A \leqslant B$ if there exists $A \to B$. If $A \in \mathscr{S}pt\,F$ and $A \leqslant B$, then $B \in \mathscr{S}pt\,F$, that is, $\mathscr{S}pt\,F$ is a co-ideal. Conversely, any co-ideal may be realized as $\mathscr{S}pt\,F$ for some $F \subset T$. Hence $\mathscr{V}al$ is isomorphic to the poset of co-ideals in $\mathbf{A}$.

1.428. An EQUALIZER diagram is defined by

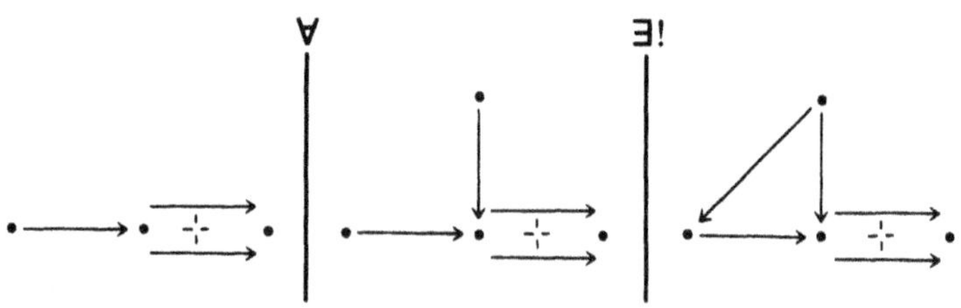

(Recall that the puncture mark removes only one equation.)

We denote equalizers thusly:

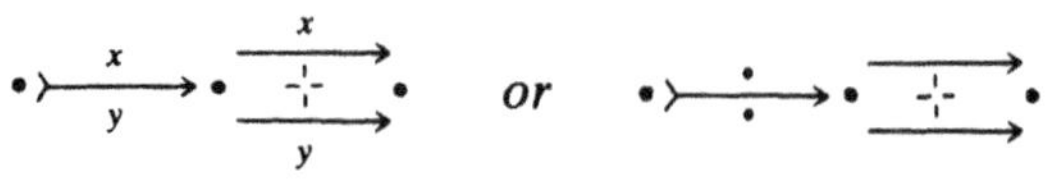

The latter is used when the diagram makes x and y unambiguous.

A category is said to *have equalizers* if

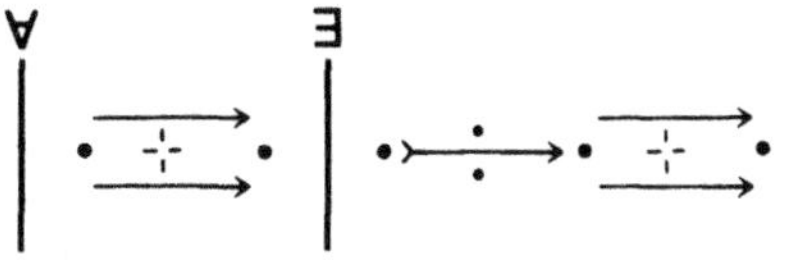

As the notation indicates, equalizers represent subobjects. As with products, they are essentially unique.

For $f, g\colon A \to B$ in the category of sets we may construct an equalizer as $\{x \mid fx = gx\}$. For transformations $\alpha, \beta\colon T_1 \to T_2$ where $T_1, T_2\colon \mathbf{A} \to \mathscr{S}$ we may construct $E \rightarrowtail T \underset{\beta}{\overset{\alpha}{\rightrightarrows}}$ in $\mathscr{S}^{\mathbf{A}}$ by $EA = \{x \in T_1A \mid \alpha_A(x) = \beta_A(x)\}$. In most examples of concrete categories one first thinks of, the equalizers are given as sets constructed this way, endowed with the additional structure. This is not always the case. In the full subcategory of the category of abelian groups, whose objects are divisible groups (for each element b, there exists an element a, and a natural number n so that $na = b$), an equalizer of $f, g\colon G \to H$ is the maximal divisible subgroup of the subgroup $\{x \in G \mid fx = gx\}$. E.g.

$$0 \rightarrowtail \xrightarrow[\text{proj}]{0} \mathbb{Q} \underset{\text{proj}}{\overset{0}{\rightrightarrows}} \mathbb{Q}/\mathbb{Z}$$

1.429. Suppose $\bullet \rightarrowtail \xrightarrow[1_A]{e} A \underset{1_A}{\overset{e}{\rightrightarrows}} A$

exists for each idempotent e. Then, given $e\colon A \to A$, $e^2 = e$, choose an equalizer $y\colon B \to A$ of e and 1_A and define $x\colon A \to B$ as the unique map such that $xy = e$ (the existence of which is insured by $ee = e1$). Then $\langle x, y\rangle$ splits e: we need show only that $yx = 1$; but $(yx)y = y(xy) = ye = y1 = 1y$ and the uniqueness condition in the definition of equalizers allows us to cancel on the right to obtain $yx = 1$. Thus, *if a category has enough equalizers then all idempotents split.*

1.43. A CARTESIAN CATEGORY is a category with finite products and equalizers. They are sometimes, but not herein, called *finitely complete categories*, sometimes *left-exact categories*. Descartes not only drew attention to finite products but to equalizers: that is, equationally defined subsets.

1.431. A PULLBACK diagram is defined by

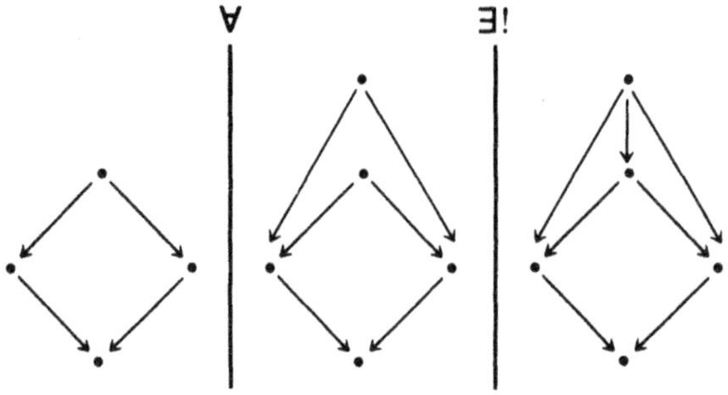

Note that the upper half of a pullback is a monic pair. The relation it represents is uniquely determined by the lower half.

We will denote pullbacks thusly:

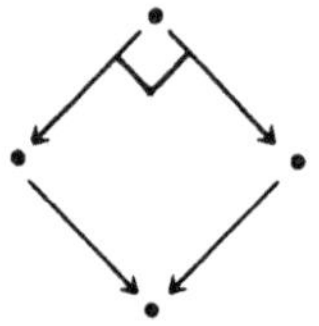

We say that a category *has pullbacks* if

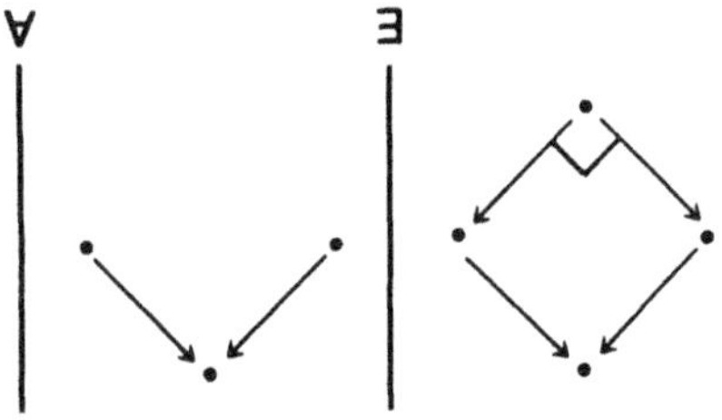

1.432. *Binary products and equalizers imply pullbacks.*

BECAUSE: Given  construct, in sequence

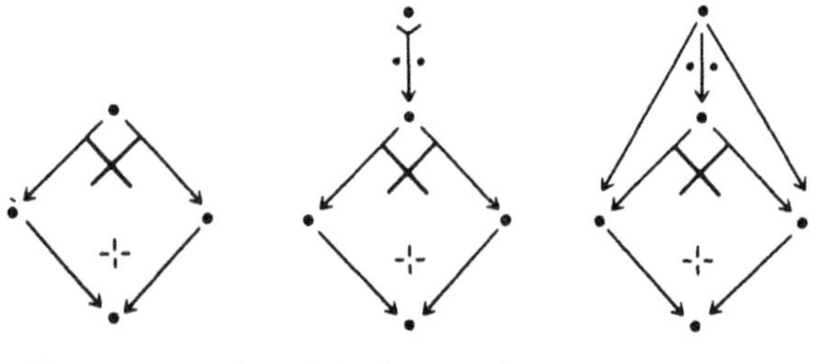

The outer diagram on the right is a pullback.

1.433. *Pullbacks and a terminator imply binary products.*

BECAUSE:

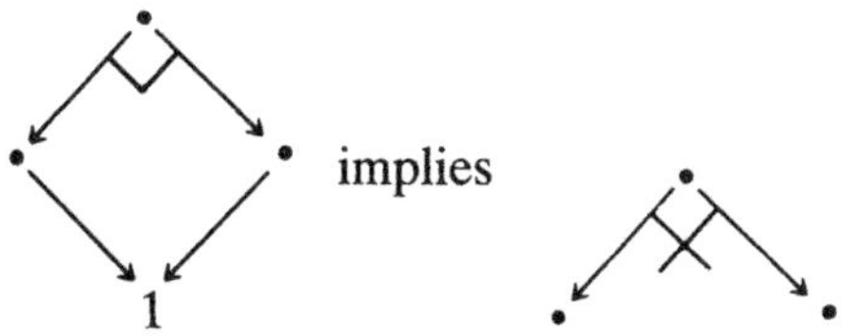

1.434. *Binary products and pullbacks imply equalizers.*

BECAUSE: Given $x, y: A \to B$ construct

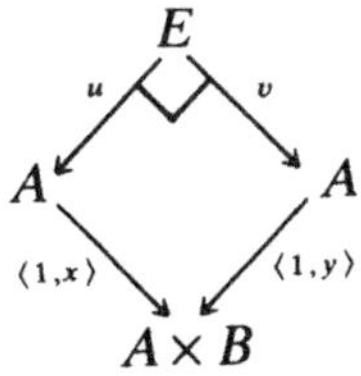

Then $u = v$ and u is an equalizer of x, y.

1.435. Combining [1.433] and [1.434]:

Pullbacks and a terminator imply cartesian.

1.436. The above three lemmas are exhaustive. In the list below, the four binary digits refer, in sequence, to the existence of *terminators*, *binary products*, *equalizers*, *pullbacks*. **A** is the monoid of all endomorphisms of an infinite set. **B** is the full subcategory of $\mathscr{S}$ obtained from an infinite set and a single-element set. The other categories are finitely presented by the indicated diagram. (Note that the 0001 example is the monoid of natural numbers.) 0111 is the empty category.

0000 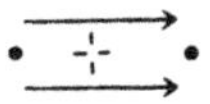

0001 

0010 

0011 • •

0100 **A**

0101 ⇒ 0111

0110 ⇒ 0111

0111

1000 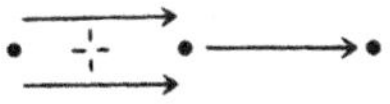

1001 ⇒1111

1010 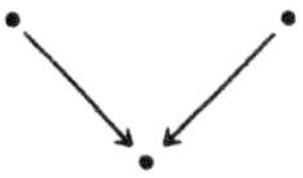

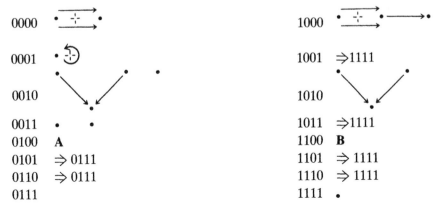

1011 ⇒1111

1100 **B**

1101 ⇒ 1111

1110 ⇒ 1111

1111 •

1.437. A functor between cartesian categories that preserves finite products and equalizers is called a REPRESENTATION OF CARTESIAN CATEGORIES. Note that the proof of [1.423] implies that representations of cartesian categories preserve pullbacks. [1.433] and [1.434] imply that a functor that preserves pullbacks and the terminator is a representation of cartesian categories. We will eventually need [1.655] the following refinement obtainable from the proof of [1.434]: a functor that preserves finite products and pullbacks of *monics* is a representation of cartesian categories.

1.438. *A functor that reflects equalizers reflects isomorphisms.*
(BECAUSE: $A \to B$ is an isomorphism iff it is an equalizer of 1_B and 1_B.)

If a category has equalizers, then any isomorphism-reflecting functor therefrom that preserves equalizers is an embedding, therefore faithful.
(BECAUSE: $\cdot \overset{x}{\underset{y}{\rightarrowtail}} \cdot$ is an isomorphism if $x = y$.)

If a category has terminators/binary products/equalizers/pullbacks then any faithful functor therefrom that preserves them also reflects them.
(BECAUSE: X is a terminator iff $X \to 1$ is an isomorphism. $A \overset{x}{\leftarrow} X \overset{y}{\to} B$ is a product diagram iff $X \xrightarrow{\langle x,y \rangle} A \times B$ is an isomorphism. The argument is similar for equalizers and pullbacks.)

1.439. *In a category with pullbacks, if* $A \overset{f}{\underset{g}{\rightrightarrows}} B \longrightarrow C$ *then there exists an equalizer for f, g.*

BECAUSE:

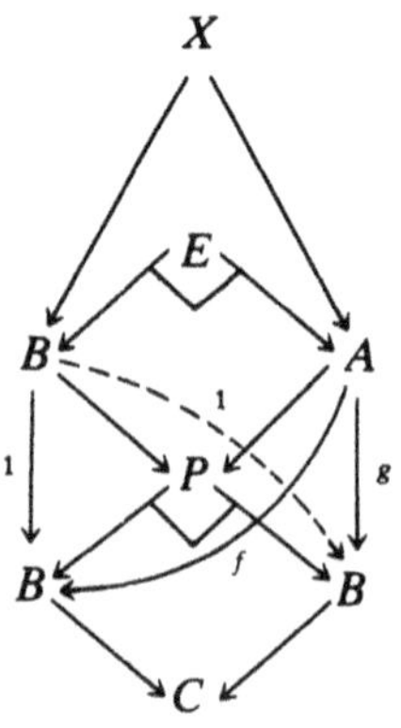

so $E \to A$ is an equalizer for f, g.
Thus:

A functor from a cartesian category that preserves pullbacks also preserves equalizers.

1.44. The forgetful functor $\Sigma\colon \mathbf{A}/B \to \mathbf{A}$ does not preserve terminators (unless B is a terminator in $\mathbf{A}$, in which case Σ is an isomorphism). $\mathbf{A}/B$ has a distinguished terminator, $B \xrightarrow{1} B$ and it is carried by Σ to B. Σ is universal in this respect:

Let $\mathbf{C}$ *be a category with a designated terminator* 1, *and let* $T\colon \mathbf{C} \to \mathbf{A}$ *be a functor such that* $T(1) = B$. *There exists unique* $T'\colon \mathbf{C} \to \mathbf{A}/B$ *such that* $T'(1) = 1_B$ *and* $T = \mathbf{C} \xrightarrow{T'} \mathbf{A}/B \xrightarrow{\Sigma} \mathbf{A}$. (Construct T' by $T'(C) = T(C) \xrightarrow{T(p_C)} T(1)$.)

As a special case, any functor $\mathbf{A} \to \mathbf{A}$ that sends each A to $B \times A$ may be factored as $\mathbf{A} \xrightarrow{\Delta} \mathbf{A}/B \xrightarrow{\Sigma} \mathbf{A}$. $\Delta\colon \mathbf{A} \to \mathbf{A}/B$ is the ubiquitous DIAGONAL FUNCTOR. $\Delta(A) = (B \times A \xrightarrow{\ell} B)$.

The uniqueness of T' is only as stated. For any automorphism $\theta\colon B \to B$ in $\mathbf{A}$ we can define T'_θ by $T'_\theta(C) = T(C) \xrightarrow{(Tp)\theta} T(1)$. T'_θ need not even by conjugate to T'. (Let $\mathbf{A} = \mathscr{S}$, $B = \{0, 1\}$, $T = \Sigma$.)

1.441. *If* $\mathbf{A}$ *has pullbacks, then* $\mathbf{A}/B$ *is cartesian and* $\Sigma\colon \mathbf{A}/B \to \mathbf{A}$ *preserves pullbacks and equalizers. As always,* Σ *is faithful.*

BECAUSE: The naive construction of pullbacks works in $\mathbf{A}/B$; the naive construction is precisely such that Σ is seen to preserve pullbacks. Since $\mathbf{A}/B$ has pullbacks and a terminator it is cartesian. Σ easily preserves equalizers.

1.442. The Cayley representation $C\colon \mathbf{A} \to \mathscr{S}$ [1.272] is easily seen to preserve and reflect pullbacks and equalizers. If $\mathbf{A}$ is cartesian we may identify $C(1)$ with $|\mathbf{A}|$. If we factor C as $\mathbf{A} \xrightarrow{C'} \mathscr{S}/|\mathbf{A}| \xrightarrow{\Sigma} \mathscr{S}$ where C' preserves the terminator we obtain a faithful representation of cartesian categories: C' is faithful since C is; C' preserves pullbacks since $C'\Sigma$ does and Σ reflects pullbacks.

The equivalence functor $\mathscr{S}/|\mathbf{A}| \to \mathscr{S}^{|\mathbf{A}|}$ of [1.261] yields:

Every small cartesian category may be faithfully represented in a power of the category of sets.

The A-th coordinate functor of C' is a REPRESENTABLE FUNCTOR: it carries $B \in |\mathbf{A}|$ to the set (A, B); it carries $f\colon B_1 \to B_2$ to the function $(A, f)\colon (A, B_1) \to (A, B_2)$ obtained by composing with f (that is, $g \in (A, B_1)$ goes to $gf \in (A, B_2)$). We may recast:

The representable functors from a cartesian category are representations of cartesian categories and are collectively faithful.

The A-th representable functor is often denoted as $(A, —)$.

1.443. Consider any universally quantified sentence in which the primitive predicates are the basic predicates of category theory together with predicates that assert that a diagram is a pullback or is an equalizer. Given a counterexample of such a statement in any category **A** we may easily find a small subcategory **A**′ in which the counterexample remains such. The Cayley representation will then yield a counterexample in $\mathscr{S}$.

1.444. Given a class of 'primitive predicates', a HORN SENTENCE is a universally quantified sentence of the form $(P_1 \wedge P_2 \wedge \cdots \wedge P_n) \Rightarrow Q$ where $P_1, P_2, \ldots, P_n, Q$ are primitive. By a Horn sentence in the theory of cartesian categories, we mean a Horn sentence in which the primitive predicates are the basic predicates of category theory together with predicates that assert that a diagram is a terminator, product or equalizer.

Any Horn sentence in the theory of cartesian categories true for the category of sets is true for all cartesian categories.

BECAUSE: Suppose there were a category **A** that contains a counterexample for a Horn sentence $(P_1 \wedge P_2 \wedge \cdots \wedge P_n) \Rightarrow Q$. The counterexample satisfies $P_1, P_2, \ldots, P_n$ but not Q. The collective faithfulness of the representable functors says that for some $A \in |\mathbf{A}|$ the functor $(A, -)$ carries the counterexample to one in $\mathscr{S}$: it continues to satisfy $P_1, P_2, \ldots, P_n$ and continues to violate Q.

1.45. *Pullbacks transfer monics*: 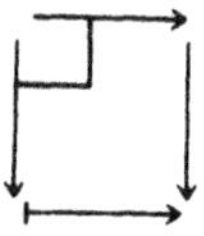 *implies* 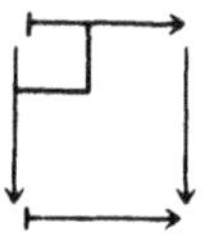 .

That is, if 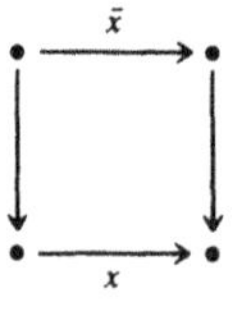

is a pullback and x is monic, then so is $\bar{x}$.

1.451. Fix $f: A \to B$. We say that $A_1 \rightarrowtail A$ is an INVERSE IMAGE of $B_1 \rightarrowtail B$ if there exists

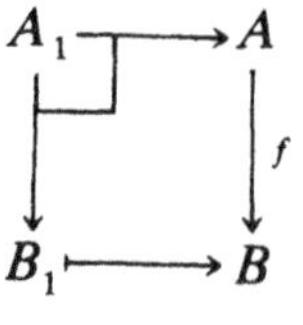

If, further, $A_2 \rightarrowtail A$ is an inverse image of $B_2 \rightarrowtail B$ and if $B_2 \rightarrowtail B$ contains $B_1 \rightarrowtail B$, then $A_2 \rightarrowtail A$ contains $A_1 \rightarrowtail A$. Hence, if two monics

represent the same subobject, then two inverse images represent the same subobject. In any category with pullbacks we obtain an order-preserving function $f^{\#}: \mathcal{S}ub(B) \to \mathcal{S}ub(A)$. $\mathcal{S}ub(-)$ may be viewed as a contravariant poset-valued functor. (In the case $\mathbf{A} = \mathcal{S}$ we called it, in [1.332], the contravariant power-set functor.)

1.452. Given monics $A_1 \rightarrowtail A$, $A_2 \rightarrowtail A$ consider

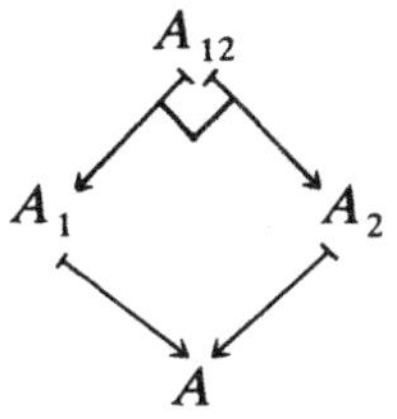

The subobject represented by $A_{12} \rightarrowtail A$ is the greatest lower bound, in $\mathcal{S}ub(A)$, of the subobjects represented by $A_1 \rightarrowtail A$ and $A_2 \rightarrowtail A$. Hence in any category with pullbacks $\mathcal{S}ub(A)$ may be viewed as a SEMI-LATTICE: that is, a poset with finite intersections. (The empty intersection is, of course, the *entire subobject* represented by $A \xrightarrow{1} A$.)

Inverse images are easily seen to preserve intersections and we may view $\mathcal{S}ub(-)$ as a contravariant functor to the category of semi-lattices.

1.453. LEMMA. *If* **A** *is cartesian and if* $T: \mathbf{A} \to \mathbf{B}$ *preserves pullbacks, then* T *is faithful iff* T *preserves properness of subobjects: that is, if* $A' \rightarrowtail A$ *is not an isomorphism, then* $TA' \rightarrowtail TA$ *is not an isomorphism.*

BECAUSE: Given $f: A \to B$ not an isomorphism in **A** we can immediately specialize to the case that f is not monic. Consider

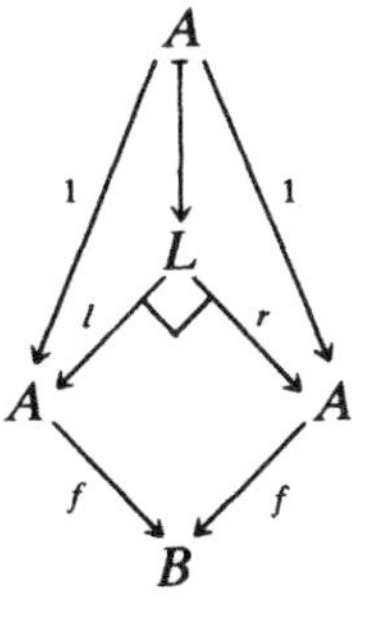

f is monic iff $A \rightarrowtail L$ is an isomorphism. Hence T reflects monics. [1.438–9]

1.454. The above pair, $A \overset{l}{\leftarrow} L \overset{r}{\rightarrow} A$, measures the extent to which f is monic and is called the LEVEL of f (elsewhere *kernel-pair*, *congruence*). The morphism $A \rightarrowtail L$ is often denoted Δ: $A \to L$ and is called the DIAGONAL morphism. It represents the *diagonal subobject*.

(1.46.) We consider the cartesian structure of the prime examples.

(1.461.) $\mathscr{LH}$ is not cartesian – it has neither a terminator nor binary products. It does have pullbacks:

Let $\mathscr{T}\!op$ be the category of spaces and all continuous maps. Given $X_1 \to Y \leftarrow X_2$ in $\mathscr{LH}$ consider

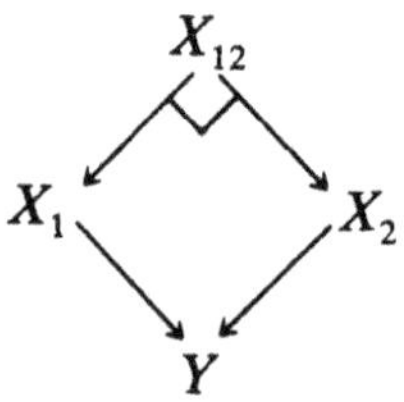

in $\mathscr{T}\!op$. Using the last sentence of [1.262] it suffices to show that $X_{12} \to Y$ is a local homeomorphism in order to show that the square is a pullback in $\mathscr{LH}$. $\mathscr{S}\!h(Y)$, therefore, is cartesian [1.441].

In any category of sheaves, $\mathscr{S}\!h(Y)$, let f_1: $X_1 \to Y$, f_2: $X_2 \to Y$ be objects. Consider their product f_{12}: $X_{12} \to Y$. For each $y \in Y$ the discrete set $f^{\#}_{12}(y)$ is a product (in $\mathscr{S}$) of the discrete sets $f^{\#}_1(y)$ and $f^{\#}_2(y)$. If inverse-images of points are called *fibers* (rather than stalks), then X_{12} is the 'fiber-wise' product of X_1, X_2. In many contexts pullbacks have been called *fiber-products*.

(1.462.) For small $\mathbf{A}$, the category of its covariant set-valued functors is cartesian [1.422, 1.424, 1.429]. The constructions show that the EVALUATION FUNCTORS, E_A: $\mathscr{S}^{\mathbf{A}} \to \mathscr{S}$ ($E_A(F) = F(A)$) are representations of cartesian categories. They are collectively faithful (such was the content of [1.274]). We may assemble them into the original forgetful functor $\mathscr{S}^{\mathbf{A}} \to \mathscr{S}^{|\mathbf{A}|}$ [1.27] to obtain a faithful representation of cartesian categories into a power of $\mathscr{S}$.

The assertion in [1.427] that $F_1 \to F_2$ is monic iff $F_1A \to F_2A$ is monic all $A \in \mathbf{A}$ is now evident.

(1.463.) When the A-th representable functor, $(A, -)$, [1.442] is viewed as an object in $\mathscr{S}^{\mathbf{A}}$ we denote it as H^A. If we view H^A as a right $\mathbf{A}$-set, it appears as a subset of the Cayley representation: $H^{\mathbf{A}}$ is the sub $\mathbf{A}$-set of $\mathbf{A}$

generated by 1_A. Given any right **A**-set X and any $x \in X$, $x\Box = A$ we obtain a unique map $H^A \to X$ that sends 1_A to x (and $y \in H^A$ to $xy \in X$). That is, $(H^A, X) \simeq \{x \in X \mid x\Box = A\}$. For the associated functor, F, of X we thus obtain $(H^A, F) \simeq F(A)$. This equivalence is natural. That is, $(H^A, -)$ *and* E_A are *conjugate functors*.

(1.464.) Since $(H^A, H^B) \simeq (B, A)$ we obtain for each $B \xrightarrow{x} A$ in **A** a transformation $H^x\colon H^A \to H^B$ and hence a contravariant full embedding $H\colon \mathbf{A} \to S^{\mathbf{A}}$ called the YONEDA REPRESENTATION.

For any category **A**, $\mathscr{H}om$ denotes the functor from $\mathbf{A}^\circ \times \mathbf{A}$ to $\mathscr{S}$ that sends a pair A, B to the set (A, B). Consider the covariant functor

$$\mathbf{A} \times \mathscr{S}^{\mathbf{A}} \xrightarrow{(H)^\circ \times 1} (\mathscr{S}^{\mathbf{A}})^\circ \times S^{\mathbf{A}} \xrightarrow{\mathscr{H}om} \mathscr{S}$$

(where $(H)^\circ$ denotes $\mathbf{A} \to \mathscr{S}^{\mathbf{A}} \to (\mathscr{S}^{\mathbf{A}})^\circ$). It is naturally equivalent to $E\colon \mathbf{A} \times \mathscr{S}^{\mathbf{A}} \to \mathscr{S}$, the functor that sends a pair A, F to $F(A)$.

(1.465.) The B-th contravariant representable functor $(-, \mathrm{B})$ when viewed as an object in $\mathscr{S}^{\mathbf{A}^\circ}$ is denoted H_B. We obtain, just as above, a natural equivalence $(H_B, F) \simeq F(B)$ and a *covariant* full embedding $\mathbf{A} \to \mathscr{S}^{\mathbf{A}^\circ}$, also called the Yoneda representation.

$\mathbf{A} \to \mathscr{S}^{\mathbf{A}^\circ}$ *preserves and reflects the cartesian predicates.*

1.47. We will say that a cartesian category **A** is *special* if every universally quantified sentence (not just the Horn sentences) in the cartesian predicates true for $\mathscr{S}$ is true for **A**. Eventually [1.646] we will show that for small **A**, special implies the existence of a faithful representation in $\mathscr{S}$.

1.471. *A special cartesian category has at most two values.*

BECAUSE: Given $V_1 \rightarrowtail V_2 \rightarrowtail 1$ in $\mathscr{S}$, then either $V_1 \rightarrowtail V_2$ or $V_2 \rightarrowtail 1$ is an isomorphism. Hence, if both V_1 and V_2 are proper subobjects of 1, then they are both isomorphic to $V_1 \cap V_2$.

1.472. Suppose that **A** is cartesian and that for every finite sequence of morphisms $x_1, x_2, \ldots, x_n$ there is a functor $T\colon \mathbf{A} \to \mathscr{S}$ that preserves and reflects the cartesian predicates when applied to $x_1, x_2, \ldots, x_n$. Then **A** is special: given any universal sentence and counterexample therein we may take $x_1, x_2, \ldots, x_n$ to be the morphisms appearing in the counterexample and apply such T to obtain a counterexample in $\mathscr{S}$.

Suppose that $T\colon \mathbf{A} \to \mathscr{S}$ is a representation of cartesian categories. By enlarging the sequence $x_1, x_2, \ldots, x_n$ we may reduce the requirement that T reflects the cartesian predicates to the requirement that it preserves properness of subobjects [1.437, 1.455]. That is, **A** is special if for every finite sequence of proper subobjects $\{A'_i \subsetneqq A_i\}_{i=1}^n$ there exists a representation of cartesian categories such that $TA'_i \subset TA_i$ is proper, $i = 1, 2, \ldots, n$.

In particular, **A** is special if for every such sequence there exists an object B such that $(B, A'_i) \rightarrowtail (B, A_i)$ is proper, $i = 1, 2, \ldots, n$. To reduce to the elementary, **A** is special if for every sequence of proper subobjects $\{A'_i \subsetneqq A_i\}_{i=1}^n$ there exists an object B and a sequence $\{f_i: B \to A_i\}_{i=1}^n$ such that $f_i^\#(A'_i) \subsetneqq B$ for all i.

One may easily check that if any such $\{f_i: B \to A_i\}_{i=1}^n$ works, then $\{\Pi_{i=1}^n A'_i \to A_i\}$ works. Finally:

$\mathbf{A}$ is special iff for any pair of proper subobjects $A' \subsetneqq A$, $B' \subsetneqq B$ it is the case that $A' \times B$ is a proper subobject of $A \times B$. (Note that if B has no proper subobjects, then $A' \times B$ can be equal to $A \times B$ in $\mathscr{S}$.)

We may restate the above condition, albeit in a non-elementary fashion:

$\mathbf{A}$ is special iff for every B with a proper subobject, the functor $(B \times -): \mathbf{A} \to \mathbf{A}$ is faithful.

1.473. *If $\mathbf{A}$ is one-valued, that is, if $|\mathscr{Val}| = 1$ (e.g. the category of groups), then $\mathbf{A}$ is special iff $B \times -$ is a faithful functor for all B.*

BECAUSE: The last section says that the condition implies that **A** is special. For the converse we need to show that in a one-valued special cartesian category, $B \times -$ is faithful, all B. Clearly $1 \times -$ is faithful. If $B \neq 1$, then $\langle 1, 1\rangle: B \to B \times B$ is proper (else $B \to 1$ is monic) and hence $(B \times B) \times -$ is faithful. That is $\mathbf{A} \xrightarrow{B\times -} A \xrightarrow{B\times -} A$ is faithful, forcing $B \times -$ to be faithful.

(The category of rings with unit is one-valued. The ring of integers $\mathbb{Z}$ has no proper subobjects. Nonetheless $\mathbb{Z} \times -$ is faithful.)

1.474. *If $\mathbf{A}$ is two-valued, that is, if $|\mathscr{Val}| = 2$ let 0 denote the unique proper subobject of 1. $\mathbf{A}$ is special iff $B \times -$ is faithful for every B not isomorphic to 0.*

BECAUSE: [1.472] says that the condition implies that **A** is special. For the converse it suffices to show that in a two-valued special cartesian category every object not isomorphic to 0 has a proper subobject.

Given $B \to V \rightarrowtail 1$ in $\mathscr{S}$, either $B \to V$ or $V \rightarrowtail 1$ is an isomorphism. Hence such is the case in a special cartesian category. In particular $\nearrow: B \times 0 \to 0$ is an isomorphism, all B. But $B \times 0 \to B \times 1$ is monic. It represents a proper subobject of B iff $B \not\simeq 0$.

1.475. Consider the category of $\mathbb{Z}$-sets, $\mathbb{Z}$ the group of integers. Let **A** be the full subcategory of those $\mathbb{Z}$-sets in which no orbit has more than three elements. **A** clearly has a terminator and equalizers. Given two objects in **A** consider first their ordinary product as $\mathbb{Z}$-sets, throw away all orbits with more than three elements, verify that what remains is a product in **A**. Then $\mathbb{Z}_2 \times \mathbb{Z}_3 = \emptyset$ and $\mathbb{Z}_2 \times -$ is not faithful. **A** is a two-valued cartesian category that is not special. $1\backslash\mathbf{A}$ is one-valued and, again, not special.

1.48. A class of monics $\mathscr{D}$ in a cartesian category **A** is said to be a DENSE class of monics if it contains all isomorphisms and it is closed under composition and under pullbacks. We denote monics in $\mathscr{D}$ by $\rightarrowtail$. Given a dense class $\mathscr{D}$ of monics, we define a RATIONAL CATEGORY $\mathbf{A}[\mathscr{D}^{-1}]$ and a functor $T_{\mathscr{D}} : \mathbf{A} \to \mathbf{A}[\mathscr{D}^{-1}]$ that sends every dense monic to an isomorphism, so that $T_{\mathscr{D}}$ is universal among all such functors from $\mathbf{A}$ to any category $\mathbf{B}$.

The objects of the rational category are those of $\mathbf{A}$, and the morphisms $A \to B$ are named by diagrams

in $\mathbf{A}$, where two such diagrams name the same morphism if they can be completed to

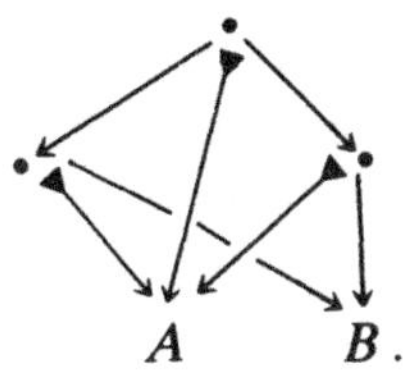

Composition is defined by

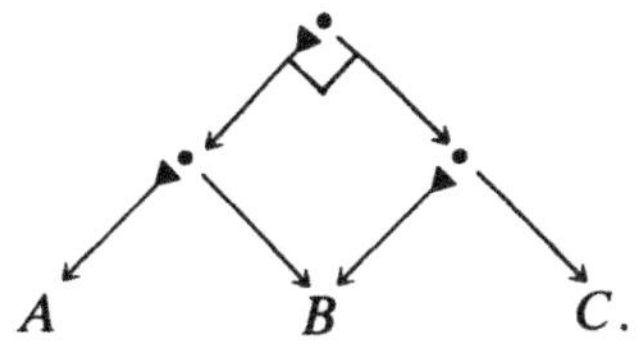

The named morphism $A \to C$ is independent of the choice of names for $A \to B$ and $B \to C$, and of the choice of pullback. Let $T_{\mathscr{D}}(A) = A$ for each $A \in |\mathbf{A}|$, and let $T_{\mathscr{D}}\ (A \xrightarrow{x} B)$ be named by

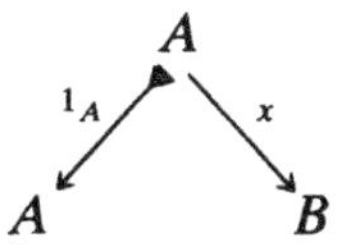

1.481. *Let* $\mathscr{D}$ *be a dense class of monics in a cartesian category. Then the rational category* $\mathbf{A}[\mathscr{D}^{-1}]$ *is cartesian and* $T_{\mathscr{D}}$ *is a representation of cartesian categories.*

BECAUSE: We can assume that morphisms $A \underset{y}{\overset{x}{\rightrightarrows}} B$ in the rational category are named by

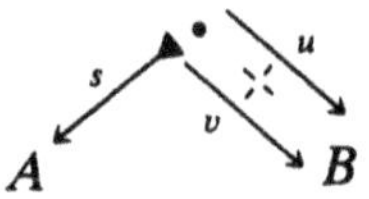

Let w be an equalizer of u, v in $\mathbf{A}$. Then $T_{\mathscr{D}}$ (ws) is an equalizer of x, y in the rational category. The argument for products is similar.

1.49. τ-Categories

For the definition of τ-categories we will need two technical definitions.

Given a table $\langle T; x_1, \ldots, x_n \rangle$ we say that x_j is a SHORT COLUMN if for every $f, g: X \to T$ such that $fx_j \neq gx_j$ there exists $i < j$ such that $fx_i \neq gx_i$ ($\langle x_1, \ldots, x_{j-1} \rangle$ is as monic as $\langle x_1, \ldots, x_{j-1}, x_j \rangle$). Borrowing from the notation conventionally used in exterior algebra, we use $\langle T; x_1, \ldots \hat{x}_j, \ldots, x_n \rangle$ to denote the family obtained by deleting x_j. If x_j is a short column, then $\langle T; x_1, \ldots \hat{x}_j, \ldots, x_n \rangle$ is still a table. But notice that just because $\langle T; x_1, \ldots, \hat{x}_j, \ldots, x_n \rangle$ is still a table does not imply that x_j is short column. (For any $f: A \to B$, $\langle A; f, 1 \rangle$ and $\langle A; \hat{f}, 1 \rangle$ are tables but f need not be short.)

Given tables $\langle T; x_1, \ldots, x_m \rangle$ and $\langle T'; y, \ldots y_n \rangle$ where $T' = x_j\Box$ we define their COMPOSITION at j as

$$\langle T; x_1, \ldots, x_m \rangle \underset{j}{\circ} \langle T'; y_1, \ldots, y_n \rangle$$
$$= \langle T; x_1, \ldots, x_{j-1}, x_j y_1, \ldots, x_j y_n, x_{j+1}, \ldots, x_m \rangle .$$

1.491. A τ-CATEGORY is a cartesian category with a distinguished class of tables, denoted here as τ such that:

τ1. Every table is isomorphic to a unique table in τ.

τ2.1. $\langle T, 1_T \rangle \in \tau$, all T.

τ2.2. If $\langle T; x_1, \ldots, x_m \rangle \in \tau$ and $\langle T'; y_1, \ldots, y_n \rangle \in \tau$ and $T' = x_j\Box$, then

$$\langle T; x_1, \ldots, x_m \rangle \underset{j}{\circ} \langle T'; y_1, \ldots, y_n \rangle \in \tau .$$

τ3. If $\langle T; x_1, \ldots, x_n \rangle \in \tau$ and x_j is a short column, then

$$\langle T; x_1, \ldots, \hat{x}_j, \ldots, x_n \rangle \in \tau .$$

Axiom 1 says that τ is a set of representatives, 2 that it is closed under 'composition', 3 that it is closed under the operations of pruning short columns.

1.492. Given a table $\langle T; x_1, \ldots, x_n \rangle$, we will say that a subsequence $i_1, \ldots, i_m$ is SUPPORTING if $\langle T; x_{i_1}, \ldots, x_{i_m} \rangle$ still satisfies the monic condition, and we will call the latter table a PRUNE of the original. Note that one can prune other than short columns.

1.493. An example of a τ-category – and as we will see, a generic example – is the category of ordinal lists, which will here be described as the category whose objects are von Neumann ordinals and whose morphisms are all functions. We define τ to be the class of tables $\langle T; f_1, \ldots, f_n \rangle$ such that for $x, y \in T$ and $j = \min\{i \mid f_i(x) \neq f_i(y)\}$ it is the case that $f_j(x) < f_j(y)$.

We can easily restrict to any initial section or ordinals closed under ordinal multiplication. Two cases will be of particular interest: **F** the finite ordinals and **P** the ω-polynomials (i.e. those less then ω^ω).

1.494. Axiom 1 says that for any table $\langle T; x_1, \ldots, x_n \rangle$ there exists a unique isomorphism $g: T' \to T$ such that $\langle T'; gx_1, \ldots, gx_n \rangle \in \tau$. We will call g the RESURFACING of $\langle T; x_1, \ldots, x_n \rangle$. A table is in τ iff its resurfacing is an identity map. A more algebraic definition of τ-structure is available by taking a sequence of partial operations $\tau_0, \tau_1, \ldots \tau_n, \ldots$ where τ_n assigns resurfacings to tables of length n.

If $f: A \to B$ is an isomorphism, then the resurfacing of $\langle A; f \rangle$ is f^{-1}. Hence the only isomorphisms that appear as one-column tables in τ are identity maps.

Axiom 3 yields its own converse: Suppose that $\langle T; x_1, \ldots, x_n \rangle$ is a table with resurfacing $g: T' \to T$ and with a short column x_j. Then $\langle T'; gx_1, \ldots, \widehat{gx_j}, \ldots, gx_n \rangle \in \tau$. Hence g is also the resurfacing of $\langle T; x_1, \ldots, \hat{x}_j, \ldots, x_n \rangle$. And thus

$$\langle T; x_1, \ldots, x_n \rangle \in \tau \quad \textit{iff} \quad \langle T; x_1, \ldots, \hat{x}_j, \ldots, x_n \rangle \in \tau .$$

A corollary is that any expansion of a τ-table is a τ-table: If $\langle T; x_1, \ldots, x_n \rangle \in \tau$, then for any x_{n+1} such that $T = \Box x_{n+1}$ we have $\langle T; x_1, \ldots, x_n, x_{n+1} \rangle \in \tau$. In particular, $\langle A; 1, f \rangle \in \tau$ for any $f: A \to B$.

1.495. DIVERSION. Axiom 3 is, of course, asymmetric. If we symmetrized it we would also obtain for any $f: A \to B$ that $\langle A; f, 1 \rangle \in \tau$. If f is an isomorphism, then $\langle A; 1, f \rangle$ and $\langle B; f^{-1}, 1 \rangle$ are isomorphic τ-tables and hence must be the same. The twist-map on $C \times C$, for any object C, is consequently the identity map and the projections are equal, therefore C is a subterminator. That is, a symmetrized axiom 3 occurs only when the category is a semi-lattice.

1.496. If T is a subterminator in a τ-category then $\langle T; f \rangle \in \tau$ for any $f: T \to T'$ (since f is short) and thus if f is an isomorphism it is an identity map. Isomorphic subterminators are equal. In particular, there is a unique terminator.

1.497. THE CANCELLATION LEMMA. *If* $\langle T; x_1, \ldots, x_m \rangle \underset{j}{\circ} \langle T'; y_1, \ldots, y_n \rangle \in \tau$ *and* $\langle T'; y_1, \ldots, y_n \rangle \in \tau$, *then* $\langle T; x_1, \ldots, x_m \rangle \in \tau$.

BECAUSE: Let $g: T'' \to T$ be the resurfacing of $\langle T; x_1, \ldots, x_m \rangle$. By axiom τ2.2, $\langle T''; gx_1, \ldots, gx_m \rangle \underset{j}{\circ} \langle T'; y_1, \ldots, y_n \rangle \in \tau$, which makes g the resurfacing of $\langle T; x_1, \ldots, x_m \rangle \underset{j}{\circ} \langle T'; y_1, \ldots, y_n \rangle \in \tau$. By assumption, the latter is a τ-table, hence $g = 1_T$.

1.498. A τ-category has a CANONICAL CARTESIAN STRUCTURE. We say the product $\langle A \times B;\ p_1, p_2 \rangle$ is canonical if it is a τ-table, and that a pullback

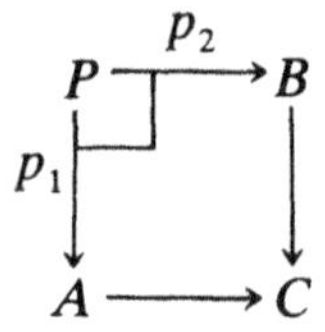

is canonical if $\langle P;\ p_1, p_2 \rangle \in \tau$.

We have already noted that there is a unique terminator.

1.499. DIVERSION. In the category of finite von Neumann ordinals let $\langle mn;\ p_1, p_2 \rangle$ be a canonical product of m and n. $p_1(x)$ is the integer part of $x \div n$ and $p_2(x)$ is the residual of $x \bmod n$ – as any computer compiler knows. It was initially a surprise to the authors that such yields a strictly associative product structure:

$$[[x \div p] \div n] = [x \div (np)]\,,$$
$$[x \div p] \bmod n = [(x \bmod (np)) \div p]\,,$$
$$x \bmod p = (x \bmod (np)) \bmod p\,.$$

The surprise evaporates, however, if we take $m = n = p = 10$ and note that $p_i(x)$ is the ith digit of x. In the general case the projections, in concert, yield the 'encode' function in APL $(n\ m\ p)\perp x$.

1.49(10). LEMMAS FOR τ-CATEGORIES. *Canonical products are strictly associative. The terminator is strictly a two-sided unit. If each square in*

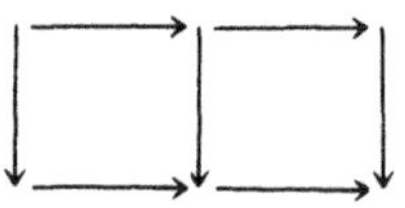

is a canonical pullback then so is the rectangle.

BECAUSE: (There is a metatheorem to come that says that it suffices to prove such theorems for **P**, the category of ω-polynomials.)

Let

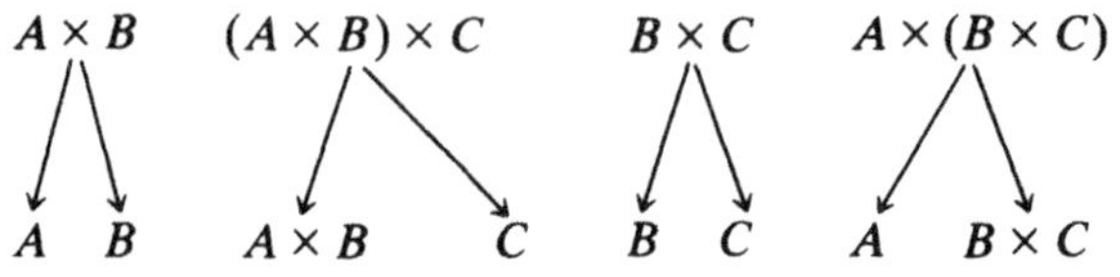

be canonical products. By 2.2, therefore, both:

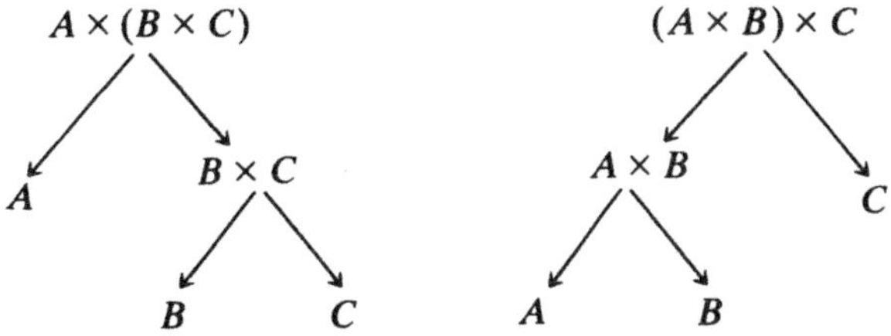

are in τ. But they are isomorphic, hence they are equal.

Let 1 denote the unique terminator. If

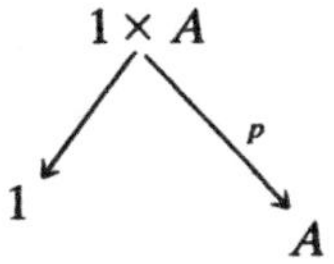

is in τ, then since $1 \times A \to 1$ is a short column we have that $\langle 1 \times \mathrm{A}; \mathrm{p} \rangle \in \tau$. But p is an isomorphism, hence $p = 1_A$. Similarly

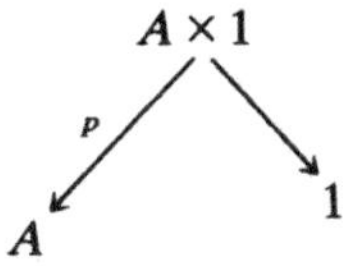

in τ implies $p = 1_A$.

Let the squares in

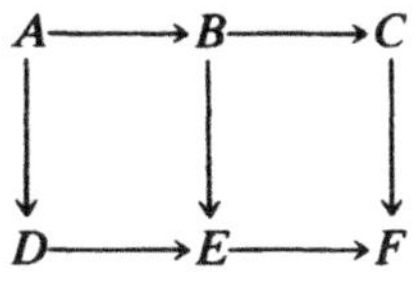

be canonical pullbacks. Then:

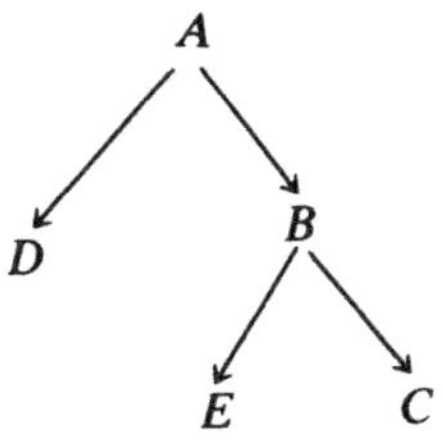

is in τ. But the middle column is a short column (since it factors through $A \to D$) and can be pruned leaving

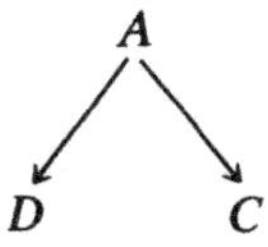

in τ.

1.49(11). A sequence $\langle T; x_1, \ldots, x_n \rangle$, monic or not, will be called AUSPICIOUS if it can be expanded to a τ-table, i.e. if there exists $\langle T; x_1, \ldots, x_n, x_{n+1}, \ldots, x_m \rangle \in \tau$. Note that in the category of von Neumann ordinals a single map is auspicious iff it preserves order.

LEMMAS FOR τ-CATEGORIES. (i) *$\langle T; x_1, \ldots, x_n \rangle$ is auspicious iff $\langle T; x_1, \ldots, x_n, 1 \rangle \in \tau$.*

(ii) *If $f\colon A \to B$ and $g\colon B \to C$ are auspicious, then so is fg.*

(iii) *If*

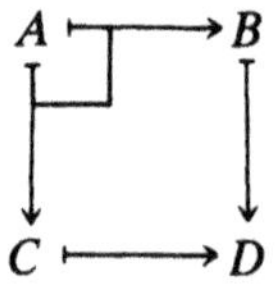

is canonical and $B \mapsto D$, $C \mapsto D$ are auspicious, then so are $A \mapsto C$, $A \mapsto B$, $A \mapsto D$.

(iv) *If the squares of*

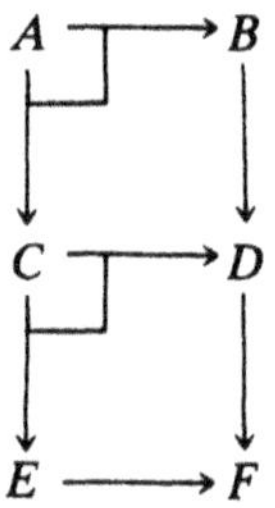

are canonical pullbacks, and if $B \to D$ is auspicious, then the rectangle is a canonical pullback.

BECAUSE: (i) Given $\langle T; x_1, \ldots, x_n, x_{n+1}, \ldots, x_m \rangle \in \tau$ the converse of axiom 3 says that $\langle T; x_1, \ldots, x_n, x_1, \ldots, x_n, x_{n+1}, \ldots, x_m \rangle \in \tau$ [1.494]. But the latter is equal to

$$\langle T; x_1, \ldots, x_n, 1 \rangle \underset{n+1}{\circ} \langle T; x_1, \ldots, x_n, x_{n+1}, \ldots, x_m \rangle$$

and the cancellation lemma [1.497] says that $\langle T; x_1, \ldots, x_n, 1 \rangle \in \tau$.

(ii) If f and g are auspicious, then $\langle f, 1 \rangle \underset{1}{\circ} \langle g, 1 \rangle = \langle fg, f, 1 \rangle \in \tau$ hence fg is auspicious.

(iii) If the displayed *square* is a canonical pullback, then $A \mapsto C$ is auspicious, hence $A \mapsto D$ is auspicious.

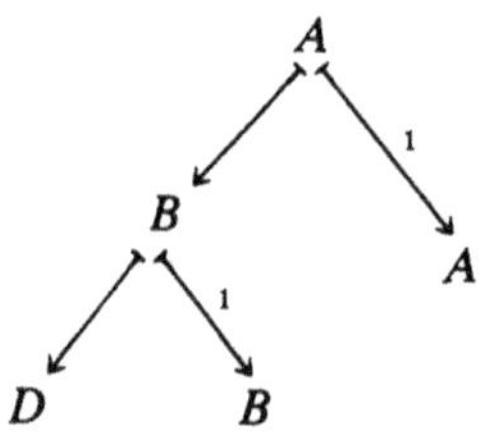

is in τ (all columns but the first are short) and the cancellation property, therefore, says $A \mapsto B$ is auspicious.

(iv) Given the hypotheses for the displayed *rectangle* we have that

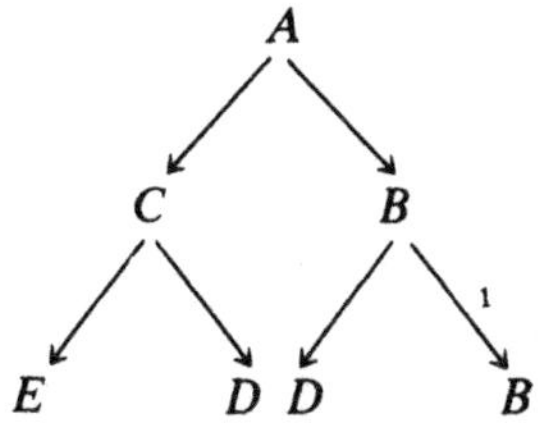

is in τ. The two middle columns are equal, hence one of them is short (the right-hand one) and we obtain that

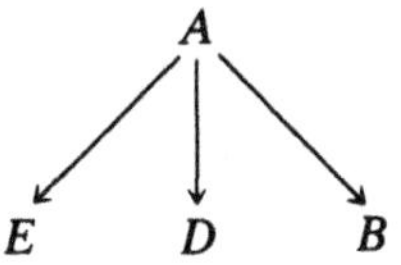

is in τ. We can display such as

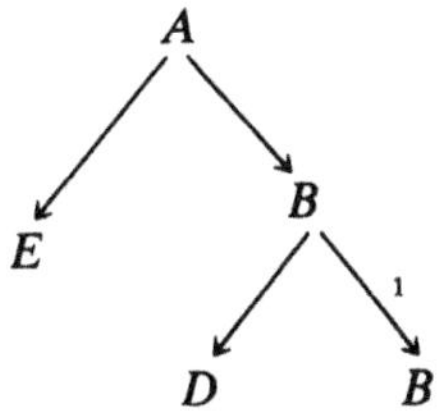

and use cancellation to obtain that

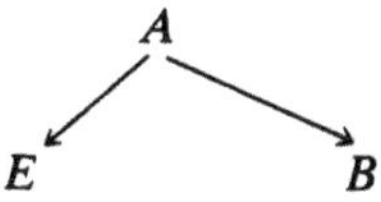

is in τ.

1.4(10). Free τ-categories

Given a cartesian category $\mathbf{A}$ we mean by a FREE τ-CATEGORY a cartesian functor $\mathbf{A} \to \mathbf{A}^\tau$ where $\mathbf{A}^\tau$ is a τ-category, such that for any cartesian functor $\mathbf{A} \to \mathbf{B}$ where $\mathbf{B}$ is a τ-category there exists a unique τ-functor $\mathbf{A}^\tau \to \mathbf{B}$ such that

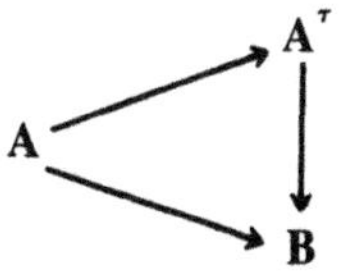

1.4(10)1. *For every small cartesian category* $\mathbf{A}$ *there exists a free* τ*-category* $\mathbf{A} \xrightarrow{F} \mathbf{A}^\tau$ *where* F *is an equivalence functor.*

N.B. We will not use the axiom of choice.

Also: F fails to be strictly one-to-one only on subterminators. (The equivalence-kernel of F is the set of identity maps and isomorphisms between subterminators.)

Given that $\mathbf{A} \to \mathbf{A}^\tau$ is an equivalence functor we are advised to search for an inflation $[P]$ with a cross-section $\mathbf{A} \to [P]$ and an equivalence-kernel $\mathscr{K} \subset [P]$ so that $A \to [P] \to [P]/\mathscr{K}$ is verifiably a free τ-category. [1.367]

We will call a table $\langle T; x_1, \ldots, x_n \rangle$ WELL-MADE if it has no short columns. Let $\mathscr{W}$ be the class of well-made tables, $P: \mathscr{W} \to |\mathbf{A}|$ the 1st coordinate function, $S: |\mathbf{A}| \to \mathscr{W}$ the function such that $SA = \langle A \rangle$ if A is a subterminator, $SA = \langle A; 1 \rangle$ otherwise. Clearly $|\mathbf{A}| \to \mathscr{W} \to |\mathbf{A}|$ is the identity functor and we obtain an inflation with cross-section $\mathbf{A} \to [P]$ [1.36].

Let $\mathscr{K} \subset [P]$ be the class of all isomorphisms of tables: $(\langle T'; x'_1, \ldots, x'_n \rangle) \xrightarrow{f} \langle T; x_1, \ldots, x_m \rangle) \in \mathscr{K}$ iff f is an isomorphism in $\mathbf{A}$, $n = m$ and $x'_i = fx_i$, $i = 1, \ldots, n$. Define $\mathbf{A} \to \mathbf{A}^\tau$ as $A \to [P] \to [P]/\mathscr{K}$. By construction, $\mathbf{A} \to \mathbf{A}^\tau$ is an equivalence functor.

We define first a class of tables τ_1 in $[P]$ (it will not satisfy the axioms). A table:

$$\{\langle T_0; x_1^0, \ldots, x_{n_0}^0 \rangle \xrightarrow{y_i} \langle T_i; x_1^i, \ldots, x_{n_i}^i \rangle\},\ i = 1, \ldots, m$$

in $[P]$ is in τ_1 iff $n_0 = n_1 + \cdots + n_m$ and:

$$\langle x_1^0, \ldots, x_{n_0}^0 \rangle = \langle y_1 x_1^1, y_1 x_2^1, \ldots, y_1 x_n^1, y_2 x_1^2, \ldots, y_{m-1} x_{n_{m-1}}^{m-1}, \\ y_m x_1^m, \ldots, y_m x_{n_m}^m \rangle .$$

τ_1 contains only well-made tables (using a cancellation lemma on such) and every well-made table is isomorphic to a τ_1 table. $\langle T_0; x_1^0, \ldots, x_{n_0}^0 \rangle$ is isomorphic in $[P]$ to $\langle T_0; yx_1^1, \ldots, y_m x_{n_m}^m \rangle$ regardless of $x_1^0, \ldots, x_{n_0}^0$. The uniqueness condition fails for τ_1 but it is the case that isomorphisms between τ_1 tables lie in $\mathscr{K}$. Hence if we define τ_2 as the class of tables in $[P]/\mathscr{K}$ with ancestors in τ_1 we gain a uniquely representative family of well-made tables. Every identity map on a non-subterminator is in τ_1 hence the same for τ_2. τ_1 satisfies axiom 2.2 hence (using finite choice) so does τ_2. We now treat $\mathbf{A}^\tau$ abstractly: it is a cartesian category with a representative class τ_2 of well-made tables satisfying axioms 2.1 for non-subterminators and axiom 2.2.

Given a table $\langle T; x_1, \ldots, x_n \rangle$ in any category we consider its WELL-MADE PART defined as the result of pruning all short columns. Define τ_3 in $\mathbf{A}^\tau$ as the class of tables whose well-made parts lie in τ_2. τ_3 easily satisfies axioms 1, 2.1, and 3. For axiom 2.2 one need only notice that the composition of well-made parts yields the well-made part of the composition. We take τ_3 for the τ-structure on $\mathbf{A}^\tau$.

Given a cartesian functor G: $\mathbf{A} \to \mathbf{B}$ where $\mathbf{B}$ is a τ-category we obtain a functor $\bar{G}: [P] \to \mathbf{B}$ as follows: Given an object $\langle T; x_1, \ldots, x_n \rangle$ in $[P]$ let $\bar{G}\langle T; x_1, \ldots, x_n \rangle \to GT$ be the resurfacing of $\langle GT; Gx_1, \ldots, Gx_n \rangle$. Given a map $\langle T'; x'_1, \ldots, x'_n \rangle \xrightarrow{y} \langle T; x_1, \ldots, x_m \rangle$ define $\bar{G}'(T' \xrightarrow{y} T)$ as the unique map such that:

$$\begin{array}{ccc} \bar{G}\langle T'; x'\rangle & \xrightarrow{\bar{G}(T' \xrightarrow{y} T)} & \bar{G}\langle T, x\rangle \\ \downarrow & & \downarrow \\ GT' & \xrightarrow[Gy]{} & GT \end{array}$$

$\bar{G}$ carries τ_1-tables into τ-tables in **B** (using axiom 2.2 on **B**), and hence clearly carries $\mathscr{K}$ into $|\mathbf{B}|$ and we obtain a factorization

$$\begin{array}{ccc} [P] & \longrightarrow & \mathbf{A}^\tau \\ & {\scriptstyle \bar{G}} \searrow & \downarrow {\scriptstyle G'} \\ & & \mathbf{B} \end{array}$$

Clearly G': $\mathrm{A}^\tau \to \mathbf{B}$ carries τ_2-tables into τ-tables.

To prove that G': $\mathbf{A}^\tau \to \mathbf{B}$ carries τ_3-tables into τ-tables we note first that a cartesian functor preserves shortness of columns. (This is the critical use of cartesianness.) Given $\langle T; x_1, \ldots, x_n\rangle$ let $E_i \rightarrowtail T \times T$ be an equalizer of ℓx_i and $r x_i$, each i. Let $F_0 \to T \times T$ be the identity map, let

$$\begin{array}{ccc} F_i & \longrightarrow & F_{i-1} \\ \downarrow & & \downarrow \\ E_i & \longrightarrow & T \times T \end{array}$$

be a pullback, $i = 1, \ldots, n$. x_j is short iff $F_j \to F_{j-1}$ is an isomorphism. Hence if x_j is short and H is a cartesian functor, then Hx_j is short.

In any τ-category a table is in τ iff its well-made part is in τ. Suppose H: $\mathbf{A}^\tau \to \mathbf{B}$ is any cartesian functor between τ-categories that preserves well-made τ-tables. Given a τ-table $\langle T; x_1, \ldots, x_n\rangle$ in $\mathbf{A}^\tau$ let $\langle T; x_{i_1}, \ldots, x_{i_m}\rangle$ be its well-made part. The well-made part of $\langle HT; Hx_{i_1}, \ldots, Hx_{i_m}\rangle$ is given, therefore, by a further subsequence of $i_1, \ldots, i_m$. Since we know that $\langle HT; Hx_{i_1}, \ldots, Hx_{i_m}\rangle$ is a τ-table we know that its well-made part is a τ-table, therefore the well-made part of $\langle HT; Hx_1, \ldots, Hx_n\rangle$ is a τ-table and hence $\langle HT; Hx_1, \ldots, Hx_n\rangle$ is a τ-table.

The uniqueness of

$$\begin{array}{ccc} & & \mathbf{A}^\tau \\ & {\scriptstyle F} \nearrow & \\ \mathbf{A} & & \downarrow {\scriptstyle G'} \\ & {\scriptstyle G} \searrow & \\ & & \mathbf{B} \end{array}$$

given G: $\mathbf{A} \to \mathbf{B}$, is obtained by noting that for every object A in $\mathbf{A}^\tau$ there exists a table $\langle T; x_1, \ldots, x_n\rangle$ in **A** and an isomorphism $A \to FT$ which is the resurfacing of $\langle FT; Fx_1, \ldots, Fx_n\rangle$. $G'A \to G'FT$ is therefore determined. Given a map

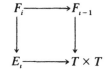

$A' \xrightarrow{y} A$ in $\mathbf{A}^\tau$, let $\langle T'; x_1', \ldots, x_n' \rangle$ be a table in $\mathbf{A}$ such that $A' \to FT'$ is a resurfacing. Since F is full and faithful there exists unique $T' \to T$ in $\mathbf{A}$ such that

$$\begin{array}{ccc} A' & \xrightarrow{y} & A \\ \downarrow & & \downarrow \\ FT' & \longrightarrow & FT \end{array}$$

and hence $G'y$ is determined.

1.4(11). Canonical slices

The last section says, in effect, that problems about cartesian functors between cartesian categories can be safely translated into problems about τ-functors between τ-categories. Since $\mathbf{A} \to \mathbf{A}^\tau$ is an equivalence functor, we can safely add any number of further conditions to the word 'cartesian', in particular any further condition described by any diagrammatic sentences and – on the functors – by diagrammatic predicates.

Given an object B in a category $\mathbf{A}$, let the proto-morphisms of $\mathbf{A}/B$ be the morphisms of $\mathbf{A}$, let objects of $\mathbf{A}/B$ be the $\mathbf{A}$-morphisms into B, and let the source-target predicate be given by $(f) \xrightarrow{x} (g)$ iff $f = xg$. If $\mathbf{A}$ is a τ-category we mean by $\mathbf{A}/B$ the full subcategory obtained by restricting to *auspicious* morphisms into B.

1.4(11)1. DIVERSION. For $\mathbf{A}$ a τ-category let $(\mathbf{A}/B)^a$ – for the purposes of this diversion — denote the full subcategory of auspicious morphisms in the bigger category $\mathbf{A}/B$. Let $\mathscr{O}$ be the class of pairs $\langle A' \xrightarrow{g} A \xrightarrow{f} B \rangle$ where g is an isomorphism and f is auspicious, and define $P: \mathscr{O} \to |(\mathbf{A}/B)^a|$ as the second coordinate function. Then $\mathbf{A}/B$ is isomorphic (and canonically so) to $[P]$. Note that P has an obvious cross-section, hence $(\mathbf{A}/B)^a$ and A/B are strongly equivalent [1.36].

1.4(11)2. We restrict attention to τ-categories $\mathbf{A}$, emphasizing the just stated convention: the objects of $\mathbf{A}/B$ are auspicious morphisms, $\mathbf{A}/B$ inherits a τ-structure from the forgetful functor $\Sigma: \mathbf{A}/B \to \mathbf{A}$. We define τ/B as $\Sigma^{-1}(\tau)$ together with all columnless tables in $\mathbf{A}/B$. (Note that $\Sigma^{-1}(\tau)$ contains $\langle (1_B); 1_B \rangle$ but does not – unlike a good τ-structure – contain the columnless table with (1_B) as top.) It is routine to verify the axioms for τ/B.

$\Sigma: \mathbf{A}/1 \to \mathbf{A}$ is a τ-isomorphism and we will rather freely confuse $\mathbf{A}$ and $\mathbf{A}/1$.

For $f: B_1 \to B_2$ in $\mathbf{A}$ we define $f^\#: \mathbf{A}/B_2 \to \mathbf{A}/B_1$ as follows: an object $(g) \in \mathbf{A}/B_2$ is sent to $(\bar{g}) \in \mathbf{A}/B_1$, where

$$\begin{array}{ccc} \cdot & \longrightarrow & \cdot \\ {\scriptstyle \bar{g}}\downarrow & & \downarrow{\scriptstyle g} \\ \cdot & \xrightarrow[f]{} & \cdot \end{array}$$

is a canonical pullback; a morphism $(g_2) \xrightarrow{x} (g_1)$ in $\mathbf{A}/B_2$ is sent to $(\bar{g}_2) \xrightarrow{\bar{x}} (\bar{g}_1)$ where the bottom square and the outer rectangle

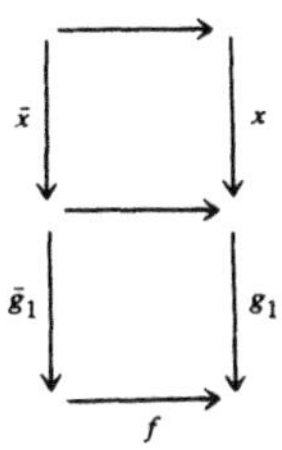

are canonical pullbacks (the other square need not be canonical). From the horizontal pasting property for canonical pullbacks we know that $(f_1 f_2)^\# = (f_2)^\#(f_1)^\#$. This *equality* (instead of an isomorphism of functors) is our motivation for τ-categories.

Δ: $\mathbf{A} \to \mathbf{A}/B$ is the functor that sends A to:

$$\begin{array}{c} B \times A \\ \Big\downarrow \ell \\ B \end{array}$$

where $\langle B \times A, \ell, r \rangle$ is the canonical product. We note that it is the unique functor such that

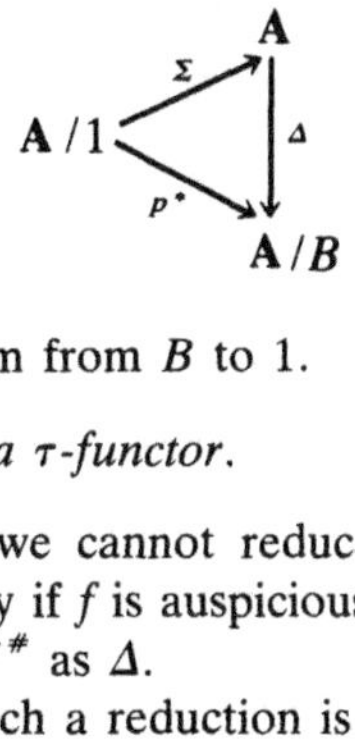

where p is the unique morphism from B to 1.

1.4(11)3. *$f^\#$: $\mathbf{A}/B_2 \to \mathbf{A}/B_1$ is a τ-functor.*

NOTE. For the τ-preservation we cannot reduce in the customary way to the special case where $B_2 = 1$. Only if f is auspicious can we make the identification $\mathbf{A}/B_1 = (\mathbf{A}/B_2)/(f)$ and view $f^\#$ as Δ.

For the cartesianness of f, such a reduction is still possible. Let

$$\begin{array}{ccc} B_1' & \xrightarrow{u} & B_1 \\ {\scriptstyle f'}\Big\downarrow & & \Big\downarrow{\scriptstyle f} \\ B_2 & \xrightarrow[1]{} & B_2 \end{array}$$

be a canonical pullback. Then $f^\# = (f')^\#(u^{-1})^\#$ and since $(u^{-1})^\#$ is an isomorphism of categories (because $(-)^\#$ is a functor) it suffices to show that $(f')^\#$ is cartesian.

$$\mathbf{A}/B_2 \xrightarrow{\Delta} (\mathbf{A}/B_2)/(f') \xrightarrow{\theta} \mathbf{A}/B_1' \;=\; (f')^\#$$

and Θ is an isomorphism, hence it suffices to show that Δ-functors are cartesian.

We recall the usual argument: $\Delta: \mathbf{A} \to \mathbf{A}/B$ clearly preserves the terminator, and $\mathbf{A} \xrightarrow{\Delta} \mathbf{A}/B \xrightarrow{\Sigma} \mathbf{A}$ preserves pullbacks (that is, $B \times -: \mathbf{A} \to \mathbf{A}$ preserves pullbacks) hence Δ preserves pullbacks because Σ reflects pullbacks.

Even if we knew that Δ is a τ-functor we cannot thus infer that $f^{\#}$ is, unless we know that $(u^{-1})^{\#}$ is a τ-functor. There seems, however, to be no special advantage in the case for isomorphisms. We do note, however, that the τ-preservation by Δ follows from the case for $(-)^{\#}$.

$$\mathbf{A} \xrightarrow{\Delta} \mathbf{A}/B \;\; = \;\; \mathbf{A} \xrightarrow{\Sigma^{-1}} \mathbf{A}/1 \xrightarrow{p^{\#}} \mathbf{A}/B$$

and Σ is a τ-functor.

That $f^{\#}: \mathbf{A}/B_2 \to \mathbf{A}/B_1$ preserves columnless tables is routine. Let $\{(g_0) \xrightarrow{x_i} (g_i)\}_{i=1,\ldots,n}$ be a τ/B_2-table,

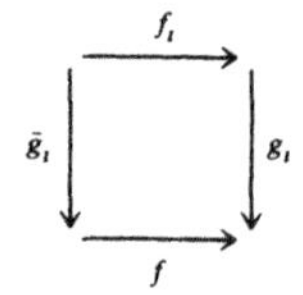

canonical pullbacks in $\mathbf{A}$, and let $\bar{x}_i$ be such that $\bar{x}_i \bar{g}_i = \bar{g}_0$, $\bar{x} \bar{f}_i = \bar{f}_0 x_i$. We wish to show that $\{(\bar{g}_0) \xrightarrow{\bar{x}_i} (\bar{g}_i)\}$ is a τ/B_1-table, equivalently that $\langle \bar{x}_1, \ldots, \bar{x}_n \rangle \in \tau$. By the cancellation lemma it suffices to show that

$$\langle \bar{x}_1 \bar{g}_1, \bar{x}_1 \bar{f}_1, \bar{x}_2 \bar{g}_2, \ldots, \bar{x}_n \bar{g}_n, \bar{x}_n \bar{f}_n \rangle \in \tau.$$

We can delete short columns $(\bar{g}_0 = \bar{x}_1 \bar{g}_1 = \bar{x}_2 \bar{g}_2 = \cdots = \bar{x}_n \bar{g}_n)$ and it suffices to show that $\langle \bar{g}_0, \bar{x}_1 \bar{f}_1, \bar{x}_2 \bar{f}_2, \ldots, \bar{x}_n \bar{f}_n \rangle \in \tau$. But $\bar{x}_i \bar{f}_i = \bar{f}_0 x_i$ and $\langle \bar{g}_0, \bar{x}_1 \bar{f}_1, \ldots, \bar{x}_n \bar{f}_n \rangle = \langle \bar{g}_0, \bar{f}_0 \rangle \underset{1}{\circ} \langle x_1, \ldots, x_n \rangle$. Since $\langle \bar{g}_0 \bar{f}_0 \rangle$ and $\langle x_1, \ldots, x_n \rangle$ are τ-tables (the latter by definition of τ/B_2) we are done.

1.4(11)4. By a POINT we mean a morphism with the terminator as source. Given $\mathbf{A}' \rightarrowtail \mathbf{A}$ we say that a point $1 \to A$ 'is in $\mathbf{A}$' if there exists $1 \to \mathbf{A}'$ such that $1 \to \mathbf{A}' = 1 \to \mathbf{A}' \rightarrowtail A$. We define the GENERIC POINT, $1 \xrightarrow{g} \Delta B$ in $\mathbf{A}/B$ as the morphism such that $(\Sigma 1 \xrightarrow{\Sigma g} \Sigma \Delta B) = (B \xrightarrow{\langle 1,1 \rangle} B \times B)$. The word 'generic' is suggested by two facts: (1) For any proper $B' \rightarrowtail B$ in $\mathbf{A}$, g fails to lie in $\Delta B' \rightarrowtail \Delta B$ and (2):

1.4(11)5. *Every object and morphism in $\mathbf{A}/B$ is obtainable by taking canonical pullbacks of the generic point and morphisms of the form Δx.*

BECAUSE: Given an object $(f) \in \mathbf{A}/B$, that is, given auspicious $f: A \to B$ in $\mathbf{A}$, note that

$$\begin{array}{ccc}
A & \xrightarrow{\;f\;} & B \\
{\scriptstyle \langle f,1 \rangle} \downarrow & & \downarrow {\scriptstyle \langle 1,1 \rangle} \\
B \times A & \xrightarrow{\;1 \times f\;} & B \times B
\end{array}$$

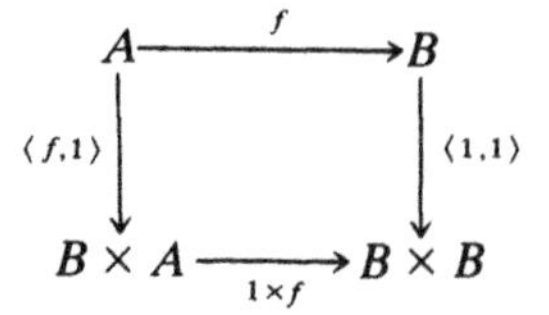

is a pullback: Suppose $A' \to A$ were its resurfacing. Then $A' \to A \xrightarrow{\langle f,1\rangle} B \times A$ would be auspicious, hence a one-column τ-table, and $A' \to A$ would be the resurfacing of $\langle A; \langle f, 1\rangle\rangle$. But since f is auspicious $\langle A; \langle f, 1\rangle\rangle$ is already in τ hence $A' \to A$ is the identity map on A. Thus the above square is a canonical pullback, and

$$\begin{array}{ccc} (f) & \longrightarrow & 1 \\ \downarrow & & \downarrow{\scriptstyle g} \\ \Delta A & \xrightarrow[\Delta f]{} & \Delta B \end{array}$$

is a canonical pullback in $\mathbf{A}/B$.

Given a morphism $(f_1) \xrightarrow{x} (f_2)$ in $\mathbf{A}/B$ we likewise show that

$$\begin{array}{ccc} (f_1) & \xrightarrow{x} & (f_2) \\ \downarrow & & \downarrow \\ \Delta A_1 & \xrightarrow[\Delta x]{} & \Delta A_2 \end{array}$$

is a canonical pullback.

1.4(11)6. *For any τ-functor $F: \mathbf{A} \to \mathbf{B}$ and point $x: 1 \to FB$ there exists a unique τ-functor $F_x: \mathbf{A}/B \to \mathbf{B}$ such that $\Delta F_x = F$ and $F_x(g) = x$.*

BECAUSE: The last lemma easily gives the uniqueness. For the existence note first that $F/B: \mathbf{A}/B \to \mathbf{B}/FB$ defined in the obvious manner $F((f) \xrightarrow{y} (h)) = (Ff) \xrightarrow{Fy} (Fh))$ is a τ-functor and that

$$\begin{array}{ccc} \mathbf{A} & \xrightarrow{\Delta} & \mathbf{A}/B \\ {\scriptstyle F}\downarrow & & \downarrow{\scriptstyle F/B} \\ \mathbf{B} & \xrightarrow[\Delta]{} & \mathbf{B}/FB \end{array}$$

F/B carries the generic point of $\mathbf{A}/B$ to the generic point of $\mathbf{B}/FB$.

$(x)^\#: \mathbf{B}/FB \to \mathbf{B}/1$ carries the generic point of $\mathbf{B}/FB$ to x. Hence $\mathbf{A}/B \xrightarrow{F/B} \mathbf{B}/FB \xrightarrow{x^\#} \mathbf{B}/1 \xrightarrow{\Sigma} \mathbf{B}$ is as required.

1.4(11)7. When we speak of set-valued functors we will not require that the category of sets $\mathscr{S}$ be a τ-category. $(B, -): A \to \mathscr{S}$ will denote the functor represented by B, and $\Gamma = (1, -)$.

1.4(11)8. DIVERSION. Is there a way to construe representable functors as τ-functors? (Except for free τ-categories, we think not.)

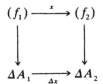

1.4(11)9. Given any τ-functor $F: \mathbf{C} \to \mathbf{B}$ there is a unique transformation $\Gamma(-) \to \Gamma(F(-)$. (It sends $x: 1 \to A$ to $Fx: 1 \to FA$.)

Given τ-functors

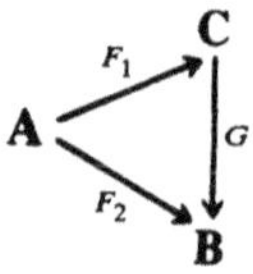

we obtain a transformation $\gamma_G: \Gamma(F_1(-)) \to \Gamma(F_2(-))$ (by sending $x: 1 \to F_1A$ to $Gx: 1 \to F_2A$). Note that $\mathbf{A} \xrightarrow{\Delta} \mathbf{A}/B \xrightarrow{\Gamma} \mathscr{S}$ is isomorphic to $(B, -)$. We may restate the last lemma as:

Given any τ-functor $\mathbf{A} \xrightarrow{F} \mathbf{B}$ and transformation $\eta: \Gamma(\Delta(-)) \to \Gamma(F(-))$ there exists unique τ-functor $G: \mathbf{A}/B \to \mathbf{B}$ such that $\gamma_G = \eta$.

1.4(12). A metatheorem for τ-categories

By a TERM we mean a sequence of morphism-variables $x_1, \ldots, x_n$ and a recipe $R(x_1, \ldots, x_n)$ for constructing a morphism using the partial operations on τ-categories: canonical products, equalizers, pullbacks, resurfacing of tables, and the terminator constant. An equation $R(x_1, \ldots, x_n) = R'(x_1, \ldots, x_n)$ will be written only when the axioms of τ-categories force $\Box R = \Box R'$ and $R\Box = R'\Box$.

Let $\mathbf{P}$ denote the τ-category of von Neumann ordinals less than ω^ω. We may explicitely describe $\mathbf{P}$ as follows: consider, first, the set W of words $\langle a_1, \ldots, a_n \rangle$ of natural numbers, subject to the condition that $a_1 \neq 0$. Well-order W by $\langle a_1, \ldots, a_n \rangle \leq \langle b_1, \ldots, b_m \rangle$ if $n < m$ or if $n = m$ then for some $i \leq n$, $a_j = b_j$ all $j < i$ and $a_i < b_i$.

We may take the objects of $\mathbf{P}$ to be the initial proper segments of W. The morphisms, recall, are all functions.

1.4(12)1. METATHEOREM. *An equation is true for all τ-categories iff it is true for* $\mathbf{P}$.

Given $R(x_1, \ldots, x_n) = R'(x_1, \ldots, x_n)$ note first that if we take $R''(x_1, \ldots, x_n)$ to be the canonical equalizer of $R(x_1, \ldots, x_n)$ and $R'(x_1, \ldots, x_n)$, then the equation is equivalent to $R''(x_1, \ldots, x_n) = (R''(x_1, \ldots, x_n))\Box$. We may then specialize to such equations.

If an equation has a counterexample in any τ-category it has a counterexample in a countable τ-category, namely the τ-subcategory generated by the counterexample (note that we do not need the axiom of choice). Suppose then that $\mathbf{A}$ is countable, $x_1, \ldots, x_n \in \mathbf{A}$ and $R(x_1, \ldots, x_n) \not\in |\mathbf{A}|$. Let $\mathbf{A} \to \mathbf{A}^\tau$ be the free τ-category of $\mathbf{A}$. We shall view $\mathbf{A}$ as a subcategory of $\mathbf{A}^\tau$ (it is not a sub-τ-category). If $R_{\mathbf{A}^\tau}(x_1, \ldots, x_n) \in |\mathbf{A}^\tau|$ then from the induced $\mathbf{A}^\tau \to \mathbf{A}$ we would obtain $R_{\mathbf{A}}(x_1, \ldots, x_n) \in |\mathbf{A}|$.

Given any ω-ordering of $\mathbf{A}$ and object $B \in \mathbf{A}$ we may interpret the values of $(B, -)$ as objects in $\mathbf{P}$ and obtain a τ-functor $\mathbf{A}^\tau \to \mathbf{P}$.

1.4(12)2. The metatheorem follows from:

If $\mathbf{A}$ *is countable, then for any morphism* $f \not\in |\mathbf{A}^\tau|$ *there exists an* ω*-ordering of* $\mathbf{A}$ *and an object* $B \in \mathbf{A}$ *such that* $(B, f) \not\in |\mathbf{P}|$.

We will find it useful to recall the construction of $\mathbf{A}^\tau$ as $[P]/\mathscr{K}$ [1.4(10)1].

Given $\langle T; x_1, \ldots, x_n\rangle \xrightarrow{f} \langle T'; y_1, \ldots, y_m\rangle$ in $[P]$ we first note that if f is not a isomorphism, then standard techniques say that either (T, f) or (T', f) fails to be an isomorphism. We assume therefore that f is an isomorphism but not in $\mathscr{K}$, that is, it is not the case that $\langle x_1, \ldots, x_n\rangle = \langle fy_1, \ldots, fy_m\rangle$. It suffices to consider the case $(\langle T; x_1, \ldots, x_n\rangle \xrightarrow{1} \langle T; z_1, \ldots, z_m\rangle) \not\in \mathscr{K}$ (where $z_i = fy_i$). For $B \in \mathbf{A}$, $(B, \langle T; x_1, \ldots, x_n\rangle) \in \mathbf{P}$ may be constructed as the set (B, T) with a particular well-ordering, namely that induced by $\langle (B, x_1), (B, x_2), \ldots, (B, x_n)\rangle$. We wish to find B and an ω-ordering on $\mathbf{A}$ such that the induced orderings on (B, T) by $\langle (B, x_1), \ldots, (B, x_n)\rangle$ and $\langle (B, z_1), \ldots, (B, z_m)\rangle$ are different.

Suppose $n \leqslant m$. If $x_i = z_i$ for all $i \leqslant n$, then since $\langle T; z_1, \ldots, z_m\rangle$ is a well-made table, $n = m$. There must exist $j \leqslant n$ such that $x_j \neq z_j$, and we may take j to be the smallest such, that is, $x_i = z_i$ for all $i < j$.

Let $B \mapsto T \times T$ be the level of $\langle x_1, \ldots, x_{j-1}\rangle$ (denoted F_{j-1} in the proof that $\mathbf{A}^\tau$ is free). That is, $p_1 x_i = p_2 x_i$ for all $i < j$, and for any $f, g\colon X \to T$ such that $fx_i = gx_i$ for all $i < j$ there exists unique:

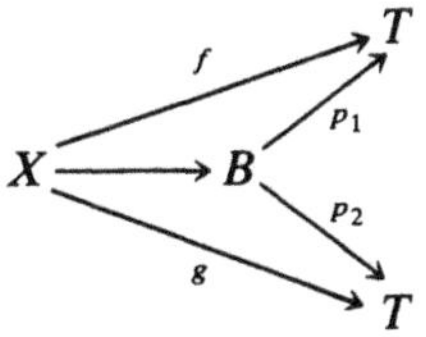

Because x_j is not a short column there exists such $f, g\colon x \to T$ with the further property: $fx_j \neq gx_j$. Hence $p_1 x_j \neq p_2 x_j$ and similarly $p_1 z_j \neq p_2 z_j$.

Let $q\colon T \to B$ be the unique morphism such that $qp_1 = qp_2$. If $p_1 x_j$ were equal to $p_1 z_j$, then $qp_1 x_j = qp_2 z_j$, contradicting our assumption on j. Similarly $p_1 x_j \neq p_2 z_j$, $p_2 x_j \neq p_1 z_j$, $p_2 x_j \neq p_2 z_j$.

$\mathbf{A}$ is countable, that is, has an ω-ordering. By a simple transposition at most, we can obtain an ω-ordering such that $(p_1 x_j) < (p_2 x_j)$ but $(p_2 z_j) < (p_1 z_j)$. The induced orderings on (B, T) therefore differ on p_1 and p_2.

1.4(12)3. A fairly obvious τ-structure exists for $\mathbf{P}^\circ$. We say that a disjoint union $\{f_i\colon A_i \mapsto B\}_{i=1,\ldots,n}$ is canonical if for $x \in \mathrm{Im}(f_i)$, $y \in \mathrm{Im}(f_j)$ and $x < y$ then either $i < j$ or $i = j$ and $x < y$ according to the induced ordering from f_i.

We say that an onto function $g\colon A \to B$ is co-auspicious if for all $x < y$ in B, $\min(g^{-1}(x)) < \min(g^{-1}(y))$. Define τ^* in $\mathbf{P}^\circ$ as the class of co-tables $\langle f_1, \ldots, f_n; T\rangle$ such that the induced map from the canonical disjoint unions of $\Box f_1, \ldots, \Box f_n$ to B is co-auspicious.

If we stick to canonical sums and products then $(A + B) \times C = (A \times C) + (B \times C)$. (But not $A \times (B + C) = (A \times B) + (A \times C)$.)

1.5. REGULAR CATEGORIES

1.51. A subobject represented by $B' \rightarrowtail B$ ALLOWS $f\colon A \to B$ iff there exists

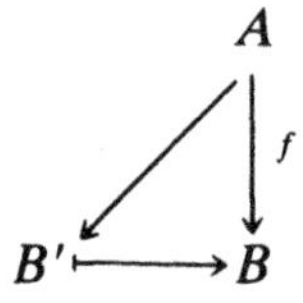

In a category with pullbacks, B' allows f iff $f^{\#}(B')$ is entire, that is, $f^{\#}(B')$ is all of A. In this case, a faithful functor which preserves pullbacks, reflects allowance.

The IMAGE of f, if it exists, is the smallest subobject that allows f. We say that a category *has images* if every morphism has an image. In that case we obtain an order-preserving function $f\colon \mathcal{S}ub(A) \to \mathcal{S}ub(B)$ that sends a subobject represented by $A' \rightarrowtail A$ to the image of $A' \rightarrowtail A \xrightarrow{f} B$.

If, further, the category has pullbacks, then we obtain our first example of an *adjoint pair of functions between posets*,

$$f(A') \subset B' \quad \textit{iff} \quad A' \subset f^{\#}(B')\,.$$

(f is called the LEFT-ADJOINT of $f^{\#}$. $f^{\#}$ is called the RIGHT-ADJOINT of f.) The adjointness relation is equivalently stated:

$$f(A') \textit{ is the minimal } B' \subset B \textit{ such that } A' \subset f^{\#}(B')$$

or:

$$f^{\#}(B') \textit{ is the maximal } A' \subset A \textit{ such that } f(A') \subset B'\,.$$

1.511. *If* $\mathbf{A}$ *has images and if* $T\colon \mathbf{A} \to \mathbf{B}$ *is faithful and preserves images, then* T *reflects images.*

1.512. $A \to B$ is a COVER if its image is entire. We shall denote such, when convenient, by $A \twoheadrightarrow B$:

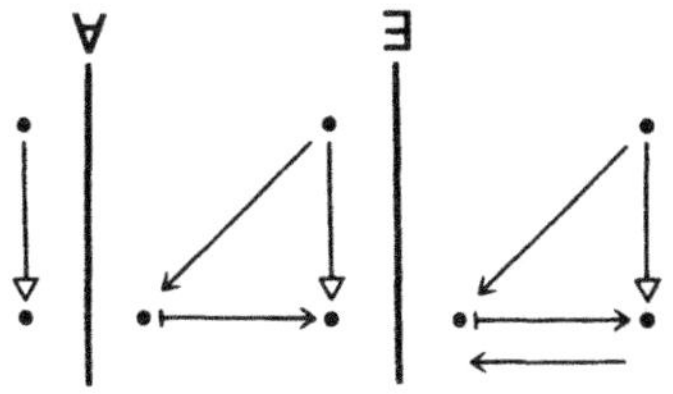

The class of covers is closed under composition and left cancellation (if $A \to B \to C$ is a cover, then so must be $B \to C$). Note that *a monic cover is an isomorphism*.

1.513. A collection of morphisms $\{A_i \to B\}$ with a common target is said to cover if no proper subobject of B allows all of them. When convenient we shall indicate a covering pair by

:

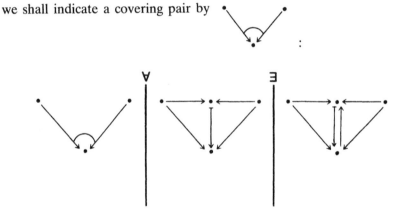

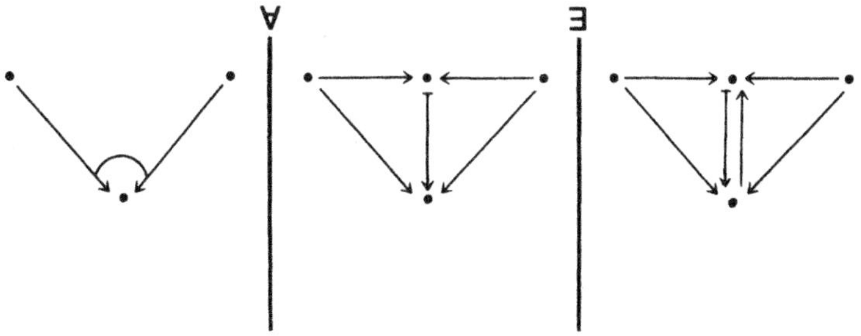

1.514. A collection $\{A_i \to B\}$ is EPIC (elsewhere, *epimorphic*,) if they collectively cancel (if it is a monic family in the opposite category). In a category with equalizers, cover implies epic. The converse, whenever it holds, is a special property. Note that in a poset viewed as a category all morphisms are epic, but the only covers are identities. In a category of equationally defined algebras (groups, rings, etc.), cover coincides with onto (in other words, the forgetful functor preserves and reflects covers.) In the category of groups, epic does imply cover (not entirely trivially), but not in the category of monoids: the inclusion of the natural numbers (under addition) into the integers is epic but not a cover.

1.52. A REGULAR CATEGORY is a cartesian category with images and in which pullbacks transfer covers. Diagrammatically

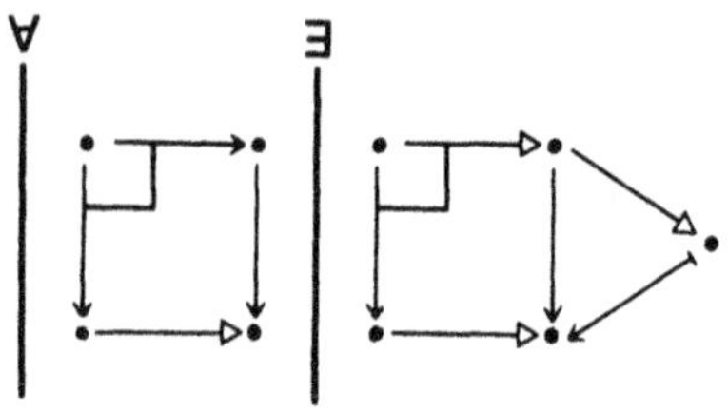

For technical reasons (that is, for reasons other than the existence of interesting examples) we will prove statements about PRE-REGULAR categories: cartesian categories with or without images in which pullbacks transfer covers.

A representation of pre-regular categories is a functor which preserves the defining structure: finite products, equalizers, covers. Note that such necessarily preserves whatever images may exist.

(1.521.) For small **A**, the category of functors, $\mathscr{S}^{\mathbf{A}}$, is regular and the evaluation functors form a collectively faithful family of representations of regular categories.

$\mathscr{LH}$ has images and pullbacks in $\mathscr{LH}$ transfer covers. The forgetful functor $\mathscr{LH} \to \mathscr{S}$ preserves and reflects pullbacks and covers.

For any space Y the category of Lazard sheaves, $\mathscr{Sh}(Y)$, is regular. Define $F\colon \mathscr{Sh}(Y) \to \mathscr{S}$ as $\mathscr{Sh}(Y) \xrightarrow{\Sigma} \mathscr{LH} \to \mathscr{S}$. F is faithful and preserves pullbacks and covers. If we factor F as $\mathscr{Sh}(Y) \xrightarrow{F'} \mathscr{S}/F(1) \xrightarrow{\Sigma} \mathscr{S}$ where F' preserves the terminator [1.44], F' is a faithful representation of regular categories. $F(1)$ is the same as $|Y|$, the underlying set of Y. We obtain a faithful representation $\mathscr{Sh}(Y) \to \mathscr{S}/|Y| \to \mathscr{S}^{|Y|}$. The y-th coordinate functor for $y \in Y$ is the y-th STALK-FUNCTOR: it sends an object $X \xrightarrow{f} Y$ to the discrete set $f^{\#}(y)$.

1.522. In a regular category the SUPPORT of an object A, denoted $\mathscr{Spt}(A)$, is the image of $A \to 1$. We say that A is WELL-SUPPORTED if $\mathscr{Spt}(A) = 1$. In a pre-regular category, an object A is well-supported if $A \to 1$ is a cover.

1.523. We say that an object A is WELL-POINTED if the collection of all morphisms from 1 to A covers A. To become elementary:

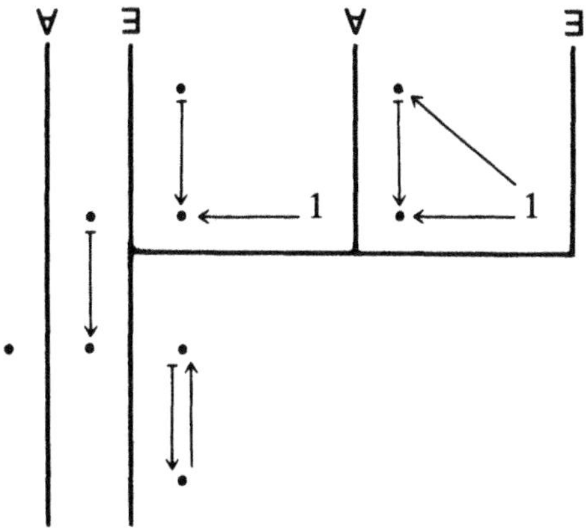

1.524. An object P is PROJECTIVE if the representable functor $(P, -)$ preserves covers. Stated elementarily:

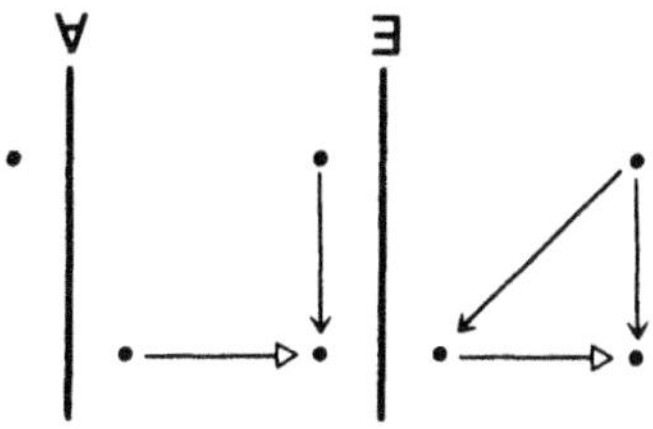

To show that P is projective it suffices to show, in a pre-regular category, that all covers targeted at P have a left inverse:

Given 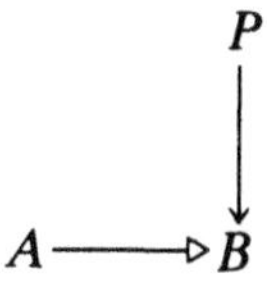

we consider 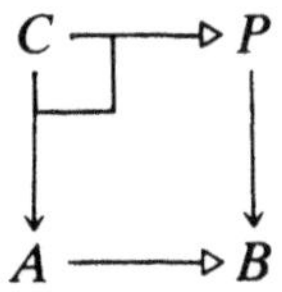

and choose a left-inverse $P \to C$. Then for $P \to A = P \to C \to A$ we have

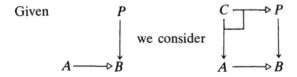

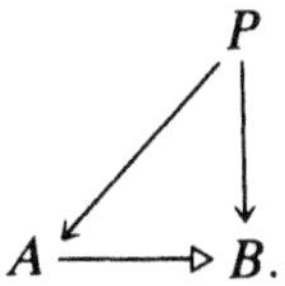

1.525. A pre-regular category is CAPITAL if every well-supported object is well-pointed.

In a capital pre-regular category the terminator is projective.

BECAUSE: Given $A \twoheadrightarrow 1$ we seek $1 \to A$, that is, we wish to show that every well-supported object has a point. Clearly if $A \twoheadrightarrow 1$ is an isomorphism we are done. Hence, if $A \twoheadrightarrow 1$ is monic we are done. Consider the level [1.454] of $A \twoheadrightarrow 1$

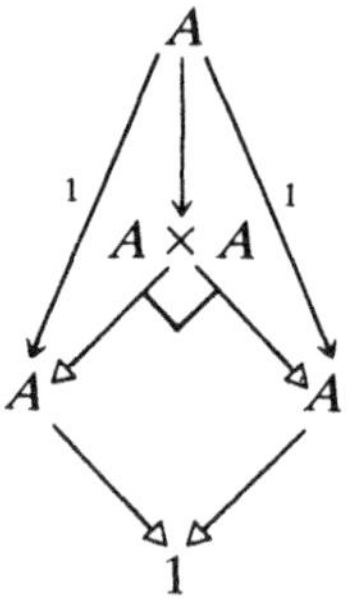

The diagonal $A \to A \times A$ is proper if $A \twoheadrightarrow 1$ is not monic. $A \times A$ is well-supported. Hence, if $A \neq 1$, $A \times A$ has a proper subobject. Thus, there exists $1 \to A \times A$. Hence there exists $1 \to$ A.

1.526. The functor represented by 1 is often denoted Γ. If $\mathbf{A}$ is a capital pre-regular category we have just said that Γ is a representation of pre-regular categories.

1.53. The SLICE LEMMA for regular categories

Given a category $\mathbf{A}$ with finite products and an object $B \in \mathbf{A}$ consider the functor $(B \times -): \mathbf{A} \to \mathbf{A}$. We may factor $(B \times -)$ as $\mathbf{A} \xrightarrow{\Delta} \mathbf{A}/B \xrightarrow{\Sigma} \mathbf{A}$ where $\Delta(1) = (B \xrightarrow{1} B)$ $\Delta: \mathbf{A} \to \mathbf{A}/B$ is the diagonal functor [1.44].

If $\mathbf{A}$ *is (pre-) regular, then so is* $\mathbf{A}/B$.
Δ is a representation of pre-regular categories.
Δ is faithful iff B is well-supported.

BECAUSE: $\mathbf{A}/B$ is cartesian [1.44]. Furthermore [1.531–4]:

1.531. $\Sigma: \mathbf{A}/B \to \mathbf{A}$ preserves and reflects covers and pullbacks which quite sufficies to show that $\mathbf{A}/B$ is pre-regular. Given a morphism

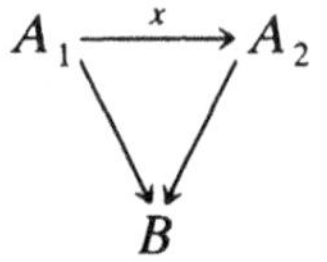

in $\mathbf{A}/B$, if x factors in $\mathbf{A}$ as $A_1 \twoheadrightarrow A_3 \rightarrowtail A_2$, then we obtain a factorization

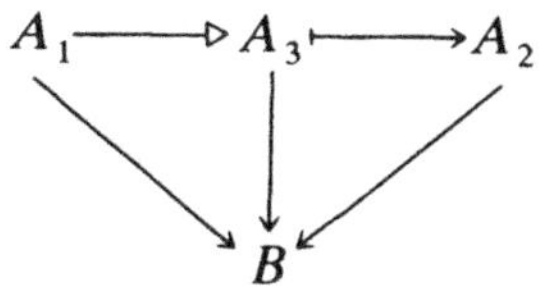

in $\mathbf{A}/B$. That is, if $\mathbf{A}$ has images then so does $\mathbf{A}/B$.

1.532. $(B \times -)$: $\mathbf{A} \to \mathbf{A}$ is easily seen to preserve pullbacks. Therefore, as in [1.442], Δ: $\mathbf{A} \to \mathbf{A}/B$ is a representation of cartesian categories. Since Σ reflects covers it suffices to show that $(B \times -)$ preserves covers in order for Δ to preserve them. In any category,

$$
\begin{array}{ccc}
B \times A_1 & \xrightarrow{1 \times x} & B \times A_2 \\
\downarrow & & \downarrow \\
A_1 & \xrightarrow{\ \ x\ \ } & A_2
\end{array}
$$

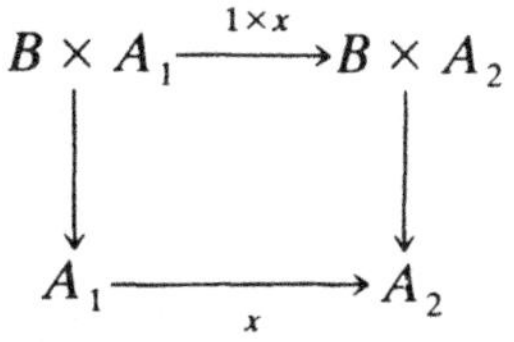

is a pullback. Hence in a pre-regular category, if x covers, then so does $B \times A_1 \xrightarrow{1 \times x} B \times A_2$.

1.533. Since Σ is faithful, Δ is faithful precisely when $(B \times -)$ is faithful. If B is well-supported then from

$$
\begin{array}{ccc}
B \times A & \to & A \\
\downarrow & & \downarrow \\
B & \to & 1
\end{array}
$$

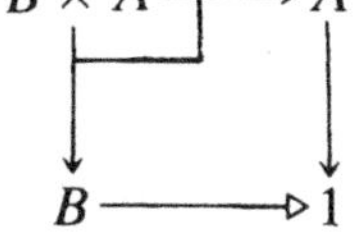

we see that $B \times A \to A$ covers. Suppose that $A' \to A$ is such that $B \times A' \to B \times A$ is an isomorphism. From

$$
\begin{array}{ccc}
B \times A' & \longrightarrow & B \times A \\
\downarrow & & \downarrow \\
A' & \rightarrowtail & A
\end{array}
$$

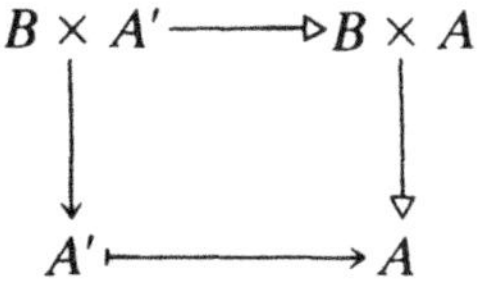

we may conclude that $A' \rightarrowtail A$ covers.

1.534. Suppose that B is not well-supported, that is, there exists $B \to U \rightarrowtail 1$ where U is a proper subobject of 1. Then $B \times U \to B \times 1$ is an isomorphism and hence $\Delta(U) \to \Delta(1)$ is an isomorphism. That is, Δ is not faithful.

1.535. The word 'diagonal' is usually used for a morphism $A \to \Pi_I A$ each of whose coordinates is the identity morphism. For $B \in \mathscr{S}$ consider $\mathscr{S} \to \mathscr{S}/B \to \mathscr{S}^B$ where the second functor is the canonical equivalence of categories. It is not 'the' diagonal functor, but it is conjugate to the standard diagonal functor.

1.54. The CAPITALIZATION LEMMA

We give two arguments, both of which depend on [1.541–5]. The second argument (beginning at [1.547]) avoids the axiom of choice and gains a naturality result, but requires a more elaborate formalization.

1.541. We shall show that every small (pre-)regular category can be faithfully represented in a capital (pre-)regular category. The same reasoning works for a significant sequence of other types of categories and hence we introduce a general setting.

Let $\mathscr{C}$ be a category whose objects are small pre-regular categories (but not, usually, all of them) and whose morphisms are faithful representations thereon (but, again, not usually all of them).

We will say that $\mathscr{C}$ satisfies the *equivalence condition* if for any $\mathbf{A} \in \mathscr{C}$ and equivalence functor $\mathbf{A} \to \mathbf{B}$ it is the case that $\mathbf{B} \in \mathscr{C}$ and $\mathbf{A} \to \mathbf{B} \in \mathscr{C}$.

$\mathscr{C}$ satisfies the *slice condition* if for every $\mathbf{A} \in \mathscr{C}$ and well-supported object $B \in \mathbf{A}$ it is the case that $\mathbf{A}/B \in \mathscr{C}$ and $\mathbf{A} \xrightarrow{\Delta} \mathbf{A}/B \in \mathscr{C}$.

$\mathscr{C}$ satisfies the *union condition* if given a category $\mathbf{A}$ and a non-empty family $\mathscr{F}$ of subcategories ordered by inclusion and such that:

(1) For any $\mathbf{A}', \mathbf{A}'' \in \mathscr{F}$ there exists $\mathbf{A}''' \in \mathscr{F}$ such that $\mathbf{A}' \subset \mathbf{A}'''$, $\mathbf{A}'' \subset \mathbf{A}'''$,

(2) $\mathbf{A} = \bigcup \mathscr{F}$,

(3) $\mathbf{A}' \in \mathscr{C}$ for all $\mathbf{A}' \in \mathscr{F}$, and

(4) $(\mathbf{A}' \subset \mathbf{A}'') \in \mathscr{C}$ for all relevant $\mathbf{A}', \mathbf{A}'' \in \mathscr{F}$,

it is then the case that $\mathbf{A} \in \mathscr{C}$ and $(\mathbf{A}' \subset \mathbf{A}) \in \mathscr{C}$ for all $\mathbf{A}' \in \mathscr{F}$.
$\mathbf{A}$ is said to be a *directed union* of $\mathscr{F}$.

1.542. If $\mathscr{C}$ is the category of all pre-regular or all regular small categories, then the equivalence condition and the union condition are easily verified. The slice condition was shown in [1.53].

1.543. *If $\mathscr{C}$ is a category of small pre-regular categories and faithful representations thereon which satisfies the three conditions defined above*:

the equivalence, slice and union conditions, then for every $\mathbf{A} \in \mathscr{C}$ *there exists capital* $\underline{\underline{\mathbf{A}}} \in \mathscr{C}$ *and faithful* $\mathbf{A} \to \underline{\underline{\mathbf{A}}} \in \mathscr{C}$.

BECAUSE: We give two arguments. The first uses the axiom of choice [1.544–6]. The second is choice-free [1.544–5 and 1.547].

1.544. For well-supported $B \in \mathbf{A}$ we shall find it convenient to consider $\mathbf{A}$ to be a subcategory of $\mathbf{A}/B$. One way to do so is to redefine $\mathbf{A}/B$ as follows:

Consider the inflation $\mathbf{A}'$ of $\mathbf{A}$ whose objects are finite sequences of objects, whose forgetful functor is obtained by choosing for each object $\langle A_1, A_2, \ldots, A_n\rangle$ a product $A_1 \times A_2 \times \cdots \times A_n$. The binary product operation on $\mathbf{A}'$ can now be taken as concatenation. We obtain a strict cancellation property: if $B \times A = B \times A'$, then $A = A'$.

We may interpret $\mathbf{A} \to \mathbf{A}/B$ as $\mathbf{A} \to \mathbf{A}' \to \mathbf{A}'/B$ where $\mathbf{A} \to \mathbf{A}'$ is the obvious cross-section, and $\mathbf{A}' \to \mathbf{A}'/B \xrightarrow{\Sigma} \mathbf{A}'$ is the concatenation operation on objects. Under this intrepretation, $\Delta: \mathbf{A} \to \mathbf{A}/B$ separates objects and, if B is well-supported, separates morphisms.

Finally we may once more reinterpret $\Delta: \mathbf{A} \to \mathbf{A}/B$ as the result of replacing the image of Δ with $\mathbf{A}$ itself.

1.545. We say that $\mathbf{A} \subset \mathbf{A}^*$ is a *relative capitalization* if for every proper $B' \rightarrowtail B$ in $\mathbf{A}$, B well-supported, there exists $x: 1 \to B$ in $\mathbf{A}^*$ so that $B' \rightarrowtail B$ does not allow x.

We shall prove that for every $\mathbf{A} \in \mathscr{C}$ there exists a relative capitalization $(\mathbf{A} \subset \mathbf{A}^*) \in \mathscr{C}$. Such will yield the capitalization lemma: by iteration we can construct a sequence of relative capitalizations $\mathbf{A} \subset \mathbf{A}^* \subset \mathbf{A}^{**} \subset \cdots$ and define $\underline{\underline{\mathbf{A}}}$ as their union. The union condition insures that $(\mathbf{A} \subset \underline{\underline{\mathbf{A}}}) \in \mathscr{C}$ and it easily verified that $\underline{\underline{\mathbf{A}}}$ is capital.

1.546. The relative capitalization is also constructed as an ascending union, but not, in general, countable.

For any faithful extension $\mathbf{A} \subset \mathbf{A}_\iota$ define $\mathscr{S}_\iota \subset |\mathbf{A}|$ as the set of objects B such that for every proper $B' \rightarrowtail B \in \mathbf{A}$ there exists $(1 \xrightarrow{x} B) \in \mathbf{A}_\iota$, such that $B' \rightarrowtail B$ does not allow x. $\mathbf{A}_i$ is a relative capitalization of $\mathbf{A}$ iff $\mathscr{S}_i$ includes all well-supported objects of $\mathbf{A}$. Note that if $\mathbf{A}_\iota \subset \mathbf{A}_j$ is a further faithful extension, then $\mathscr{S}_\iota \subset \mathscr{S}_j$.

Well-order the objects of $\mathbf{A}$. Inductively define $\mathbf{A}_\alpha$, by

(0) $\mathbf{A}_0 = \mathbf{A}$.

(1) Given $\mathbf{A}_\alpha$, terminate the construction if $\mathbf{A}_\alpha$ is a relative capitalization of $\mathbf{A}$, otherwise let $B \in \mathbf{A}$ be the first well-supported object *not* in

$\mathscr{G}_\alpha$ (first with respect to the well-ordering on $|\mathbf{A}|$) and take $(\mathbf{A}_\alpha \subset \mathbf{A}_{\alpha+1}) = (\mathbf{A}_\alpha \xrightarrow{\Delta} \mathbf{A}_\alpha/B)$ as described above [1.554].

(2) Given a limit ordinal β and given $\mathbf{A}_\alpha$ for all $\alpha < \beta$ define $\mathbf{A}_\beta$ as $\bigcup \mathbf{A}_\alpha$.

It suffices to show that in step (1) we add B to $\mathscr{G}_{\alpha+1}$ (because $\mathscr{G}_\alpha$ must then eventually contain all well-supported objects). $\Delta(B) \in \mathbf{A}_\alpha/B$ has a generic point:

$$\begin{array}{ccc} B & \xrightarrow{\langle 1,1\rangle} & B \times B \\ & {}_1\searrow \quad \swarrow & \\ & B & \end{array}$$

(recall that $B \xrightarrow{1} B$ is the terminator in $\mathbf{A}_\alpha/B$). For any proper $B' \rightarrowtail B$ in $\mathbf{A}_\alpha$ it is the case that $\Delta B' \rightarrowtail \Delta B$ does not allow the generic point.

1.547. The choice-free construction of the relative capitalization uses the theory of rational categories [1.48]. First consider an intermediate category $\hat{\mathbf{A}}$ whose objects are pairs $\langle A, F\rangle$, where $A \in |\mathbf{A}|$, and F is a finite set of morphisms in $\mathbf{A}$ from A to distinct, well-supported targets. We denote the finite set of targets by $F\Box$. A morphism $A_1 \to A_2$ in $\mathbf{A}$ is construed as a morphism $\langle A_1, F_1\rangle \to \langle A_2, F_2\rangle$ in $\hat{\mathbf{A}}$ if

$$F_2\Box \subset F_1\Box \quad \text{and} \quad \begin{array}{ccc} A_1 & \longrightarrow & A_2 \\ & \searrow \quad \swarrow & \\ & B & \end{array} \quad \text{for all } B \in F_2\Box .$$

A full embedding of $\mathbf{A}$ into $\hat{\mathbf{A}}$ is obtained by sending A to $\langle A, \emptyset\rangle$.

Note that $\hat{\mathbf{A}}$ is cartesian, e.g.

$$\begin{array}{ccc} \langle D, K\rangle & \xrightarrow{wv} & \langle A_2, F_2\rangle \\ {\scriptstyle wu}\downarrow & \lrcorner & \downarrow{\scriptstyle x} \\ \langle A_1, F_1\rangle & \xrightarrow{y} & \langle B, G\rangle \end{array} \quad \text{in } \hat{\mathbf{A}} \quad \text{if} \quad \begin{array}{ccc} C & \xrightarrow{v} & A_2 \\ {\scriptstyle u}\downarrow & \lrcorner & \downarrow{\scriptstyle x} \\ A_1 & \xrightarrow{y} & B \end{array} \quad \text{in } \mathbf{A},$$

where $H = H_1 \cup H_2$ for $H_1 = \{uf \mid f \in F_1\}$, $H_2 = \{vf \mid f \in F_2, f\Box \not\in G\Box\}$, and $w: D \rightarrowtail C$ is the maximal subobject that equalizes all morphisms in H with equal targets, and $K = \{wh \mid h \in H\}$.

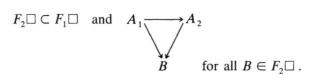

We consider $x: \langle A_1, F_1\rangle \to \langle A_2, F_2\rangle$ dense in $\hat{\mathbf{A}}$ if x together with $\{f \in F_1 \mid f\Box \not\in F_2\Box\}$ forms a product diagram in $\mathbf{A}$ [1.425]. It is readily shown that every dense morphism is monic, that every isomorphism is dense, and that dense monics are closed under composition and under pullback. The rational category [1.48] is then a relative capitalization $(\mathbf{A} \subset \mathbf{A}^*) \in \mathscr{C}$, because it is equivalent to a directed union of slices, as follows.

Given a finite set U of well-supported objects of $\mathbf{A}$, let $\mathbf{A}^* \mid U$ be the full subcategory of the rational category whose objects $\langle A, F \rangle$ satisfy $F\square \subset U$. Every such object is isomorphic to $\langle A', F' \rangle$, with $F'\square = U$, so the obvious functor $\mathbf{A}/\Pi U \to A^*|U$ is an equivalence of categories. For $V \subset U$ we have

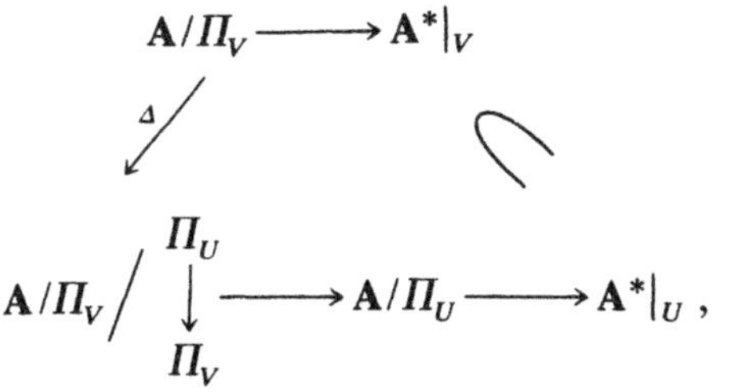

where all horizontal functors are equivalences. $\mathbf{A}^*$ is a directed union of $\mathbf{A}^* \mid U$'s. For any proper $B' \rightarrowtail B$ in $\mathbf{A}$, B well-supported, $\Delta B' \rightarrowtail \Delta B$ does not allow the generic point in $\mathbf{A}/B$.

1.55. THE HENKIN–LUBKIN THEOREM. *Every small pre-regular category may be faithfully represented in a power of the category of sets.*

BECAUSE: For $B \in \mathbf{A}$ define T_B: $\mathbf{A} \to \mathscr{S}$ as $\mathbf{A} \xrightarrow{\Delta} \mathbf{A}/B \to \underline{\underline{\mathbf{A}/B}} \xrightarrow{\Gamma} \mathscr{S}$ where $\mathbf{A}/B \to \underline{\underline{\mathbf{A}/B}}$ is a capitalization. T_B is a representation of pre-regular categories [1.543, 1.526]. Given proper $B' \rightarrowtail B$ we obtain proper $\Delta B' \rightarrowtail \Delta B$ in $\mathbf{A}/B$ and ΔB is well-supported. Such remains the case in $\underline{\underline{\mathbf{A}/B}}$ and hence $\Gamma(\Delta B') \rightarrowtail \Gamma(\Delta B)$ is proper in $\mathscr{S}$.

If $\mathbf{A}$ is regular, then one may verify that $\{T_U\}_{U \subset 1}$ is collectively faithful. (Given proper $B' \rightarrowtail B$ then $\Delta B' \rightarrowtail \Delta B$ is proper in $\mathbf{A}/\mathscr{Spt}(B)$ and ΔB is well-supported.)

1.551. As in [1.444] we have the immediate consequence:

Every Horn sentence in the defining predicates of regular categories, true for the category of sets, is true for every regular category.

1.552. We will say that a pre-regular category $\mathbf{A}$ is *special* if every universally quantified sentence in the relevant predicates true for $\mathscr{S}$ is true for $\mathbf{A}$. Since such a category is special as a cartesian category, every necessary condition discovered in [1.47] remains necessary. In fact, the situation is much simpler.

A pre-regular category is special iff $A \to U \rightarrowtail 1$ *implies that either* $A \to U$ *or* $U \rightarrowtail 1$ *is an isomorphism.*

BECAUSE: The condition above says immediately that $\mathbf{A} \to \underline{\underline{\mathbf{A}}} \xrightarrow{\Gamma} \mathscr{S}$ is faithful, where $\mathbf{A} \to \underline{\underline{\mathbf{A}}}$ is a capitalization.

We may, as before, separate the two cases:

A pre-regular category is special iff
either it is one-valued ($|\mathscr{Val}| = 1$)
or it is two-valued ($|\mathscr{Val}| = 2$)
and every object is either well-supported or isomorphic to 0, *the unique proper subterminator.*

1.56. We define composition of binary relations as follows:

Given monic pairs

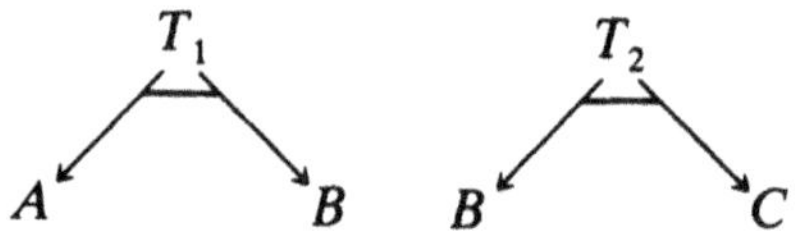

say that

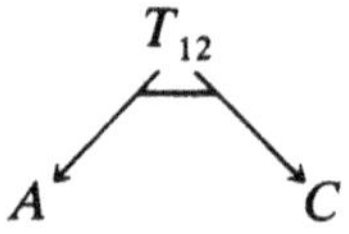

is a composition if there exists

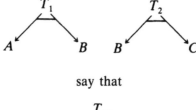

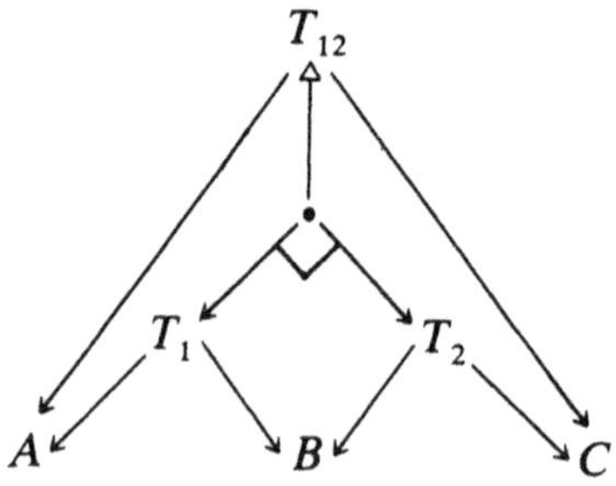

Given such T_1 and T_2 in a cartesian category with images we may construct a composition:

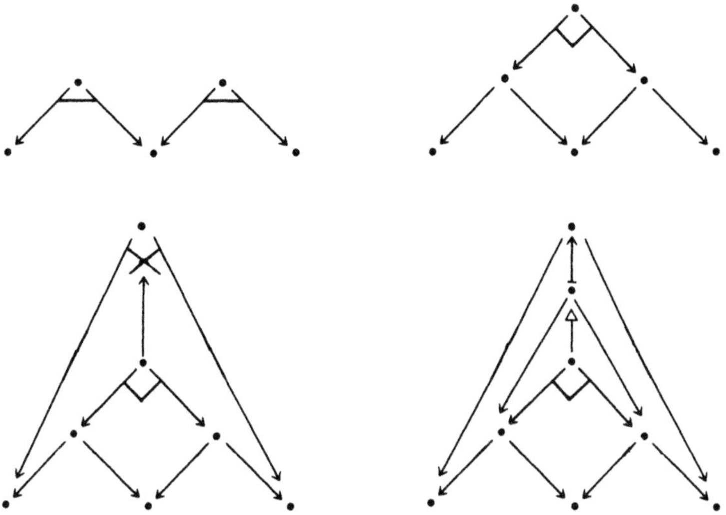

If T_1' is another table on A, B contained in T_1 and if T_{12}' is a composition of T_1' and T_2, then T_{12}' is contained in T_{12}.

Similarly, if T_2' is a table on B, C contained in T_2 and if T_{12}' is a composition of T_1 and T_2, then T_{12}' is contained in T_{12}. We obtain in this manner an order-preserving function $\mathscr{Rel}(A, B) \times \mathscr{Rel}(B, C) \to \mathscr{Rel}(A, C)$.

CAUTION. Composition of relations need not be associative [1.569].

1.561. The RECIPROCAL of a relation, R, from A to B is the relation R° from B to A obtained by twisting columns.

$$R^{\circ\circ} = R, \qquad (RS)^\circ = S^\circ R^\circ .$$

Note that reciprocation reverses composition but preserves the ordering.

1.562. In a cartesian category the natural order-isomorphism $\mathscr{Rel}(A, B) \simeq \mathscr{Sub}(A \times B)$ yields a semi-lattice structure on $\mathscr{Rel}(A, B)$.

$$(R \cap S)T \subset RT \cap ST , \qquad (R \cap S)^\circ = S^\circ \cap R^\circ .$$

1.563. *If* **A** *and* **B** *are cartesian categories with images and if* $F: \mathbf{A} \to \mathbf{B}$ *preserves the cartesian structure and images, then the induced functions* $\mathscr{Rel}(A, B) \to \mathscr{Rel}(FA, FB)$ *preserve composition, reciprocation and in-*

tersection. If F is faithful, then it reflects composition, reciprocation and intersection.

Hence,

If **A** *is a regular category, then any Horn sentence on maps and relations in the predicates of regular categories for morphisms and the operations of composition, reciprocation and intersection for relations which holds for binary relations between sets holds for* **A**.

In particular,

If **A** *is regular, then*

$$R(ST) = (RS)T , \qquad RS \cap T \subset (R \cap TS^{\circ})S .$$

The latter containment is called, for reasons to become apparent, the MODULAR IDENTITY. Bear in mind that any containment is equivalent to an equation:

$$R \subset S \quad \text{iff} \quad R = R \cap S .$$

1.564. For **A** a regular category $\mathscr{Rel}(\mathbf{A})$ will denote the category of relations. For a morphism $x: A \to B$ in **A** we obtain a relation, called its GRAPH, in $\mathscr{Rel}(\mathbf{A})$, to wit: the relation tabulated by

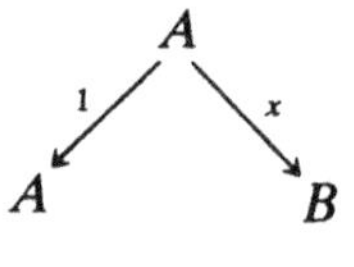

Such yields a functor, $\mathbf{A} \to \mathscr{Rel}(\mathbf{A})$.

We shall notationally treat **A** as a subcategory of $\mathscr{Rel}(\mathbf{A})$. In this context we will use the word MAP to indicate that a morphism of $\mathscr{Rel}(\mathbf{A})$ is in **A**.

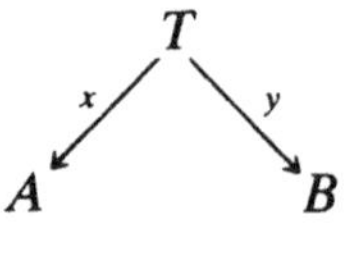

tabulates a map iff x is an isomorphism in **A**.

Given $R \in \mathscr{Rel}(\mathbf{A})$, we say that R is ENTIRE iff $1 \subset RR^{\circ}$, and that R is SIMPLE iff $R^{\circ}R \subset 1$. If R is tabulated by x, y, we can readily check that

x is a cover iff R is entire,
x is monic iff R simple.

Hence,

R is a map iff it is entire and simple.

As a consequence, given $R \in \mathscr{Rel}(\mathbf{A})$ such that for all representations of regular categories $F\colon \mathbf{A} \to \mathscr{S}$ it is the case that $F(R)$ is a map, then R is a map.

1.565. It is surprising how many existential theorems can now be inferred from the representations of regular categories.

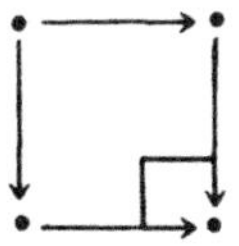

denotes a PUSHOUT, that is, a pullback in the opposite category. (Or, if desired, that which is defined by the Q-sequence obtained from the defining Q-sequence for pullbacks by reversing all the arrows.)

In any regular category a pullback of covers is a pushout, that is

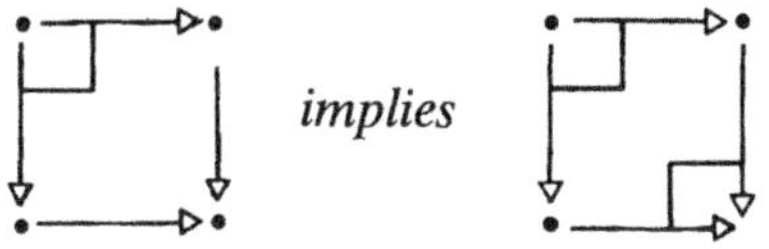

BECAUSE: Given

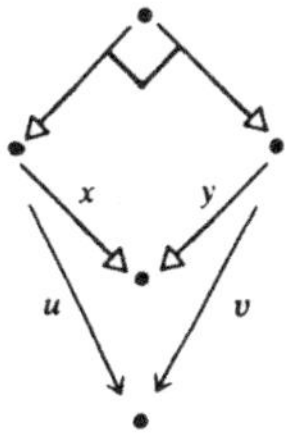

consider the relation $R = (x^\circ u) \cap (y^\circ v)$. Verify that in $\mathscr{S}$ such is a map. By the last section it is therefore a map in an arbitrary regular category. It now suffices to verify in $\mathscr{S}$ that $xR = u$ and $yR = v$. (The uniqueness condition is immediate.)

1.566. In any category a COEQUALIZER (that is, an equalizer in the opposite category) is a cover.

In a regular category every cover is a coequalizer.

BECAUSE: Given $x: A \twoheadrightarrow B$ consider its level

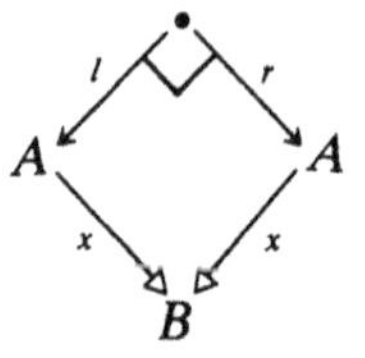

The last section implies that x is a coequalizer of l, r.

1.567. An endo-relation $E \in \mathscr{R}el(\mathbf{A})$ is an EQUIVALENCE RELATION if

$$1 \subset E, \qquad E^{\circ} \subset E, \qquad EE \subset E\,.$$

Levels are equivalence relations:

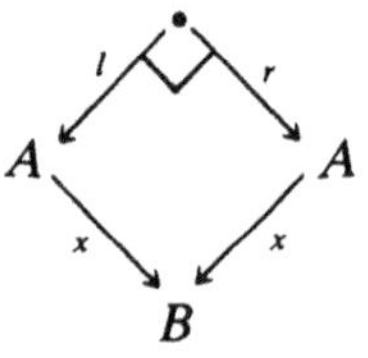

implies that $\langle l, r\rangle$ tabulates an equivalence relation.

An equivalence relation is EFFECTIVE if it is the level of some map. An EFFECTIVE REGULAR CATEGORY is one in which all equivalence relations are effective.

The metatheorem of [1.551] can easily be extended to effective categories. (Because the category of sets is effective.)

1.568. There is a preorder of covers with a given object A as source, where we let $f \leqslant g$ iff f factors through g

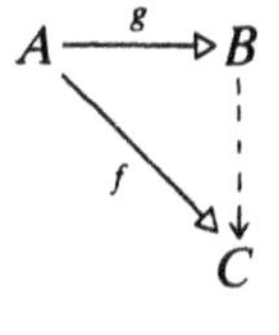

Its associated poset [1.246] consists of QUOTIENT-OBJECTS of A, and is called $\mathscr{Q}uot(A)$. Note that this is not dual to the definition of $\mathscr{S}ub(A)$ [1.412].

[1.566] yields a faithful order-reversing functor from $\mathcal{Q}uot(A)$ to the poset of equivalence relations on A.

1.569. *Let* **A** *be a cartesian category with images. Composition of relations is associative iff* **A** *is regular.*

BECAUSE: Viewing **A** as a substructure of $\mathcal{R}el(\mathbf{A})$ we note that $x: A \to B$ is a cover iff $x^\circ x = 1_B$. Given $y: C \to B$ and cover $x: A \twoheadrightarrow B$ we thus have $y(x^\circ x) = y$. Consider

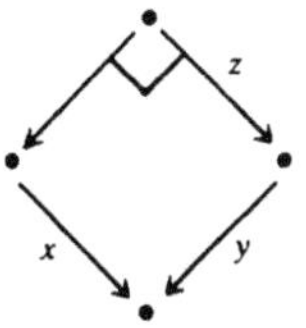

If z is not a cover, then $(yx^\circ)x$ is not even a map.

1.56(10). As an application of the metatheorem for regular categories we can obtain:

If $x: A \to B$ *is a* CONSTANT MORPHISM (*that is, for all* $y, y': C \to A$ *it is the case that* $yx = y'x$), *then its image is a subterminator.*

The non-regular category described in [1.475] provides a counterexample: $\mathbb{Z}_2 + \mathbb{Z}_3 \to 1 + 1$ is constant. (Taking $+$ as disjoint union.) With reference to the last section, note that

$$p_{\mathbb{Z}_3}(p^{\circ}_{\mathbb{Z}_2} p_{\mathbb{Z}_2}) = p_{\mathbb{Z}_3}, \qquad (p_{\mathbb{Z}_3} p^{\circ}_{\mathbb{Z}_2}) p_{\mathbb{Z}_2} = \emptyset .$$

1.56(11). *In a regular category, an object is projective iff every entire relation from it contains a map.*

BECAUSE: For an entire relation

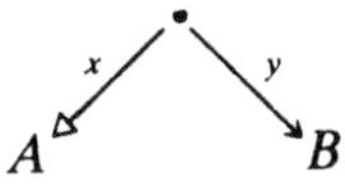

let z be a left inverse for x. Then $zy \subset x^\circ y$. In the other direction, if $C \to A$ is a cover and $z \subset x^\circ$, then $zx \subset x^\circ x \subset 1$.

1.57. An object in a regular category is CHOICE if every entire relation targeted at it contains a map.

Clearly, every object is projective iff every object is choice. An AC REGULAR CATEGORY is a regular category in which either (both) of these conditions hold. (AC stands for the axiom of choice. In particular, the axiom of choice asserts that the regular category $\mathscr{S}$ is an AC regular category.) We may alternatively define an AC regular category as a cartesian category in which every morphism factors as a left-invertible followed by a monic (left-invertibles are always covers and are always transferred by pullbacks).

Although this condition almost never holds, it is important to know it if it does. In a multiplicative monoid underlying a ring, this is the question whether left-cancellable implies left-invertible. For a group G, the category of G-sets, $\mathscr{S}^G$, [1.271] is AC iff G is trivial.

1.571. Suppose **A** is cartesian and that for every $x: A \to B$ there exists $e: A \to A$ such that $e^2 = e$, $ex = x$.

e and x have the same level.

Let $C \to A$ be an equalizer of $1, e$. let $(A \to C \to A) = e$. Then $x = (A \to C \to A \to B)$. $A \to C$ has a left-inverse $(C \to A)$ and $C \to A \to B$ is monic. Such an **A** is an AC regular category. (The image of x has been constructed, somehow, as a subobject of A.)

1.572. The *category composed of recursive functions* $\underline{\mathbf{R}}$ has the extended natural numbers $0, 1, 2, \ldots, \omega$ as objects and recursive functions as morphisms (using the von Neumann convention that $\alpha = \{\beta \mid \beta < \alpha\}$). Any function from a finite natural number is understood to be recursive.

$\underline{\mathbf{R}}$ is an AC regular category: given $x: \alpha \to \beta$ we define $e: \alpha \to \alpha$ by $e(n) = \min\{i \leqslant n \mid x(i) = x(n)\}$ and use the last section. $\underline{\mathbf{R}}$ is not effective: if E is an equivalence relation on ω which appears as a level of $\omega \twoheadrightarrow \alpha$, then a left-inverse $\alpha \to \omega$ chooses a set of representatives for E, forcing E to be not just recursively enumerable, but recursive. Finitely presented groups do not live in $\underline{\mathbf{R}}$ unless they yield solvable word problems. When we construct [2.169] effective closures of regular categories we will gain a category in which finitely presented algebraic systems can live but we will necessarily lose the axiom of choice.

1.573. If we replace the word 'recursive' with 'primitive recursive' the resulting category is not cartesian because equalizers do not always exist. However, given $x, y; \omega \to \alpha$ let $n \in \omega$ be such that $x(n) = y(n)$ (if there is no such n, then 0 is the equalizer of x, y) and define $e: \omega \to \omega$ by

$$e(i) = \begin{cases} i & \text{if } x(i) = y(i)\,, \\ n & \text{if } x(i) \neq y(i)\,. \end{cases}$$

Then $e^2 = e$, $ex = ey$ and for any $z: \beta \to \omega$ such that $zx = zy$ it is the case that $ze = e$.

Such is precisely what is needed to show that x, y have an equalizer if we split all idempotents. We define $\underline{\mathbf{P}}$ as the result of doing so. (It is equivalent to the category whose objects are taken to be all primitively recursive subsets of ω.)

1.574. The forgetful functor $\underline{\mathbf{R}} \to \mathscr{S}$ is faithful. Not so for $\underline{\mathbf{P}}$. There exist one-to-one onto $x: \omega \to \omega$ in $\underline{\mathbf{P}}$ such that $x^{-1} \not\in \underline{\mathbf{P}}$ (Indeed $\underline{\mathbf{R}}$ is obtainable as a 'category of fractions' of $\underline{\mathbf{P}}$: that is, if all monic-epics in $\underline{\mathbf{P}}$ are formally inverted and the result is equivalent to $\underline{\mathbf{R}}$.) The functor $\underline{\mathbf{P}} \to \mathscr{S}$ represented by ω is a faithful set-valued representation of regular categories.

(1.58.) A BICARTESIAN CATEGORY is one that is both cartesian and COCARTESIAN, the latter meaning that the opposite category is cartesian. We have already mentioned that pushouts are dual to pullbacks, coequalizers to equalizers. A COTERMINATOR (elsewhere: *initial* or *coterminal object*), the notion dual to terminator, is usually denoted 0. A COPRODUCT, the notion dual to product, is usually denoted $A + B$.

Our previous use of 0 is consistent [1.474, 1.552]. In any cartesian category with an object 0 such that all morphisms targeted at 0 are isomorphisms, then 0 is a coterminator. This extra property is said to make 0 a STRICT COTERMINATOR. (To obtain $0 \to A$ invert $A \times 0 \to 0$ and compose with $A \times 0 \to A$. Since 0 can have no proper subobjects, all equalizers in 0 are entire.)

(1.581.) If $\mathbf{A}$ and $\mathbf{B}$ are regular and cocartesian and if $F: \mathbf{A} \to \mathbf{B}$ is a representation of bicartesian categories, then F is a representation of regular categories: F preserves coequalizers, hence preserves covers [1.566].

(1.582.) If $\mathbf{A}$ is bicartesian, then its regularity is a pair of Horn sentences in the bicartesian predicates, because images in a bicartesian regular category may be constructed as follows: given $x: A \to B$ consider its level $A \xleftarrow{l} L \xrightarrow{r} A$ and then the coequalizer $A \to C$ of l, r. The unique morphism $C \to B$ such that $x = (A \twoheadrightarrow C \to B)$ is monic.

(1.583.) If $\mathbf{A}$ is bicartesian and regular, then its effectiveness is a Horn sentence in the bicartesian predicates. If

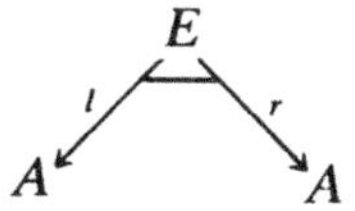

tabulates an equivalence relation consider the coequalizer $x: A \to C$ of l, r. E is effective iff

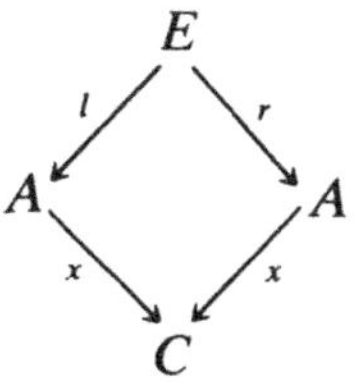

is a pullback.

(1.584.) *If* **A** *is cocartesian, then so is* $\mathbf{A}/B$ *and* $\Sigma: \mathbf{A}/B \to \mathbf{A}$ *is a faithful representation of cocartesian categories.*

The second part of the sentence tells how to prove the first part.

(1.585.) $\mathcal{LH}$ has coproducts (disjoint unions) but is not co-cartesian: let θ be an irrational rotation of the circle $\mathbb{R}/\mathbb{Z}$; 1 and θ cannot have a coequalizer, indeed there exists no local homeomorphism $f: \mathbb{R}/\mathbb{Z} \to X$ such that $f = \theta f$. $\mathcal{LH}$ does have, however, enough coequalizers to make $\mathcal{SH}(Y)$ cocartesian. Given

$$X_1 \underset{y}{\overset{x}{\rightrightarrows}} X_2 \longrightarrow Y$$

in $\mathcal{LH}$ let $z: X_2 \to X_3$ be a coequalizer of x, y in $\mathcal{T}op$, that is, the quotient space of X_2 obtained from the equivalence relation generated by $x^{\circ}y$. Using the fact that x and y are local homeomorphisms one may verify for any open $U \subset X_2$ that $\tilde{z}(U)$ is open, hence that z is an open map. Since z may be continued to a local homeomorphism $X_2 \xrightarrow{z} X_3 \xrightarrow{u} Y$ and since z is open we can now obtain that z is a local homeomorphism. Finally, the ontoness of z forces u to be a local homeomorphism.

The stalk-functors $\mathcal{SH}(Y) \to \mathcal{S}$ are a collectively faithful family of representations of bicartesian categories. As a consequence $\mathcal{SH}(Y)$ is effective regular.

(1.586.) *For small* **A**, *the functor category* $\mathcal{S}^{\mathbf{A}}$ *is cocartesian and the evaluation functors are a collectively faithful family of representations of cocartesian categories.*

The second part of the sentence tells how to prove the first part.

In particular, $\mathcal{S}^{\mathbf{A}}$ is effective regular.

1.587. A natural question arises: which bicartesian categories satisfy all Horn sentences in the bicartesian predicates that hold for the category of sets. There can be no elementary (in the technical sense) answer:

Let $\mathbb{N}$ be the set of natural numbers, $0: 1 \to \mathbb{N}$ the inclusion map of $\{0\}$, $s: \mathbb{N} \to \mathbb{N}$ the successor function. Then

$$1 + \mathbb{N} \xrightarrow{\binom{0}{s}} \mathbb{N}$$

is an isomorphism and $\mathbb{N} \to 1$ is a coequalizer of $1, s$. Such yields a characterization in $\mathcal{S}$ (up to isomorphism) of the triple $\mathbb{N}, 0, \mathrm{s}$.

Let $a: \mathbb{N} \times \mathbb{N} \to \mathbb{N}$ be the addition map. It is characterized by

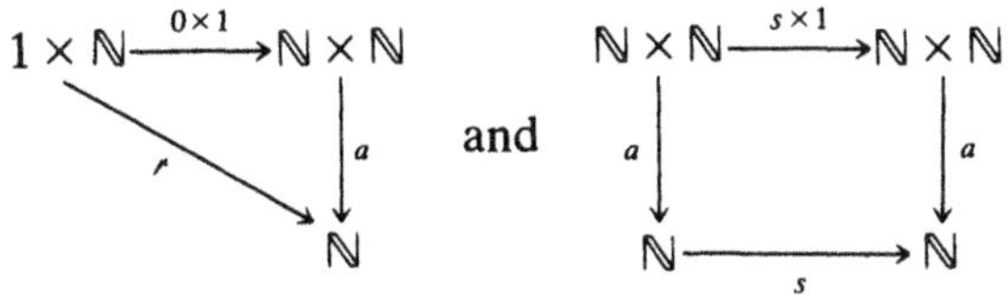

Let $m\colon \mathbb{N} \times \mathbb{N} \to \mathbb{N}$ be the multiplication map. It is characterized by

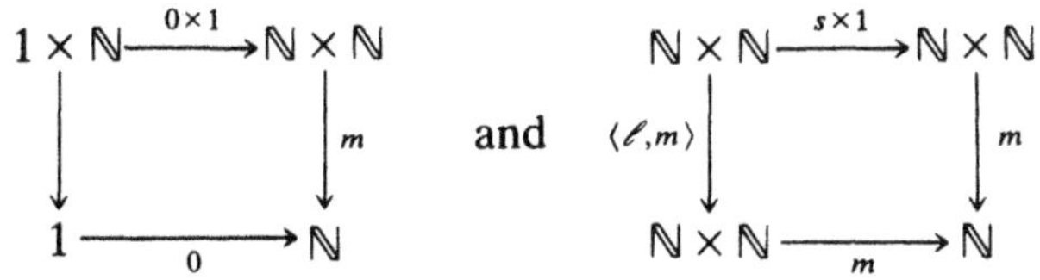

Given two polynomials on natural numbers we may form their equalizer. It is empty (that is, is a coterminator) iff there is no solution to the equation. Hence, any diophantine problem is equivalent to a Horn sentence in the bicartesian predicates. The set of such Horn sentences true for the category of sets is not recursively enumerable.

1.59. An ABELIAN CATEGORY is a bicartesian category which satisfies all Horn sentences in the bicartesian predicates which hold for $\mathscr{Ab}$, the category of abelian groups. We shall find remarkably simple characterizations for abelian categories. We shall also find that abelianness implies much more than initially might be expected.

1.591. $0 \to 1$ is an isomorphism in $\mathscr{Ab}$, hence in any abelian category. When a terminator and coterminator coincide it is called a ZERO OBJECT and denoted, naturally, as 0. In any such category we obtain for every pair A, B a distinguished ZERO MORPHISM $A \to 0 \to B$. It is the unique morphism that factors through 0. The set of zero morphisms is a two-sided ideal.

In any bicartesian category a morphism from a coproduct $A_1 + A_2$ to a product $B_1 \times B_2$ is uniquely described by a 2×2 matrix, the (i, j)-th entry of which is an element of (A_i, B_j). In an abelian category

$$\begin{pmatrix} 1 & 0 \\ 0 & 1 \end{pmatrix} : A_1 + A_2 \to A_1 \times A_2$$

is an isomorphism. That is, finite products and coproducts coincide. In such categories the symbol $\oplus$ is conventionally used to denote (co-) products. The top half of the diagram below is a coproduct diagram. The bottom half is a product diagram.

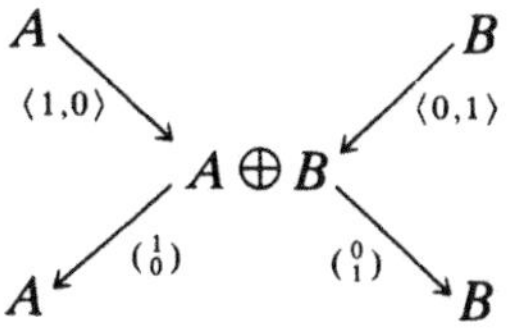

In any *category with zero* (that is, any category with a zero-object) in which products and coproducts coincide we may define binary operations on (A, B) as follows:

$$x \underset{L}{+} y = A \xrightarrow{\langle 1,1\rangle} A \oplus A \xrightarrow{\binom{x}{y}} B ,$$

$$x \underset{R}{+} y = A \xrightarrow{\langle x,y\rangle} B \oplus B \xrightarrow{\binom{1}{1}} B .$$

The following equations are easily verified.

$$x \underset{L}{+} 0 = x = 0 \underset{L}{+} x , \qquad x \underset{R}{+} 0 = x = 0 \underset{R}{+} x ,$$

$$(x \underset{L}{+} y)z = xz \underset{L}{+} yz , \qquad z(x \underset{R}{+} y) = zx \underset{R}{+} zy .$$

For any $u, v, x, y \in (A, B)$ we have

$$(u \underset{R}{+} v) \underset{L}{+} (x \underset{R}{+} y) = (u \underset{L}{+} x) \underset{R}{+} (v \underset{L}{+} y) .$$

Because,

$$\langle 1,1\rangle\left(\begin{pmatrix} u & v \\ x & y\end{pmatrix}\begin{pmatrix}1\\1\end{pmatrix}\right) = \langle 1,1\rangle\left(\begin{pmatrix}u\\x\end{pmatrix} \underset{R}{+} \begin{pmatrix}v\\y\end{pmatrix}\right) = \langle 1,1\rangle\begin{pmatrix}u\\x\end{pmatrix} \underset{R}{+} \langle 1,1\rangle\begin{pmatrix}v\\y\end{pmatrix}$$
$$= (u \underset{L}{+} x) \underset{R}{+} (v \underset{L}{+} y) ,$$

$$\left(\langle 1,1\rangle\begin{pmatrix} u & v \\ x & y\end{pmatrix}\right)\begin{pmatrix}1\\1\end{pmatrix} = \left(\langle u,v\rangle \underset{L}{+} \langle x,y\rangle\right)\begin{pmatrix}1\\1\end{pmatrix}$$
$$= \langle u,v\rangle\begin{pmatrix}1\\1\end{pmatrix} \underset{L}{+} \langle x,y\rangle\begin{pmatrix}1\\1\end{pmatrix} = (u \underset{R}{+} v) \underset{L}{+} (x \underset{R}{+} y) .$$

If we let $v = x = 0$ we obtain $u \underset{L}{+} y = u \underset{R}{+} y$. The binary operations are the same. Erasing the subscripts we obtain the *middle-two interchange law*:

$$(u + v) + (x + y) = (u + x) + (v + y) .$$

Letting $v = 0$ we obtain the associativity of $+$. Letting $u = y = 0$ we obtain its commutativity. Any category in which finite products and coproducts coincide has, therefore, a binary partial operation $x + y$ defined iff $\Box x = \Box y$ and $x\Box = y\Box$ and satisfying:

$$\Box(x + y) \succeq \Box x, \qquad (x + y)\Box \succeq y\Box ,$$

$$x + 0 = x, \qquad x + y = y + x, \qquad x + (y + z) = (x + y) + z ,$$

$$z(x + y) = zx + zy, \qquad (x + y)z = xz + yz .$$

Such is called a HALF-ADDITIVE CATEGORY.

Conversely, in a half-additive category products and coproducts must coincide. If

$$\begin{array}{ccc} & A \times B & \\ {}^{\ell}\swarrow & & \searrow{}^{r} \\ A & & B \end{array}$$

is a product and

$$\begin{array}{ccc} A & & B \\ {}_{\ell'}\searrow & & \swarrow{}_{r'} \\ & A + B & \end{array}$$

a coproduct, then $\ell\ell' + rr'$ is an inverse for

$$\begin{pmatrix} 1 & 0 \\ 0 & 1 \end{pmatrix} : A + B \to A \times B\,.$$

An ADDITIVE CATEGORY is a half-additive category in which (A, B) is an abelian group, all A, B, equivalently, in which $\left(\begin{smallmatrix} 1 & x \\ 0 & 1 \end{smallmatrix}\right)$ is an isomorphism, all $x\colon A \to B$. (If $\left(\begin{smallmatrix} u & v \\ x & y \end{smallmatrix}\right)$ is its inverse one may show first that $y = 1$ and then that $v + x = 0$.) Such is the case in $\mathscr{Ab}$, hence in any abelian category.

1.592. In a category with zero we define a KERNEL of x, denoted $\mathscr{Ker}(x)$, as an equalizer of $x, 0$. A COKERNEL, denoted $\mathscr{Cok}(x)$, is a coequalizer of $x, 0$. In an additive category a (co-) equalizer of x, y reduces to a (co-) kernel, namely of $x - y$.

As for any category of equationally defined algebras, $\mathscr{Ab}$ is effective regular, hence any abelian category is such [1.582, 1.583]. Note that the image of x is constructible as $\mathscr{Ker}(\mathscr{Cok}(x))$.

Any small abelian category **A** *may be faithfully representation as a bicartesian category in the category of abelian groups.*

BECAUSE: Since **A** is one-valued we obtain a faithful representation $T\colon \mathbf{A} \to \mathscr{S}$ of regular categories [1.552]. For every $A \in \mathbf{A}$, $T(0) \to T(A)$, $T\left(\begin{smallmatrix} 1 \\ 1 \end{smallmatrix}\right)$: $T(A \times A) \to T(A)$ are easily verified to give an abelian group structure on $T(A)$. For $x\colon A \to B$, $T(x)$ is easily verified to be a homomorphism with respect to the structures. That is, we may factor T as $\mathbf{A} \xrightarrow{T'} \mathscr{Ab} \xrightarrow{U} \mathscr{S}$ where U is the forgetful functor. Since U is a faithful representation of regular categories, T' is. We need only show that T' preserves the cocartesian structure. Since **A** is half-additive, T' preserves coproducts. Given $x\colon A \to B$ we know that $B \to \mathscr{Cok}(x)$ is a cover in **A** whose kernel is the image of x. Such characterizes cokernels in $\mathscr{Ab}$.

1.593. Inspection of the argument above leads to a number of characterizations of abelian categories. A subobject is a NORMAL SUBOBJECT if it is the kernel of some morphism.

A *is abelian iff it is a regular additive category in which every subobject is normal.*

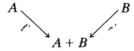

BECAUSE: As above we obtain (for small $\mathbf{A}$) a faithful representation of regular categories $T: \mathbf{A} \to \mathscr{Ab}$. We must show that $\mathbf{A}$ has cokernels and that T preserves them. Using just the existence of images (and their preservation by T) it clearly suffices to concentrate on cokernels of monic morphisms. Given $x: A \rightarrowtail B$ in $\mathbf{A}$, let $y: B \to D$ be such that x is a kernel of y. We may factor y as $B \xrightarrow{z} C \rightarrowtail D$. Clearly x is a kernel of z. It suffices now to show, in any regular additive category with zero, that if a monic is a kernel of a cover then the cover is a cokernel of the monic. Since the cover, z, is a coequalizer, say of $u, v: E \to B$ [1.566] it is a cokernel, namely of $u - v$. We thus obtain

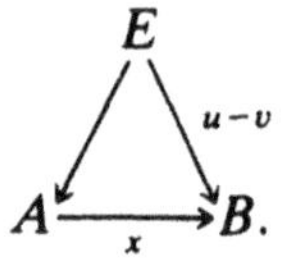

Given $w: B \to X$ such that $xw = 0$ then necessarily $(u - v)w = 0$ and there exists

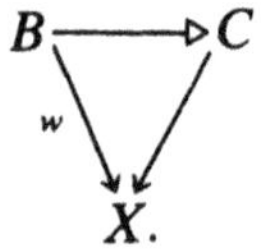

1.594. $\mathbf{A}$ *is abelian iff it is an effective regular additive category.*

BECAUSE: The first part of the argument in [1.592] says that any regular additive category may be faithfully represented as a regular additive category in $\mathscr{Ab}$. The calculus of relations that holds for $\mathscr{Ab}$ therefore holds for $\mathbf{A}$. In particular, if an endo-relation R is reflexive ($1 \subset R$), then it is an equivalence relation.

Given monic $x: A \rightarrowtail B$ in $\mathbf{A}$ consider the monic pair

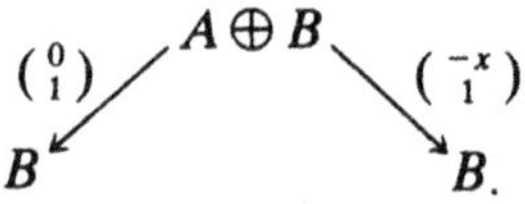

The relation it tabulates is clearly reflexive. Let $z: B \to C$ be such that

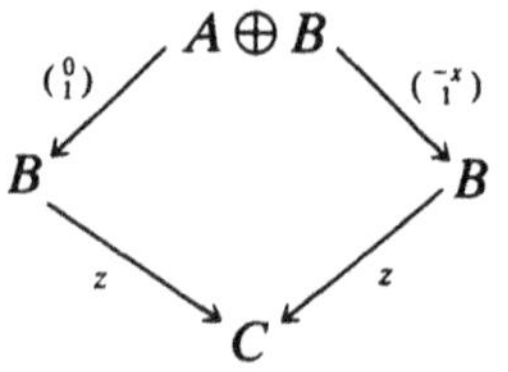

is a pullback. Verify that x is a kernel of z.

1.595. In any category $\mathbf{A}$ with finite products we may define an *abelian group object* as an object A together with morphisms $0: 1 \to A$, $n: A \to A$, $a: A \times A \to A$ such that

$$\begin{array}{ccc} 1\times A & \xrightarrow{0\times 1} & A\times A \\ & \searrow^{r} & \downarrow{\scriptstyle a} \\ & & A \end{array} \qquad\qquad \begin{array}{ccc} A\times A & \xrightarrow{n\times 1} & A\times A \\ \downarrow & & \downarrow{\scriptstyle a} \\ 1 & \xrightarrow[0]{} & A \end{array}$$

$$\begin{array}{ccc} A\times A\times A & \xrightarrow{1\times a} & A\times A \\ {\scriptstyle a\times 1}\downarrow & & \downarrow{\scriptstyle a} \\ A\times A & \xrightarrow[a]{} & A \end{array} \qquad\qquad \begin{array}{ccc} A\times A & \xrightarrow{\langle r,\ell\rangle} & A\times A \\ & {\scriptstyle a}\searrow \swarrow{\scriptstyle a} & \\ & A & \end{array}$$

Given another abelian group object B we say that $x: A \to B$ is a *homomorphism* if

$$\begin{array}{ccc} A\times A & \xrightarrow{a} & A \\ {\scriptstyle x\times x}\downarrow & & \downarrow{\scriptstyle x} \\ B\times B & \xrightarrow[a]{} & B. \end{array}$$

$\mathscr{Ab}(\mathbf{A})$ denotes the category founded on $\mathbf{A}$ whose objects are abelian group objects and whose source-target predicate is the definition of homomorphism.

If $\mathbf{A}$ *is effective regular, then so is* $\mathscr{Ab}(\mathbf{A})$ *and the forgetful functor* $\mathscr{Ab}(\mathbf{A}) \to \mathbf{A}$ *is a faithful representation of regular categories.*

The second part of the sentence tells how to prove the first part.

Consequently,

For any effective regular category $\mathbf{A}$ *the category of abelian group objects* $\mathscr{Ab}(\mathbf{A})$ *is an abelian category.*

1.596. For any small $\mathbf{A}$, $\mathscr{Ab}(\mathscr{S}^{\mathbf{A}})$ is isomorphic to $(\mathscr{Ab}(\mathscr{S}))^{\mathbf{A}}$ that is, the category of functors from $\mathbf{A}$ to $\mathscr{Ab}$.

$\mathscr{Ab}(\mathscr{Sh}(Y))$ is not so easily dispatched. Indeed, it is the historical motivation for the abstract theory of abelian categories, and consequently, of much of category theory.

An abelian group structure on $X \to Y$ in $\mathscr{Sh}(Y)$ yields an abelian group structure on each stalk. The objects of $\mathscr{Ab}(\mathscr{Sh}(Y))$ are often called "sheaves of abelian groups over Y". Since stalks are discrete there is no restriction of the individual groups structures – but the structures must continuously vary from stalk to stalk. (An abelian Lazard sheaf is often described as being vertically algebraic, horizontally geometric.)

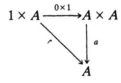

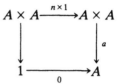

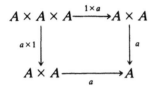

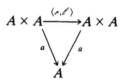

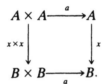

1.597. We may further simplify the characterization of abelian categories. Let **A** be a category with zero, kernels and cokernels. For any $x\colon A \to B$ and any choice of kernels and cokernels there is a unique morphism θ such that

$$x = A \to \mathscr{C}ok\,(\mathscr{K}er(x)) \xrightarrow{\theta} \mathscr{K}er(\mathscr{C}ok\,(x)) \to B\,.$$

We will say that **A** is an EXACT CATEGORY if θ is always an isomorphism.

An alternative description of an exact category is a category with zero in which each morphism $A \to B$ factors as $A \to I \to B$ where $A \to I$ is a cokernel (of something) and $I \to B$ is a kernel. To show that such implies the existence of cokernels let $A \to B$ be arbitrary and $A \to I \to B$ a factorization as above. It clearly suffices to show that $I \to B$ has a cokernel. Let $I \to B$ be a kernel of $B \to D$. We may factor $B \to D$ as $B \to C \to D$ where $B \to C$ is a cokernel and $C \to D$ a kernel. Clearly $I \to B$ is a kernel of $B \to C$. Now, imitating the argument in [1.593] we let $B \to C$ be a cokernel of $E \to B$. There exists

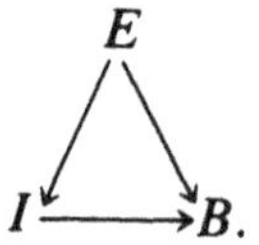

Given any $B \to X$ such that $I \to B \to X = 0$ it is necessarily the case that $E \to B \to X = 0$. Hence there exists

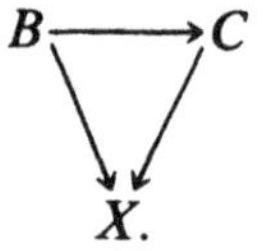

Since this description of exact categories is self-dual, we know that the dual construction works for kernels.

A *is abelian iff it is an exact additive category.*

Indeed,

A *is abelian iff it is an exact category with either binary products or coproducts.*

BECAUSE: We first show that the existence of binary products yields an additive structure. We shall repeatedly use the following immediate consequence (and its dual) of the definition of exact category.

If $A \rightarrowtail B$ *is monic and* $B \twoheadrightarrow C$ *its cokernel, then* $A \rightarrowtail B$ *is a kernel of* $B \twoheadrightarrow C$.

For any object A, let $y\colon A \times A \to C$ be a cokernel of $\langle 1,1\rangle\colon A \to A \times A$. Let $\theta = A \xrightarrow{\langle 1,0\rangle} A \times A \xrightarrow{y} C$. Consider $x\colon \mathscr{K}er(\theta) \to A$. Since $(x\langle 1,0\rangle)y = 0$ there must exist $x'\langle 1,1\rangle = x\langle 1,0\rangle$. Hence $\langle x', x'\rangle = \langle x, 0\rangle$ and $x = 0$. Thus $\mathscr{K}er(\theta) = 0$.

Consider $x: C \to \mathscr{C}ok(\theta)$. Since $\langle 1,0\rangle: A \to A \times A$ is a kernel of $\mathscr{r}$: $A \times A \to A$ it is the case that $\mathscr{r}$ is a cokernel of $\langle 1,0\rangle$ and there must exist $\mathscr{r}z' = yz$. But $z' = 1 \cdot z' = \langle 1,1\rangle \mathscr{r}z' = \langle 1,1\rangle yz = 0 \cdot z = 0$ and hence $z = 0$ since $yz = 0$ and y is monic.

The definition of exact categories forces y to be an isomorphism. Define $s_A = A \times A \xrightarrow{y\theta^{-1}} A$. s_A is still a cokernel of $\langle 1,1\rangle$ (which together with the equation $\langle 1,0\rangle s_A = 1$ characterizes it).

$\{s_A\}_A$ yields a transformation from the cartesian squaring functor to the identity functor: given $x: A \to B$ we note first that $\langle 1,1\rangle(x \times x)s_B = x\langle 1,1\rangle s_B = 0$, hence there is a morphism

$$\begin{array}{ccc} A \times A & \xrightarrow{s_A} & A \\ {\scriptstyle x\times x}\downarrow & & \downarrow{\scriptstyle x'} \\ B \times B & \xrightarrow[s_B]{} & B. \end{array}$$

But $x' = 1 \cdot x' = \langle 1,0\rangle s_A x' = \langle 1,0\rangle(x \times x)s_B = x\langle 1,0\rangle s_B = x \cdot 1 = x$.

We obtain a binary operation on (A, B) for each pair of objects A, B:

$$(x - y) = A \xrightarrow{\langle x,\, y\rangle} B \times B \xrightarrow{s_B} B\,.$$

We obtain the following equations

$$\begin{aligned} &x - 0 = x\langle 1,0\rangle s_B = x\,,\\ &x - x = x\langle 1,1\rangle s_B = 0\,,\\ &z(x - y) = z\langle x, y\rangle s_B = \langle zx, zy\rangle s_B = zx - zy\,,\\ &(x - y)z = \langle x, y\rangle s_B z = \langle x, y\rangle(z \times z)s_C = \langle xz, yz\rangle s_C = xz - yz\,. \end{aligned}$$

For $a, b \in (A, B)$ and $c, d \in (B, C)$ we have the equation

$$(ac - ad) - (bc - bd) = (ac - bc) - (ad - bd)$$

because

$$\begin{aligned} (ac - ad) - (bc - bd) &= a(c - d) - b(c - d) = (a - b)(c - d)\\ &= (a - b)c - (a - b)d\\ &= (ac - bc) - (ca - bd)\,. \end{aligned}$$

Given $u, v, x, y \in (A, B)$ let $a = \langle u, v\rangle$, $b = \langle x, y\rangle$, $c = \ell$, $d = \mathscr{r}$ and apply the above equation to obtain the middle-two interchange law

$$(u - v) - (x - y) = (u - x) - (v - y).$$

Define $x + y = x - (0 - y)$.
We obtain the equations

$$x + 0 = x - (0 - 0) = x - 0 = x\,,$$

$$\begin{aligned} 0 + x &= 0 - (0 - x) = (x - x) - (0 - x) \\ &= (x - 0) - (x - x) = (x - 0) - 0 = x\,, \end{aligned}$$

$$\begin{aligned} (u + v) + (x + y) &= (u - (0 - v)) - (0 - (x - (0 - y))) \\ &= (u - 0) - ((0 - v) - (x - (0 - y))) \\ &= (u - 0) - ((0 - x) - (v - (0 - y))) \\ &= (u - (0 - x)) - (0 - (v - (0 - y))) \\ &= (u + x) + (v + y)\,. \end{aligned}$$

As already noted [1.591] the above equations imply that + is associative and commutative. (A, B) is an abelian group because

$$(0 - x) + x = (0 - x) - (0 - x) = 0.$$

Clearly + distributes on both the right and left. **A** is an additive category. Since kernels and cokernels exist, **A** is a bicartesian category. The definition of exact category immediately implies that **A** has images (cokernels are necessarily covers). We need only show that **A** is regular to apply [1.593].

Given a square

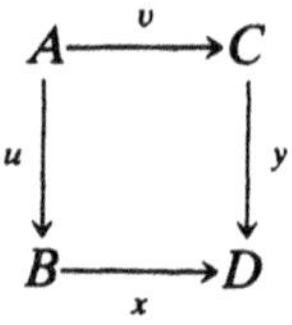

consider the morphisms

$$\langle u, v\rangle\colon A \to B \oplus C \quad \text{and} \quad \begin{pmatrix} x \\ -y \end{pmatrix}\colon B \oplus C \to D.$$

The square commutes iff $\langle u, v\rangle\begin{pmatrix} x \\ -y \end{pmatrix} = 0.$

The square is a pullback iff $\langle u, v\rangle$ *is a kernel of* $\begin{pmatrix} x \\ -y \end{pmatrix}$.

The square is a pullback iff $\begin{pmatrix} x \\ -y \end{pmatrix}$ *is a cokernel of* $\langle u, v\rangle$.

If the square is a pullback and if x is a cover then necessarily $\binom{x}{-y}$ is a cover and $\binom{x}{-y}$ is a cokernel of $\langle u, v\rangle$. That is, such a square is also a pushout. Consider $C \to \mathscr{Cok}(v)$. Since the square is a pushout there exists

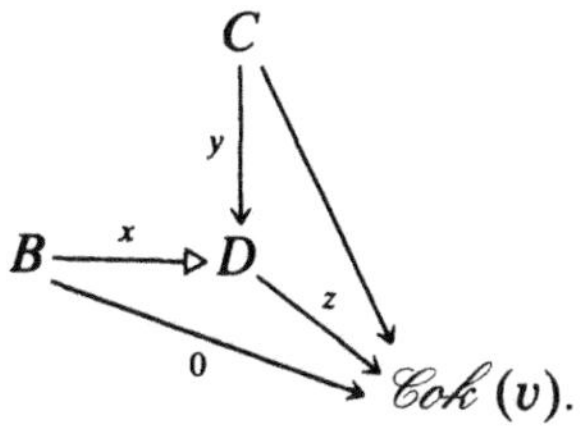

Since x is epic, $z = 0$. Hence $yz = 0$ and $\mathscr{C}ok(v) = 0$. The definition of exact category now immediately implies that v is cover.

1.598. Say that a category with zero is *left-normal* if every subobject is normal, and that it is *right-normal* if the dual condition holds (every comonic is a cokernel). A *normal category* is one that is both left and right-normal. (The category of all groups, abelian of not, is right-normal but not left-normal. The category of pointed compact Hausdorff spaces is left-normal but not right-normal. Both categories are, of course, bicartesian.) The first book on the subject defined abelian categories as normal categories with kernels, cokernels, binary products and coproducts. We are thus obliged to prove:

A *is abelian iff it is a normal category with kernels, cokernels and either binary products or coproducts.*

BECAUSE: We will suppose that products exist. Given

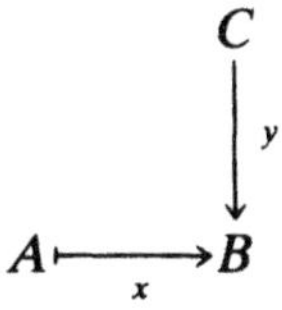

let $z: B \to D$ be such that x is a kernel of z. There exists

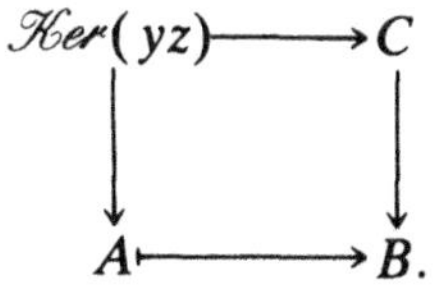

It is readily verified to be a pullback. In particular we can construct pullbacks of pairs of monics. By the argument of [1.434] equalizers exist. Every cover, therefore, is epic.

Given $x: A \to B$ in any category with kernels and cokernels, $\mathscr{K}er(\mathscr{C}ok(x))$ is easily verified to be the minimal normal subobject that allows x. Since every subobject is normal, it is the image of x. We can factor $A \to B$ as $A \twoheadrightarrow C \rightarrowtail B$. Since $A \twoheadrightarrow C$ is epic it is a cokernel. **A** is exact.

(Consider the monoid with three elements $1, e, 0$ where $e^2 = e$. Viewing such as a category, formally adjoin a zero-object to obtain a normal category with kernels and cokernels. It is not an exact category.)

1.599. The consequences of abelianness are many. We say that

$$\cdots \to A_{-1} \to A_0 \to A_1 \to A_2 \to \cdots$$

is an EXACT SEQUENCE if for every n the image of $A_{n-1} \to A_n$ is a kernel of $A_n \to A_{n+1}$. (This is, in fact, a self-dual condition.) Algebraic topologists invented the following, called the *five lemma*, for abelian groups:

Consider

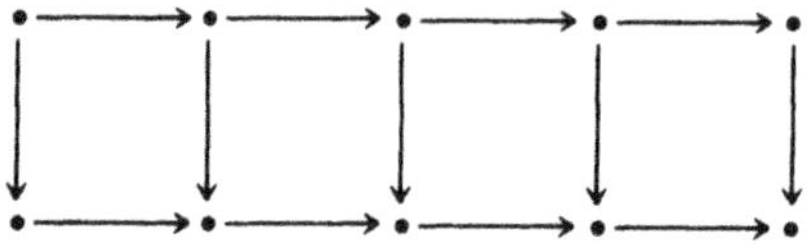

If the rows are exact and if the outer four verticals are isomorphisms, then so is the middle vertical.

The lemma is a Horn sentence is bicartesian predicates hence holds in any abelian category if it does in $\mathscr{Ab}$.

It is fairly easy to show in $\mathscr{Ab}$ that the center vertical has zero kernel. The exact-category definition of abelian category is self-dual. Hence this half of the five lemma is true in $\mathscr{Ab}^{\circ}$. But that means that the center vertical has zero cokernel, hence is an isomorphism.

The most remarkable short lemma true in abelian categories is the *snake lemma* (it is used to construct connecting homorphisms in homological algebra):

Consider

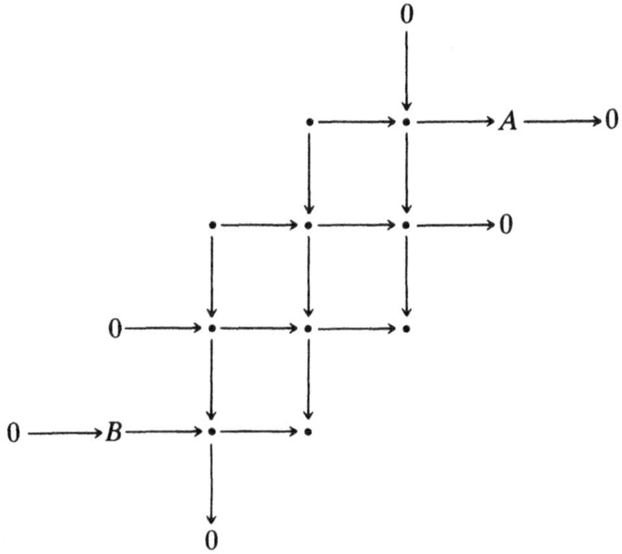

If all rows and columns are exact, then A is isomorphic to B.

The isomorphism is not fortuitous. It is constructed as a relation, namely that which is obtained by composing the reciprocal of the horizontal morphism into A with the relevant vertical morphism then with the reciprocal of the relevant horizontal morphism, and alternating in this way until B is reached.

If suffices to verify in $\mathscr{Ab}$ that such is an isomorphism from A to B.

1.59(10). Given $f\colon A \to B$ in an abelian category, the adjoint pair of functions $f\colon \mathscr{Sub}(A) \to \mathscr{Sub}(B)$, $f^{\#}\colon \mathscr{Sub}(B) \to \mathscr{Sub}(A)$ satisfies the further equations

$$f(A' \cap f^{\#}(B')) = (fA') \cap B', \qquad f^{\#}(f(A') \cup B') = A' \cup f^{\#}B' .$$

For the special case of $f\colon A \rightarrowtail B$ we have that $f^{\#}(B') = A \cap B'$ and the second of the two equations implies that $\mathscr{Sub}(B)$ is a modular lattice.

Let **M** be the category of modular lattices and order-preserving maps each of which has a right-adjoint satisfying the equations above. Then **M** is an exact category. It cannot be faithfully represented in an abelian category.

1.6. PRE-LOGOI

A PRE-LOGOS is a regular category in which $\mathscr{S}\!ub(A)$ is a lattice (not just a semi-lattice) for each A and in which $f^{\#}: \mathscr{S}\!ub(B) \to \mathscr{S}\!ub(A)$ is a lattice homomorphism for each $f: A \to B$.

We may, of course, make the definition elementary by requiring each pair of subobjects to have a union preserved under inverse image.

An equivalent definition: a cartesian category with images in which pullbacks transfer finite covers; that is, given a cover $\{B_i \to B\}_{i=1}^{n}$, a map $A \to B$, there exist

$$\begin{array}{ccc} A_i & \longrightarrow & A \\ \downarrow & & \downarrow \\ B_i & \longrightarrow & B, \end{array} \qquad i = 1, 2, \ldots, n\,,$$

where $\{A_i \to A\}_{i=1}^{n}$ is a cover. Note that the existence of such squares implies that the pullback squares must also be such.

1.61. Let 0 denote the minimal subobject of 1. For any $p: A \to 1$ in a pre-logos, $p^{\#}(0)$ is the minimal subobject of A. If there exists $A \to 0$ then $p_A^{\#}(0) = A$, that is, A has no proper subobjects. The same must be true for $A \times A$. Hence $\langle 1, 1\rangle: \mathrm{A} \to \mathrm{A} \times \mathrm{A}$ is entire and A is a subterminator. Since it is contained in 0 it is isomorphic to 0:

Any morphism to 0 *is an isomorphism.*

Since $p_A^{\#}(0)$ is isomorphic to 0 there exists for each A a morphism $0 \to A$. Since 0 has no proper subobjects there cannot be another morphism to A (else their equalizer would be proper). That is, 0 is a coterminator.

Note that a pre-logos is degenerate iff it is one-valued, that is if $0 \simeq 1$. (If any map $1 \to 0$ exists, then $0 \simeq 1$.)

1.611. We may thus define a pre-logos as a cartesian category with images such that

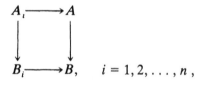

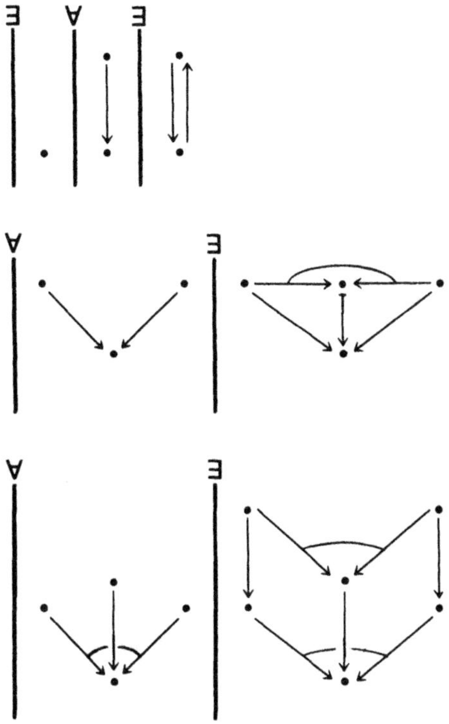

1.612. For the special case of monic $f\colon A \rightarrowtail B$ the inverse-image operation $f^{\#}\colon \mathcal{Sub}(B) \to \mathcal{Sub}(A)$ sends $B' \subset B$ to $A \cap B'$. $f^{\#}$ preserves binary unions for all monic f targeted at B iff $\mathcal{Sub}(B)$ is a DISTRIBUTIVE LATTICE:

$$A \cap (B_1 \cup B_2) = (A \cap B_1) \cup (A \cap B_2)\,.$$

1.613. A poset when viewed as a category is cartesian iff it is a semi-lattice (intersections give products, equalizers exist by default). A semi-lattice is a regular category (the only covers are identity morphisms). A semi-lattice is a pre-logos iff it is a distributive lattice.

1.614. By a REPRESENTATION OF PRE-LOGOI we mean a functor between pre-logoi that preserves the cartesian structure, images, and finite unions (including empty unions).

1.615. Given $x_1\colon A_1 \rightarrowtail A$, $x_2\colon A_2 \to A$ in a bicartesian category with images, their union is the image of $\binom{x_1}{x_2}\colon A_1 + A_2 \to A$. Building on [1.582] we can find a finite set of Horn sentences in bicartesian predicates that hold for a given bicartesian category iff it is a pre-logos.

Since our prime examples $\mathscr{S}^{\mathbf{A}}$ and $\mathscr{H}(Y)$ may be faithfully represented as bicartesian categories in a power of the category of sets [1.585–6], and since the category of sets is a pre-logos, it follows that the prime examples are pre-logoi.

1.616. In a pre-logos $\mathscr{R}el(A, B)$ is a distributive lattice (since it is isomorphic to $\mathscr{S}ub(A \times B)$). We obtain a binary operation $R \cup S$ on relations:

$$R \cap (S \cup T) = (R \cap S) \cup (R \cap T)\,,$$

$$(R \cup S)^\circ = S^\circ \cup R^\circ\,.$$

Viewing S, T as subobjects of $A \times B$, note that for any $y\colon C \to A$ it is the case that yS, when viewed as a subobject of $C \times B$, is $(y \times 1)^{\#}(S)$. Since inverse images preserve unions we thus have

$$y(S \cup T) = (yS) \cup (yT)\,.$$

Direct images always preserve unions. Given $x\colon C \to D$ the composition $x^\circ(yS)$ viewed as a subobject of $D \times B$ is the direct image $(x \times 1)(yS)$. Since any relation is of the form $x^\circ y$ we thus have

$$R(S \cup T) = (RS) \cup (RT)\,.$$

Using reciprocation we can further obtain

$$(S \cup T)R' = (SR') \cup (TR')\,.$$

1.62. PASTING LEMMA. *Suppose that A_1, A_2 are subobjects of A in a prelogos. Then*

$$\begin{array}{ccc} A_1 \cap A_2 & \longrightarrow & A_2 \\ \downarrow & & \downarrow \\ A_1 & \longrightarrow & A_1 \cup A_2 \end{array}$$

is a pushout. Restated:

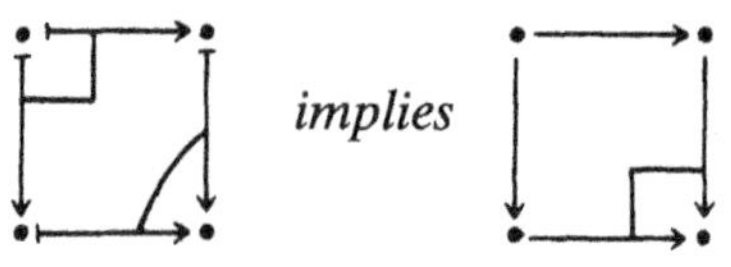

BECAUSE: Given

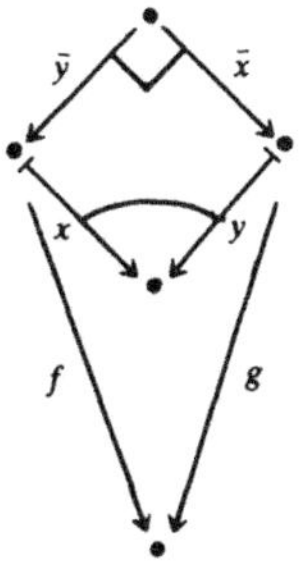

we note first, using just that we are in a regular category, that

$$xx^\circ = 1\,, \qquad yy^\circ = 1\,, \qquad xy^\circ = \bar{y}^\circ\bar{x}\,,$$

$$1 \subset ff^\circ\,, \qquad 1 \subset gg^\circ\,, \qquad f^\circ f \subset 1\,, \qquad g^\circ g \subset 1\,,$$

$$\bar{x}g = \bar{y}f\,, \qquad (\bar{y}f)^\circ \bar{y}f \subset 1\,.$$

A monic morphism z with the same target as x, y contains $\mathrm{Im}(x)$ and $\mathrm{Im}(y)$ (the image of x and the image of y) iff

$$x^\circ x \subset z^\circ z \quad \text{and} \quad y^\circ y \subset z^\circ z\,.$$

Hence, in a pre-logos, we have $x^\circ x \cup y^\circ y = 1$.

Define $R = x^\circ f \cup y^\circ g$. Then

$$\begin{aligned} 1 &\subset x^\circ 1 x \cup y^\circ 1 y \subset x^\circ ff^\circ x \cup y^\circ gg^\circ y \\ &\subset (x^\circ f \cup y^\circ g)(f^\circ x \cup g^\circ y) \subset RR^\circ\,. \end{aligned}$$

$$\begin{aligned} R^\circ R &\subset (f^\circ x \cup g^\circ y)(x^\circ f \cup y^\circ g) \subset f^\circ xx^\circ f \cup f^\circ xy^\circ g \cup g^\circ yx^\circ f \\ &\quad \cup g^\circ yy^\circ g \\ &\subset f^\circ f \cup f^\circ \bar{y}^\circ \bar{x} g \cup g^\circ \bar{x}^\circ \bar{y} f \cup g^\circ g \subset 1 \cup (\bar{y}f)^\circ \bar{x}g \\ &\quad \cup (\bar{x}g)^\circ \bar{y}f \cup 1 \\ &\subset 1 \cup (\bar{y}f)^\circ \bar{y}f \cup (\bar{y}f)^\circ \bar{y}f \subset 1\,. \end{aligned}$$

Since $1 \subset \mathrm{RR}^\circ$ and $R^\circ R \subset 1$, R is a map [1.564].

$$\begin{aligned} xR &= x(x^\circ f \cup y^\circ g) = xx^\circ f \cup xy^\circ g = f \cup \bar{y}^\circ \bar{x} g = f \cup \bar{y}^\circ \bar{y} f \\ &= (1 \cup \bar{y}^\circ \bar{y}) f = f\,. \end{aligned}$$

Similarly, $yR = g$.

The uniqueness of R is a consequence of the fact that x, y cover.

1.621. As a special case,

If A_1, A_2 are subobjects of A in a pre-logos and such that $A_1 \cap A_2 = 0$ and $A_1 \cup A_2 = A$ then A is a coproduct of A_1, A_2.

1.622. It is a curious fact that the pasting lemma holds in abelian categories, but for entirely different reasons (unions of relations do not distribute with composition in any non-degenerate additive regular category). It does not hold in the category of all groups: consider $\mathbb{Z}_2$ and $\mathbb{Z}_3$ as subgroups of the third symmetric group S_3.

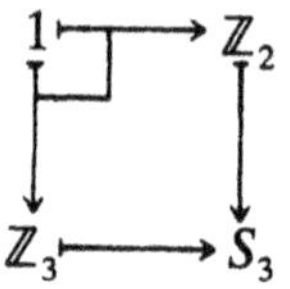

is not a pushout.

1.623. A POSITIVE PRE-LOGOS is a pre-logos in which for every pair of objects A, B there exists

$$\begin{array}{ccc} 0 & \longrightarrow & B \\ \downarrow & & \downarrow \\ A & \longrightarrow & C. \end{array}$$

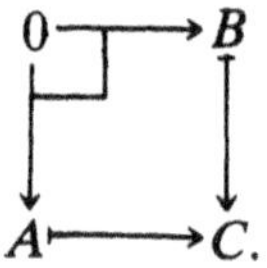

We can, of course, choose C so that $A \cup B = C$, hence a positive pre-logos has coproducts. The prime examples, $\mathscr{S}^A$ and $\mathscr{SR}(Y)$, are easily seen to be positive.

1.624. *In a positive pre-logos any morphism $f: A \to B_1 + B_2$ yields a decomposition*

$$\begin{array}{ccc} A & \xrightarrow{\ \sim\ } & A_1 + A_2 \\ & {}_{f}\searrow & \downarrow {}^{f_1 + f_2} \\ & & B_1 + B_2 \end{array}$$

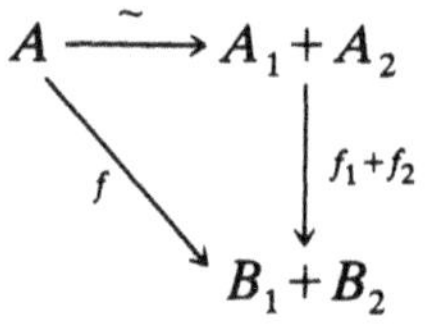

where $A_i = f^{\#}(B_i)$.

1.625. *Suppose $\mathbf{A}$ and $\mathbf{B}$ are positive pre-logoi and that $T: \mathbf{A} \to \mathbf{B}$ is a representation of regular categories. Then T is a representation of pre-logoi iff it preserves disjoint unions.*

BECAUSE: The union of A_1, $A_2 \subset A$ is the image of $A_1 + A_2 \to A$.

1.626. Coproducts can exist without positivity. Any distributive lattice, viewed as a category is a pre-logos with coproducts. It is positive iff it is degenerate.

We shall show in chapter 2 that every pre-logos may be faithfully represented in a positive pre-logos [2.217]. The result of the construction when applied to the two-element lattice is the category of finite sets, the lattice appearing as the lattice of subterminators. Note that the inclusion does not preserve coproducts.

1.63. For any category **A** and object $B \in \mathbf{A}$ the functor $\Sigma: \mathbf{A}/B \to \mathbf{A}$ yields an isomorphism $\mathcal{S}ub_{\mathbf{A}/B}(-) \simeq \mathcal{S}ub_{\mathbf{A}}(\Sigma(-))$. If **A** has pullbacks then these isomorphisms respect inverse images. If **A** has unions, then so does **A**/B. Hence if **A** is a pre-logos then so is $\mathbf{A}/B$ and $\Delta: \mathbf{A} \to \mathbf{A}/B$ is a representation of pre-logoi.

If **A** *is a (positive) pre-logos then so is* $\mathbf{A}/B$.

The equivalence and union conditions needed for the capitalization lemma [1.543] are easily verified. Hence:

Any (positive) pre-logos is faithfully representable in a capital (positive) pre-logos.

1.631. In a pre-logos we say that $A_1 \subset A$ is a COMPLEMENTED SUBOBJECT if there exists $A_2 \subset A$ such that

$$A_1 \cap A_2 = 0\,, \qquad A_1 \cup A_2 = A\,.$$

The distributivity of $\mathcal{S}ub(\mathbf{A})$ implies that A_2, if it exists, is unique.

In a positive pre-logos a complemented subobject of a projective object is projective.

BECAUSE: If P appears as a complemented subobject of some projective, then there exists P' such that $P + P'$ is projective. Given a cover $x: A \twoheadrightarrow P$, use the projectiveness of $P + P'$ to obtain

$$\begin{array}{ccc} & A + P' & \\ {}^{y+1}\nearrow & & \searrow^{x+1} \\ P + P' & \xrightarrow[1+1]{} & P + P'. \end{array}$$

Clearly, $yx = 1$.

Using [1.525] we thus obtain:

The COMPLEMENTED SUBTERMINATORS (*i.e. the complemented subobjects of* 1) *in a capital pre-logos are projective.*

1.632. For any category **A**, a set of objects $\mathscr{G} \subset |\mathbf{A}|$ is a GENERATING SET if the representable functors $\{(G, -)\}_{G \in \mathscr{G}}$ collectively yield an embedding. A set $\mathscr{B} \subset |\mathbf{A}|$ is a BASIS if $\{(B, -)\}_{B \in \mathscr{B}}$ is collectively faithful. Both definitions can be made elementary. Note that in a poset viewed as a category, any set of elements is a generating set, even the empty set. A basis, on the other hand, is a set such that for every element x it is the case that x is the least upper bound of the basis-elements that it bounds. A basis for a topological space, therefore, is precisely the same as a basis, as defined here, for its lattice of open sets.

If **A** is cartesian then $\mathscr{B}$ is a basis iff for every proper $A' \mapsto A$ there exists

$$\begin{array}{ccc} B' & \longrightarrow & B \\ \downarrow & & \downarrow \\ A' & \longmapsto & A \end{array}$$

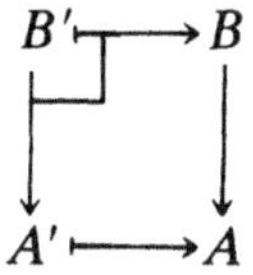

where $B \in \mathscr{B}$ and $B' \mapsto B$ is proper [1.442].

1.633. *A positive pre-logos is capital iff the complemented subterminators are projective and form a basis.*

BECAUSE: We have already seen that in a capital positive pre-logos the complemented subterminators are projective. To see that they form a basis, let $A' \mapsto A$ be proper. Then $A' + 1 \mapsto A + 1$ is still proper and since $A + 1$ is well-supported there exists

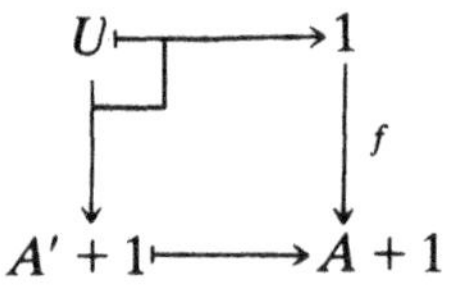

$$\begin{array}{ccc} U & \longrightarrow & 1 \\ \downarrow & & \downarrow f \\ A' + 1 & \longrightarrow & A + 1 \end{array}$$

where $U \mapsto 1$ is proper. f decomposes as $f_1 + f_2 \colon V_1 + V_2 \to A + 1$ [1.631]. For $U_i = U \cap V_i$ we obtain

$$\begin{array}{ccc} U_1 + U_2 & \longrightarrow & V_1 + V_2 \\ \downarrow & & \downarrow \\ A' + 1 & \longrightarrow & A + 1. \end{array}$$

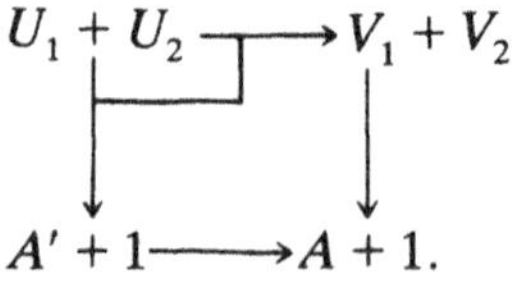

One may easily verify that

$$\begin{array}{ccc} U_1 & \longrightarrow & V_1 \\ \downarrow & & \downarrow \\ A_1' & \longrightarrow & A \end{array} \quad \text{and} \quad \begin{array}{ccc} U_2 & \longrightarrow & V_2 \\ \downarrow & & \downarrow \\ 1 & \longrightarrow & 1 \end{array} \text{ are pullbacks .}$$

Necessarily $U_2 = V_2$. Since U is proper in 1 it must be the case that U_1 is proper in V_1. V_1 is a complemented subterminator.

For the converse, we suppose that the complemented subterminators are projective and form a basis. Given proper $A' \rightarrowtail A$ where A is well-supported, choose $V_1 + V_2 = 1$,

$$\begin{array}{ccc} U_1 & \rightarrowtail & V_1 \\ \downarrow & & \downarrow \\ A' & \rightarrowtail & A, \end{array}$$

$U_1 \rightarrowtail V_1$ proper. Since V_2 is projective, and, $A \twoheadrightarrow 1$ a cover, there exists $V_2 \rightarrow A$. We thus obtain

$$\begin{array}{ccc} U_1 + U_2 & \longrightarrow & V_1 + V_2 \\ \downarrow & & \downarrow \\ A' & \longrightarrow & A. \end{array}$$

$U_1 + U_2$ is proper in $V_1 + V_2 = 1$.

1.634. A subset $\mathscr{F}$ of a poset **P** is a PRE-FILTER if it is non-empty and if for all $x, y \in \mathscr{F}$ there exists $z \in \mathscr{F}$ such that $z \leq x, z \leq y$. It is a FILTER if it is also an updeal.

For an arbitrary category **A** and $\mathscr{F}$ a pre-filter in $\mathscr{V}al_{\mathbf{A}}$, we define a set-valued functor $\Gamma_{\mathscr{F}}$ as follows: $\Gamma_{\mathscr{F}}(A)$ is the set whose elements are named by morphisms of the form $x: U \rightarrow A$, $U \in \mathscr{F}$. $x: U \rightarrow A$, $y: V \rightarrow A$ name the same element if there exists $W \subset U$, $W \subset V$, $W \in \mathscr{F}$ such that

$$\begin{array}{ccc} W & \longrightarrow & U \\ \downarrow & & \downarrow{\scriptstyle x} \\ V & \underset{y}{\longrightarrow} & A. \end{array}$$

$\Gamma_{\mathscr{F}}$ preserves finite products and equalizers. If the elements of $\mathscr{F}$ are projective then $\Gamma_{\mathscr{F}}$ preserves covers.

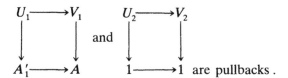

If **A** *is a pre-logos then* $\Gamma_{\mathscr{F}}$ *preserves disjoint unions iff*

$$0 \notin \mathscr{F}\,,$$

$$U_1 + U_2 \in \mathscr{F} \quad \textit{implies} \quad U_1 \in \mathscr{F} \textit{ or } U_2 \in \mathscr{F}\,.$$

BECAUSE: $0 \notin \mathscr{F}$ is equivalent with $\Gamma_{\mathscr{F}}(0) = \emptyset$. Given an element of $\Gamma_{\mathscr{F}}(A_1 + A_2)$ we wish to show that it comes from either $\Gamma_{\mathscr{F}}(A_1)$ or $\Gamma_{\mathscr{F}}(A_2)$. (Since $\Gamma_{\mathscr{F}}$ preserves pullbacks and $\Gamma_{\mathscr{F}}(0) = \emptyset$ it cannot come from both.) Suppose the element is named by $U \to A_1 + A_2$. We obtain a decomposition $f_1 + f_2$: $U_1 + U_2 \to A_1 + A_2$. If $U_i \in \mathscr{F}$, then the given element comes from the element in $\Gamma_{\mathscr{F}}(A_i)$ named by f_i: $U_i \to A_i$.

1.635. THE REPRESENTATION THEOREM FOR PRE-LOGOI

We shall show in chapter 2 that every pre-logos is faithfully representable in a positive pre-logos [2.217]. Hence the word 'positive' will be removable from:

Every small positive pre-logos is faithfully representable in a power of the category of sets.

BECAUSE: We may concentrate on a capital positive pre-logos **A** [1.63]. Let $\mathscr{B}$ be a representative set of complemented subterminators. $\mathscr{B}$ will be viewed as a sublattice of $\mathscr{V}\!al$. It is distributive and every element has a complement, hence it is a BOOLEAN ALGEBRA. Any pre-filter $\mathscr{F} \subset \mathscr{B}$ is a pre-filter in $\mathscr{V}\!al$ and we obtain a representation $\Gamma_{\mathscr{F}}$: $\mathbf{A} \to \mathscr{S}$ of regular categories (since the elements of $\mathscr{F}$ are projective).

We recall the standard theorems: if $\mathscr{F}$ is a pre-filter in a boolean algebra, $\mathscr{B}$, we may use the axiom of choice to find an ULTRA-FILTER $\mathscr{F} \subset \bar{\mathscr{F}}$, that is, a maximal pre-filter such that $0 \notin \bar{\mathscr{F}}$.

If $\bar{\mathscr{F}}$ *is an ultra-filter in* $\mathscr{B}$ *then* $\Gamma_{\bar{\mathscr{F}}}$ *is a representation of pre-logoi.*

To prove the above assertion, it suffices to show that $\Gamma_{\bar{\mathscr{F}}}$ preserves disjoint unions [1.625] and for that it suffices to show that if U, V are disjoint and if $U \cup V \in \bar{\mathscr{F}}$ then either $U \in \bar{\mathscr{F}}$ or $V \in \bar{\mathscr{F}}$ [1.634]. Consider the set $\mathscr{F}' = \bar{\mathscr{F}} \cup \{U \cap W \mid W \in \bar{\mathscr{F}}\}$, easily seen to be a pre-filter. If $\mathscr{F}' = \bar{\mathscr{F}}$ then $U \in \bar{\mathscr{F}}$. If $\mathscr{F}' \neq \bar{\mathscr{F}}$ then the maximality of $\bar{\mathscr{F}}$ says that $0 \in \mathscr{F}'$. There must exist $W \in \mathscr{F}'$ disjoint from U. An ultra-filter is easily seen to be a filter, hence closed under intersection. Hence $V \cap W = (U \cup V) \cap W \in \bar{\mathscr{F}}$. Since it is a filter, $V \in \bar{\mathscr{F}}$.

We must show that the collection of such representations is collectively faithful, that is, given proper $A' \rightarrowtail A$ we must show that there is an ultra-filter $\bar{\mathscr{F}}$ such that $\Gamma_{\bar{\mathscr{F}}}(A') \rightarrowtail \Gamma_{\bar{\mathscr{F}}}(A)$ is proper. Using the fact that $\mathscr{B}$ is a basis we choose $U \in \mathscr{B}$,

$$\begin{array}{ccc} U' & \rightarrowtail & U \\ \downarrow & & \downarrow \\ A' & \rightarrowtail & A \end{array}$$

such that $U' \rightarrowtail U$ is proper. It suffices to find an ultra-filter $\bar{\mathscr{F}}$ such that $\Gamma_{\bar{\mathscr{F}}}(U') \rightarrowtail \Gamma_{\bar{\mathscr{F}}}(U)$ is proper. There is only one way in which that can occur:

$$\Gamma_{\bar{\mathscr{F}}}(U') = \emptyset\,, \qquad \Gamma_{\bar{\mathscr{F}}}(U) \neq \emptyset\,.$$

The first condition is equivalent to: $U \in \bar{\mathscr{F}}$.

The second condition is equivalent to: $V \subset U'$, $V \in \mathscr{B}$ implies $V \not\in \bar{\mathscr{F}}$.

Let $\mathscr{F}$ be the set of all $W \in \mathscr{B}$ such that $U \subset U' \cup W$. Since $\mathscr{V}\!al$ is distributive, $\mathscr{F}$ is a pre-filter. Let $\bar{\mathscr{F}} \subset \mathscr{B}$ be an ultra-filter containing $\mathscr{F}$. The first condition is immediate. For the second condition suppose that $V \subset U'$, $V \in \mathscr{B}$. Then the complement V' of V is in $\mathscr{F}$, hence in $\bar{\mathscr{F}}$. If V were in $\bar{\mathscr{F}}$ then $0 = V \cap V' \in \bar{\mathscr{F}}$.

1.636. Consequently

Any Horn sentence in the predicates of pre-logoi true for the category of sets, is true for all positive pre-logoi.

In chapter 2 we shall be able to remove the word 'positive' [2.217].

1.637. By a *special pre-logos* we mean a pre-logos which satisfies every universal sentence in the predicates of pre-logoi satisfied by the category of sets. A pre-logos is a special regular category iff it is two-valued.

A pre-logos is special iff for every pair of proper subobjects $A' \subset A$, $B' \subset B$ it is the case that $(A' \times B) \cup (A \times B')$ is a proper subobject of $A \times B$.

BECAUSE: The condition is equivalent to a universal sentence that holds for the category of sets, hence is a necessary condition for special pre-logoi.

The same argument used in [1.472] may be used here to show that it suffices to construct a set-valued representation of pre-logoi that preserves any given finite set of proper subobjects $\{A'_i \subset A_i\}_{i=1}^{n}$. The condition guarantees that the union B of the subobjects of $A_1 \times A_2 \times \cdots \times A_n$ of the form $A_1 \times A_2 \times \cdots \times A'_i \times \cdots \times A_n$ is proper. If T is a representation of pre-logoi that preserves the properness of B in $A_1 \times A_2 \times \cdots \times A_n$ then it preserves the properness of each $A'_i \subset A_i$.

1.638. For small **A** the functor category $\mathscr{S}^{\mathbf{A}}$ is a special regular category iff **A** is strongly connected, that is, if for every ordered pair of objects $\langle A, B\rangle$ there exists $A \to B$ [1.427, 1.552].

$\mathscr{S}^{\mathbf{A}}$ is a special pre-logos iff it is *well-joined*: for every pair of objects A, B there exists

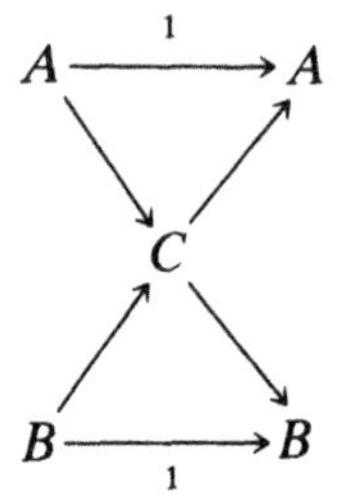

(e.g. any strongly connected category with products).

1.639. The category of recursive functions $\underline{\mathbf{R}}$ [1.572] and the category of primitive recursive functions $\underline{\mathbf{P}}$ [1.573] are easily seen to be positive pre-logoi. The forgetful functor from $\underline{\mathbf{R}}$ to $\mathscr{S}$ is a faithful representation of pre-logoi. There is no such representation for $\underline{\mathbf{P}}$ because $\underline{\mathbf{P}}$ is not a special pre-logos:

Let A, A' be complementary primitive recursive infinite sets that are not isomorphic to the natural numbers ω. We can assume $0 \in \mathrm{A}$. Define $B = A' \cup \{0\}$.

Let $f: \omega \to \omega$ be the function defined by

$$f(0) = 0\,,$$
$$f(n+1) = \begin{cases} f(n) & \text{if } n+1 \notin A\,, \\ f(n)+1 & \text{if } n+1 \in A\,. \end{cases}$$

Let $g: \omega \to \omega$ be the function similarly defined using B in place of A. Note that $f(n) + g(n) = n$.

$A \rightarrowtail \omega \xrightarrow{f} \omega$ is monic. It must be a proper subobject (else $A \simeq \omega$). Similarly $B \rightarrowtail \omega \xrightarrow{g} \omega$ represents a proper subobject. The specialness of $\underline{\mathbf{P}}$ will be obstructed by the fact that $(A \times \omega) \cup (\omega \times B) = (\omega \times \omega)$.

Define $h: \omega + \omega \to \omega$ by

$$h\langle n, m\rangle = \min\{i \leqslant n + m \mid f(i) = n \text{ or } g(i) = m\}.$$

Let $C \subset \omega \times \omega$ be the primitive recursive subset defined as $\{\langle n, m\rangle \mid fh\langle n, m\rangle = n\}$.

Define $f': C \to A \times \omega$ as the function that sends $\langle n, m\rangle$ to $\langle h\langle n, m\rangle, m\rangle$.

Since $C \xrightarrow{f'} A \times \omega \xrightarrow{f\times 1} \omega \times \omega$ is the inclusion map of C we have that C is contained in $A \times \omega$.

Similarly, let C' be the complement of C and define $g: C' \to \omega \times B$ as the function that sends $\langle n, m\rangle$ to $\langle n, h\langle n, m\rangle\rangle$. C' is contained in $\omega \times B'$.

$$\omega \times \omega = C \cup C' \subset (A \times \omega) \cup (\omega \times B).$$

(1.64.) A BOOLEAN PRE-LOGOS is a pre-logos in which $\mathcal{S}ub(A)$ is a boolean algebra for all A. Because $\mathcal{S}ub(A)$ is distributive we need only require the existence of complements (which are necessarily unique)

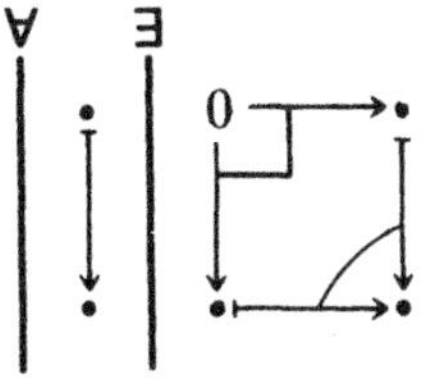

(1.641.) If $\mathbf{A}$ and $\mathbf{B}$ are boolean pre-logoi and if $T: \mathbf{A} \to \mathbf{B}$ is a representation of pre-logoi then T automatically preserves the boolean structure: that is, $\mathcal{S}ub(A) \to \mathcal{S}ub(TA)$ preserves complements and is a map of boolean algebras, for all $A \in \mathbf{A}$.

(1.642.) *For $\mathbf{A}$ a small category, $\mathcal{S}^{\mathbf{A}}$ is boolean iff $\mathbf{A}$ is a groupoid.*

BECAUSE: If $\mathbf{A}$ is a groupoid it is equivalent to a disjoint union of groups $\{G_i\}_I$ (namely, its skeleton), hence $\mathcal{S}^{\mathbf{A}}$ is equivalent to $\Pi_I \mathcal{S}^{G_i}$. Products of boolean pre-logoi are boolean. The category of G-sets, for any group G, is easily seen to be boolean.

Conversely, if $\mathcal{S}^{\mathbf{A}}$ is boolean, consider for any $A \in \mathbf{A}$ the maximal proper subfunctor $M \subset H^A$. $M(B)$ is the set of morphisms $A \to B$ which are *not* right-invertible. If M has a complement it is necessary for $M(A)$ to be empty. Hence every morphism in $\mathbf{A}$ is right-invertible, hence $\mathbf{A}$ is a groupoid [1.17].

(1.643.) Assuming that all spaces are T_0:

$\mathcal{S}h(Y)$ *is boolean iff Y is discrete.*

BECAUSE: If Y is discrete then $\mathcal{S}h(Y) = \mathcal{S}/Y$.

Conversely, if $\mathcal{S}h(Y)$ is boolean then the lattice of open subsets of Y is boolean. Every open set has an open complement, hence every open set is closed. Every closed set is open. T_0 now implies T_2. Every subset is a union of closed subsets, hence a union of open subsets, hence open.

(1.644.) Any power of $\mathcal{S}$ is a boolean capital positive pre-logos. We shall notationally confuse $\mathcal{V}al_{\mathcal{S}^I}$ with the power set of I. Given an ultra-filter $\mathcal{F} \subset \mathcal{V}al_{\mathcal{S}^I}$ the functor $\Gamma_{\mathcal{F}}: \mathcal{S}^I \to \mathcal{S}$ is called an ULTRA-PRODUCT FUNCTOR, and $\mathcal{S} \xrightarrow{\Delta} \mathcal{S}^I \xrightarrow{\Gamma_{\mathcal{F}}} \mathcal{S}$ is called an ULTRA-POWER FUNCTOR. Both are representations of boolean pre-logoi.

(1.645.) For a representation of boolean pre-logoi $T: \mathbf{A} \to \mathbf{B}$ we define $\mathcal{K}er(T)$ as the set of values killed by T: $\mathcal{K}er(T) = \{U \subset 1 \mid T(U) = 0\}$.

Given a subobject $A' \subset A$ in $\mathbf{A}$ we define $\mathit{prop}(A' \subset A)$, its *properness*, as the support of the complement of A'. $A' \subset A$ is proper iff $\mathit{prop}(A' \subset A) \neq 0$. $TA' \subset TA$ is proper iff $\mathit{prop}(A' \subset A) \not\in \mathcal{K}er(T)$. T is faithful iff $\mathcal{K}er(T) = 0$.

$1 \in \mathcal{K}er(T)$ iff $\mathbf{B}$ is degenerate. If $\mathbf{A}$ is two-valued and $\mathbf{B}$ non-degenerate then T is necessarily faithful. Every ultra-power functor, in particular, is faithful.

For any ultra-filter $\mathcal{F} \subset \mathcal{V}al_{\mathbf{A}}$, $\mathcal{K}er(\Gamma_{\mathcal{F}})$ is the complement of $\mathcal{F}$ in $\mathcal{V}al_{\mathbf{A}}$.

1.646. Two propositions with the same proof:

Every small special cartesian category is faithfully representable in the category of sets.

Every small special positive pre-logos is faithfully representable in the category of sets.

BECAUSE: We have shown in each case [1.472, 1.637] for every finite set S of proper subobjects $\{A'_i \subsetneqq A_i\}_{i=1}^{n}$ that there is a representation $T_S: \mathbf{A} \to \mathcal{S}$ such that $TA'_i \subsetneqq TA_i$, $i = 1, 2, \ldots, n$.

Let I be the set of such finite sets and choose T_S for each $S \in I$ to obtain $T: \mathbf{A} \to \mathcal{S}^I$. Let $\mathcal{F}$ be the pre-filter of principal co-ideals in I: that is, $I' \in \mathcal{F}$ iff there exists $S \in I$ such that $I' = \{S' \mid S \subset S'\}$. Let $\bar{\mathcal{F}}$ be an ultra-filter which contains $\mathcal{F}$. We wish to show that $\mathbf{A} \xrightarrow{T} \mathcal{S}^I \xrightarrow{\Gamma_{\bar{\mathcal{F}}}} \mathcal{S}$ is faithful.

Note that $\mathit{prop}(T_S A' \subset T_S A) = 1$ if $(A' \subset A) \in S$. Regarding $\mathit{prop}(TA' \subset TA)$ as a subset of I we see that it contains an element of $\mathcal{F}$. It is therefore an element of $\bar{\mathcal{F}}$. It is not an element of $\mathcal{K}er(\Gamma_{\bar{\mathcal{F}}})$.

1.647. Let A_i be a proper subobject of B_i $(i = 1, 2)$ in a boolean pre-logos, and let A'_i be the complement of A_i in B_i. The complement of $(A_1 \times B_2) \cup (B_1 \times A_2)$ in $B_1 \times B_2$ is $A'_1 \times A'_2$. Hence, [1.637],

A boolean pre-logos is special iff it is two-valued.

1.648. When is an ultra-power functor $T = \mathcal{S} \xrightarrow{\Delta} \mathcal{S}^I \xrightarrow{\Gamma_{\mathcal{F}}} \mathcal{S}$ [1.644] a representation of bicartesian categories? Let $\mathbb{N}, 0, s$ be as described in [1.587]. For any natural number n we will denote $1 \xrightarrow{0} \mathbb{N} \xrightarrow{s^n} \mathbb{N}$ as $1 \xrightarrow{n} \mathbb{N}$. Since $T(\mathbb{N}), T(0), T(s)$ is isomorphic to the natural numbers [1.587] we know that for any $x: T(1) \to T(\mathbb{N})$ there exists n such that $x = T(1 \xrightarrow{n} \mathbb{N})$. Let $\{A_n\}_{n=0}^{\infty}$ be a countable partioning of I. Consider the element $x: \Delta 1 \to \Delta\mathbb{N}$ in $\mathcal{S}^I$ whose i-th coordinate is $1 \xrightarrow{n} \mathbb{N}$ if $i \in A_n$. Let n be such that $\Gamma_{\mathcal{F}}(x) = T(1 \xrightarrow{n} \mathbb{N})$. Let E be the equalizer of x and Δn. Regarding E as a value and values as subsets of I, E is the subset A_n. Since $\Gamma_{\mathcal{F}}(E) \neq \emptyset$ it must be the case that $A_n \in \mathcal{F}$.

We have shown that if an ultra-power functor preserves coequalizers (indeed, just one particular coequalizer) then the ultra-filter has the property that it meets every countable partitioning of I. Such an ultra-filter is called a COMPLETE MEASURE. Every acceptable axiom in set theory is known to be consistent with the assertion that all complete measures are ATOMIC MEASURES, that is, of the form $\{j \mid i \leq j\}$, for some i. The last sentence is a tautology. If $\mathscr{F}$ is an atomic measure then $\Gamma_{\mathscr{F}}$ is conjugate to a projection and T is conjugate to the identity functor.

1.65. Pre-topoi

A PRE-TOPOS is an effective positive pre-logos.

In chapter 2, we will show that every pre-logos may be faithfully represented in a pre-topos [2.217].

1.651. The AMALGAMATION LEMMA. *Given a pair of monics with a common source $x: A \rightarrowtail B$, $y: A \rightarrowtail C$ in a pre-topos there exists*

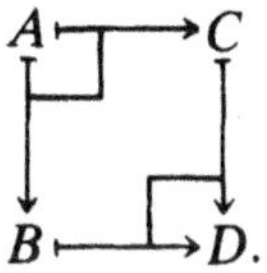

BECAUSE: Let $B \xrightarrow{l} B+C \xleftarrow{r} C$ be a coproduct and consider the endorelation $E = l^\circ x^\circ yr \cup 1 \cup r^\circ y^\circ xl$ on $B + C$. As with regular categories, we may rely on the category of sets for Horn sentences involving the calculus of relations, with the union operation now included. E is easily verified to be an equivalence relation. Let $\binom{\bar{y}}{\bar{x}}: B + C \twoheadrightarrow D$ be such that E is its level. It suffices to show, in $\mathscr{S}$, that $\bar{x}$, $\bar{y}$ are monic and that

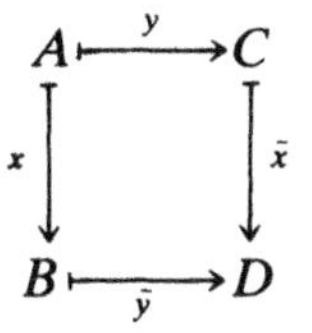

is a pullback. The pasting lemma [1.62] says that it is also a pushout.

1.652. *In a pre-topos, covers coincide with epics and monics coincide with cocovers.*

BECAUSE: Given monic $x: A \rightarrowtail B$ consider

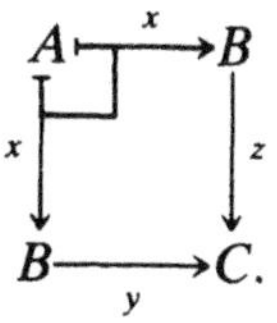

x is an equalizer of y, z, hence a cocover.

Since every subobject now appears as an equalizer, the image of an epic must be entire. That is, epics are covers.

1.653. *Given a pair of morphisms in a pre-topos with a common source, one of which is monic, $A \to B$, $y: A \mapsto C$, there exists*

$$\begin{array}{ccc} A & \rightarrowtail & C \\ \downarrow & & \downarrow \\ B & \longmapsto & D. \end{array}$$

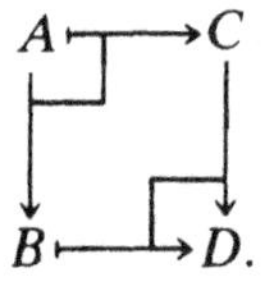

BECAUSE: Factor $A \to B$ as $A \overset{x}{\to} I \mapsto B$. It suffices to show that there exists

$$\begin{array}{ccc} A & \overset{y}{\longmapsto} & C \\ x\downarrow & & \downarrow \\ I & \longmapsto & D' \end{array}$$

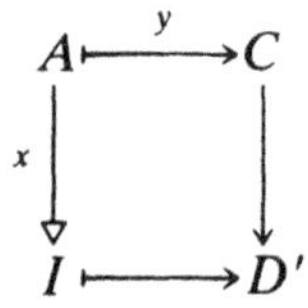

because [1.651] can be applied to $I \mapsto B$, $I \mapsto D'$ and the squares can be stacked together.

Let $E = xx^\circ$ be the level of x. $E' = 1 \cup y^\circ E y$ is easily checked to be an equivalence relation on C. Let $z: C \to D'$ be such that E' is its level. Define $R \in \mathscr{Rel}(I, D')$ as $x^\circ yz$. Verify (in $\mathscr{S}$) that R is a monic map and that the resulting square is a pullback.

Given

$$\begin{array}{ccc} A & \overset{y}{\longmapsto} & C \\ x\downarrow & & \downarrow f \\ I & \underset{g}{\longrightarrow} & X \end{array}$$

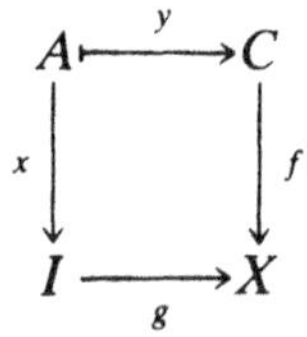

define $H \in \mathscr{Rel}(D', X)$ as $z^\circ f$. Verify (in $\mathscr{S}$) that H is a map and that $RH = g$, $zH = f$.

1.654. A pre-topos need not be cocartesian. If it is, however, the last two sections show that its opposite category is regular.

(1.655.) *If $\mathbf{A}$ and $\mathbf{B}$ are pre-topoi and $T: \mathbf{A} \to \mathbf{B}$ a functor which preserves 0, pushouts, finite products and monics then T is a bicartesian representation.*

BECAUSE: Given a pair of monics in **A** with a common target $A_1 \rightarrowtail A$, $A_2 \rightarrowtail A$ consider their pullback

$$\begin{array}{ccc} A_{12} & \rightarrowtail & A_2 \\ \downarrow & \lrcorner & \downarrow \\ A_1 & \rightarrowtail & A. \end{array}$$

To see that

$$\begin{array}{ccc} TA_{12} & \longrightarrow & TA_2 \\ \downarrow & & \downarrow \\ TA_1 & \longrightarrow & TA \end{array} \quad \text{is a pullback}$$

it suffices to show that

$$\begin{array}{ccc} TA_{12} & \longrightarrow & TA_2 \\ \downarrow & & \downarrow \\ TA_1 & \longrightarrow & T(A_1 \cup A_2) \end{array} \quad \text{is a pullback}\,.$$

By the pasting lemma [1.62]

$$\begin{array}{ccc} A_{12} & \longrightarrow & A_2 \\ \downarrow & & \downarrow \\ A_1 & \longrightarrow & A_1 \cup A_2 \end{array} \quad \text{is a pushout}\,.$$

Hence by hypothesis on T,

$$\begin{array}{ccc} TA_{12} & \longrightarrow & TA_2 \\ \downarrow & & \downarrow \\ TA_1 & \longrightarrow & T(A_1 \cup A_2) \end{array} \quad \text{is a pushout}\,.$$

By [1.651], the last square is a pullback.

Since T preserves finite products the proof of [1.434] now says that it preserves equalizers.

Since covers are coequalizers and by hypothesis T preserves pushouts and 0 it is the case that T preserves co-equalizers and therefore covers.

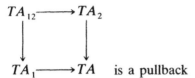

1.656. For functors between abelian categories the analogous theorem holds: preservation of the cocartesian structure and monics implies preservation of the cartesian structure. And for abelian categories the dual theorem holds. Not so for pre-topoi. We have seen that there must be examples [1.648] using the axiom of choice. We will later find an example independent of the axiom of choice [1.853].

(1.657.) Let **A** be an effective regular category, R an endo-relation on an object A, tabulated by

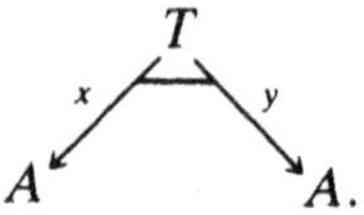

Suppose that $f\colon A \to B$ is a coequalizer of x, y. The effectiveness of **A** implies that ff° is the minimal equivalence relation that contains R.

Suppose, conversely, that for every R there were a minimal equivalence relation $\bar{R}$ containing R. Then for $x, y\colon C \to A$ we can choose $A \to B$ so that the level of $A \to B$ is $\overline{x^\circ y}$ and verify that $A \to B$ is a coequalizer of x, y. Hence,

A pre-topos is cocartesian iff for every endo-relation R there is a minimal equivalence relation that contains R. A functor between bicartesian pre-topoi preserves the bicartesian structure iff it is a representation of pre-logoi that preserves minimal equivalence relations.

1.658. An object A in a pre-logos is DECIDABLE if the diagonal $\langle 1, 1\rangle\colon A \rightarrowtail A \times A$ has a complement.

Every object in a pre-topos is decidable iff the pre-topos is boolean.

BECAUSE: Given $A' \rightarrowtail A$ consider

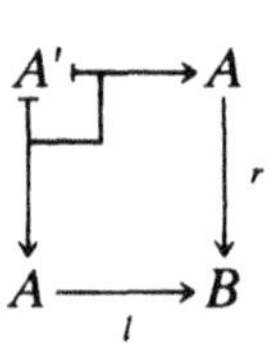

and

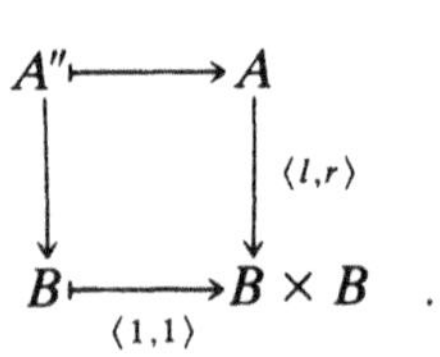

A' is the equalizer of l, r. So is A''. Pullbacks of complemented subobjects are complemented.

1.659. $T \in \mathscr{S}^{\mathbf{A}}$ is decidable iff $T(x)$ is a monic map for all $x\colon A \to B \in \mathbf{A}$.

$X \to Y \in \mathscr{H}(Y)$ is decidable iff for any $x, x' \in X$ lying in the same stalk, there exist disjoint neighborhoods. Hence if Y is Hausdorff, then $X \to Y$ is decidable iff X is Hausdorff.

Given connected open $U \subset Y$ and a pair of maps $f, g\colon U \to X$ into a decidable sheaf, then the equalizer of f, g is either all of U or nothing. For Y a reasonable space (the reals, for example) the sheaf of germs of continuous functions is not decidable. (The sheaf of germs of analytic functions *is* decidable.)

1.66. We study choice objects [1.57] in a pre-topos. First, in any regular category it is immediate that

A subobject of a choice object is choice.

Also,

A quotient of a choice object is choice.

BECAUSE: A quotient [1.568] $x: A \twoheadrightarrow B$ is also a subobject of A via a map contained in x°.

1.661. *In a regular category, finite products of choice objects are choice.*

BECAUSE: Any entire relation targeted at a terminator is already a map. For binary products, let B_1, B_2 be choice, and let R be an entire relation targeted at $B_1 \times B_2$. Then $R\ell$ is entire and targeted at B_1, so it contains a map f_1. In $\mathscr{S}$ets, it follows that $R \cap f_1\ell^\circ$ is entire, and thus it follows in any regular category [1.551, 1.563]. Then $(R \cap f_1\ell^\circ)r$ is an entire relation targeted at B_2, so it contains a map f_2. $\langle f_1, f_2\rangle \subset R$.

1.662. *In a pre-topos, the following are equivalent:*
(1) *Binary coproducts of choice objects are choice.*
(2) $1 + 1$ *is choice.*
(3) *The pre-topos is boolean.*

BECAUSE: (1) implies (2) trivially. From (2) to (3), we pass through an intermediate condition:

(2a) In every $\mathscr{S}ub(B)$, if $X \cup Y = B$, then there exists X', Y' such that $X' \subset X$, $Y' \subset Y$, $X' \cup Y' = B$, and $X' \cap Y' = 0$.
The statement (2a) is just a restatement of (2) because $\mathscr{R}el(B, 1 + 1) \simeq \mathscr{S}ub(B) \times \mathscr{S}ub(B)$. In particular, an entire relation from B to $1 + 1$ is a pair of subobjects of B that cover B. $\mathscr{M}ap(B, 1 + 1)$ consists of partitions of B.

Note that (2a) is inherited by slices. Thus it suffices to show that $\mathscr{S}ub(1)$ is a boolean algebra, because $\mathscr{S}ub_{\mathbf{A}}(B) \simeq \mathscr{S}ub_{\mathbf{A}/B}(1)$. Let $U \in \mathscr{S}ub(1)$. We show that it is complemented. Look at

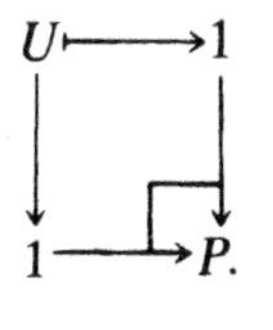

P is a quotient of $1 + 1$, and we are assuming that $1 + 1$ is choice, so P is a subobject of $1 + 1$. $1 + 1$ is decidable [1.658], and so is P. Now

$$\begin{array}{ccc} U & \rightarrowtail & 1 \\ \downarrow & \lrcorner & \downarrow \\ P & \underset{\langle 1,1\rangle}{\rightarrowtail} & P \times P, \end{array}$$

so U is complemented as a pullback of a complemented object.

(3) Implies (1): Given a relation R from an object, let $\mathscr{D}om(R)$ be the subobject given by $1 \cap RR^\circ$. If S is an entire relation from A to $B_1 + B_2$, and if u_i: $B_i \rightarrowtail B_1 + B_2$ ($i = 1, 2$) are the canonical monics, the restriction of $\mathrm{S}u_1^\circ$ to $\mathscr{D}om(\mathrm{S}u_1^\circ)$ is entire, so it contains a map f_1, because B_1 is choice. The restriction of S to the complement of $\mathscr{D}om(Su_1^\circ)$ is entire, so it contains a map, because B_2 is choice. This map factors as $f_2 u_2$. Because the pre-topos is boolean, $f_1 + f_2$ is a map from A to $B_1 + B_2$ contained in S.

Diaconescu was the first to show, by a different argument, that an AC topos is boolean.

Not every boolean pre-topos is AC. Recall that a boolean pre-topos of G-sets, G a group, is AC iff G is trivial [1.57, 1.642].

1.7. LOGOI

A LOGOS is a regular category in which $\mathcal{S}ub(A)$ is a lattice for each object A, and in which the inverse-image operation $f^{\#}: \mathcal{S}ub(B) \longrightarrow \mathcal{S}ub(A)$ has a right adjoint $f^{\#\#}: \mathcal{S}ub(A) \longrightarrow \mathcal{S}ub(B)$ for each morphism $f: A \longrightarrow B$. That is,

$$f^{\#}(B') \subset A' \quad \textit{iff} \quad B' \subset f^{\#\#}(A') .$$

The existence of $f^{\#\#}(A')$ may be restated as:

There is a maximal subobject B' such that $f^{\#}(B') \subset A'$.

1.71. In the category of sets,

$$f^{\#\#}(A') = \{b \in B \mid \forall_a [f(a) = b \Rightarrow a \in A']\}.$$

In any boolean pre-logos, $f^{\#\#}(A')$ may be constructed as $\neg(f(\neg A'))$, where $\neg$ denotes complements: $B' \subset \neg f(\neg A')$ iff $f(\neg A') \subset \neg B'$ iff $\neg A' \subset f^{\#}(\neg B')$ iff $\neg f^{\#}(\neg B') \subset A'$. Now use the fact that $f^{\#}$ preserves complements.

1.711. A logos is a pre-logos. Indeed, in any logos, $f^{\#}$ preserves all unions that may exist. Suppose that $\{B_i \subset B\}$ is a family of subobjects with a least upper bound: $f^{\#}(\bigcup B_i) \subset A'$ iff $\bigcup B_i \subset f^{\#\#}(A')$ iff $B_i \subset f^{\#\#}(A')$ for all i, iff $f^{\#}(B_i) \subset A'$ for all i.

1.712. We say that a category is LOCALLY COMPLETE if for every object A the poset $\mathcal{S}ub(A)$ is a complete lattice. Suppose that $\mathbf{A}$ is a locally complete regular category in which $f^{\#}$ preserves *all* unions for every morphism $f: A \longrightarrow B$. Then $\mathbf{A}$ is a logos: given $A' \subset A$ let $\{B_i\}$ be the family of subobjects such that $f^{\#}(B_i) \subset A'$. Then $f^{\#\#}(A')$ may be constructed as $\bigcup B_i$.

1.713. Both $\mathcal{S}^{\mathbf{A}}$ and $\mathcal{S}h(Y)$ are easily verified to be locally complete. The evaluation functors $\mathcal{S}^{\mathbf{A}} \longrightarrow \mathcal{S}$ preserve inverse-images and arbitrary unions. Because they are collectively faithful, $\mathcal{S}^{\mathbf{A}}$ is seen to satisfy the condition of the last section. For $\mathcal{S}h(Y)$, use the stalk functors.

1.714. A logos may be defined as a pre-logos in which

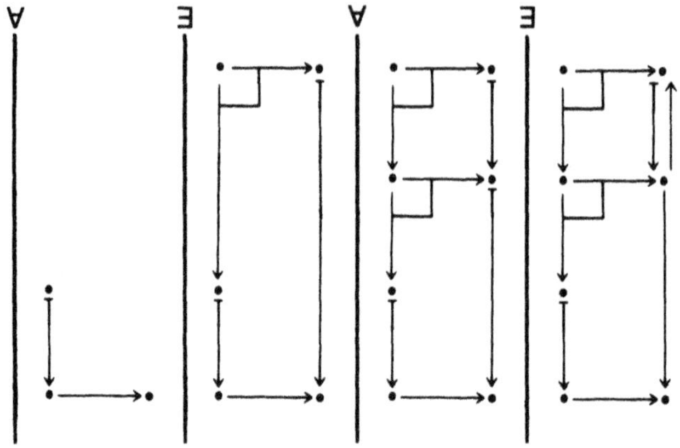

1.72. A HEYTING ALGEBRA is a lattice with a binary operation denoted $x \to y$, satisfying the double-Horn sentence:

$$z \leqslant x \to y \quad \textit{iff} \quad z \wedge x \leqslant y \,.$$

($x \to y$ is the largest element whose intersection with x is contained in y. If we fix x, then $x \to (\_)$ is the right adjoint of $x \wedge (\_)$.)

A Heyting algebra viewed as a category is a logos. The notation presents a problem. Given $x \leqslant y$, we let f denote the unique morphism from x to y. Given a subobject u of y (i.e. $u \leqslant y$) then, of course, $f^{\#}(u) = x \wedge u$. Given a subobject v of x, we have $f^{\#\#}(v) = (x \to v) \wedge y$.

1.721. Let A be an object in a logos. $\mathscr{S}\!\mathit{ub}(A)$ is a Heyting algebra: given $A_1, A_2 \subset A$ let f be the inclusion map of A_1. $(A_1 \to A_2)$ is constructible as $f^{\#\#}(A_1 \cap A_2)$.

1.722. Putting the last two sections together, we have:

A poset, when viewed as a category, is a logos iff it is a Heyting algebra (more precisely, iff it is the poset underying a [necessarily unique] Heyting algebra).

1.723. A LOCALE is a complete lattice in which finite intersections distribute over arbitrary unions:

$$x \wedge \bigvee y_i = \bigvee (x \wedge y_i) \,.$$

A complete lattice, viewed as a category is, of course, locally complete.

The distributivity condition is equivalent to the condition that inverse-images preserve arbitrary unions. Thus, using [1.712] and [1.722], a locale is a Heyting algebra.

(The phrases 'category of complete Heyting algebras' and 'the category of locales' do not usually refer to the same category. The objects are the same. The morphisms in the second category are usually taken to be the lattice homomorphisms that preserve arbitrary unions. They need not preserve the arrow operation.)

1.724. In a Heyting algebra define $x \leftrightarrow y$ as $(x \to y) \wedge (y \to x)$. $x \leftrightarrow y$ is characterized by the double-Horn sentence:

$$z \leqslant x \leftrightarrow y \quad \textit{iff} \quad z \wedge x = z \wedge y$$

($x \leftrightarrow y$ is the largest element that meets x and y in the same way.)

Given such a 'double-arrow' operation we can define $x \to y$ as $x \leftrightarrow (x \wedge y)$ and verify the double-Horn sentence that characterizes the single-arrow operation.

$\leftrightarrow$ is a commutative operation with a unit, 1, and in which every element is its own inverse: $x \leftrightarrow x = 1$. Note that $x \wedge (x \leftrightarrow y) = x \wedge y$ (it suffices to verify that $x \wedge (x \leftrightarrow y) \leqslant y$ and $x \wedge y \leqslant x \leftrightarrow y$; the first is a consequence of $x \wedge (x \leftrightarrow y) = y \wedge (x \leftrightarrow y)$; the second is a consequence of $(x \wedge y) \wedge x = (x \wedge y) \wedge y$).

$\leftrightarrow$ almost distributes with $\wedge$:

$$z \wedge (x \leftrightarrow y) = z \wedge [(z \wedge x) \leftrightarrow (z \wedge y)] .$$

(Because the equality reduces to two inequalities:

$$z \wedge (x \leftrightarrow y) \leqslant (z \wedge x) \leftrightarrow (z \wedge y) ,$$

$$z \wedge [(z \wedge x) \leftrightarrow (z \wedge y)] \leqslant x \leftrightarrow y .$$

Which in turn are equivalent to two equalities,

$$z \wedge (x \leftrightarrow y) \wedge (z \wedge x) = z \wedge (x \leftrightarrow y) \wedge (z \wedge y) ,$$

$$z \wedge [(z \wedge x) \leftrightarrow (z \wedge y)] \wedge x = z \wedge [(z \wedge x) \leftrightarrow (z \wedge y)] \wedge y .$$

Both are easy consequences of $x \wedge (x \leftrightarrow y) = x \wedge y$.)

1.725. Heyting algebras are given by an equational theory obtained by adding to the lattice equations:

$$x \leftrightarrow 1 = x = 1 \leftrightarrow x\,,$$

$$x \leftrightarrow x = 1\,,$$

$$z \wedge (x \leftrightarrow y) = [(z \wedge x) \leftrightarrow (z \wedge y)]\,.$$

We showed in the last section that these three equations are consequences of the double-Horn characterization. For the other direction suppose we are given z such that $z \wedge x = z \wedge y$. Then we have

$$z \wedge (x \leftrightarrow y) = z \wedge ((z \wedge x) \leftrightarrow (z \wedge y)) = z \wedge 1 = z, \quad \text{hence}$$

$$z \leqslant x \leftrightarrow y\,.$$

Note that

$$\begin{aligned} x \wedge (x \leftrightarrow y) &= x \wedge [(x \wedge x) \leftrightarrow (x \wedge y)] \\ &= x \wedge [(x \wedge 1) \leftrightarrow (x \wedge y)] \\ &= x \wedge [1 \leftrightarrow y] = x \wedge y\,. \end{aligned}$$

Similarly $y \wedge (x \leftrightarrow y) = x \wedge y$.

Given $z \leqslant x \leftrightarrow y$ we have $x \wedge z \leqslant x \wedge (x \leftrightarrow y) \leqslant x \wedge y \leqslant y$ and similarly $y \wedge z \leqslant x$, hence $z \wedge x = z \wedge y$.

1.726. The following equations are easily verified from the double-Horn characterization of $\rightarrow$:

$$x \rightarrow (y \rightarrow z) = (x \wedge y) \rightarrow z\,,$$

$$(x \vee y) \rightarrow z = (x \rightarrow z) \wedge (y \rightarrow z)\,,$$

$$x \rightarrow (y \wedge z) = (x \rightarrow y) \wedge (x \rightarrow z)\,,$$

$$x \rightarrow x = 1, \qquad 1 \rightarrow x = x, \qquad x \rightarrow 1 = 1\,,$$

$$x \wedge (x \rightarrow y) = x \wedge y, \qquad x \leqslant (x \rightarrow y) \rightarrow y\,.$$

As a special case of [1.711] we further have

$$x \wedge (y \vee z) = (x \wedge y) \vee (x \wedge z)$$

Note that $x \rightarrow y$ is covariant in y, contravariant in x.

1.727. Define $\neg x$, the NEGATION of x, as $x \to 0$ (or $x \leftrightarrow 0$). $\neg x$ is characterized by

$$z \leqslant \neg x \quad \textit{iff} \quad z \wedge x = 0 \,.$$

($\neg x$ is the largest element disjoint from x.)

In the lattice of open subsets of a topological space the negation of U is the interior of the complement of U.

As special cases of the last section:

$$\neg(x \vee y) = \neg x \wedge \neg y \,,$$

$$\neg 1 = 0, \qquad \neg 0 = 1,$$

$$x \leqslant \neg\neg x \,.$$

Negation is contravariant, that is, $x \leqslant y$ implies $\neg y \leqslant \neg x$.

We further have

$$\neg x = \neg\neg\neg x$$

since $x \leqslant \neg\neg x$ implies $\neg(\neg\neg x) \leqslant \neg x$ and since $(\neg x) \leqslant \neg\neg(\neg x)$.

If $\bar{x}$ denotes the double-negation, $\neg\neg x$, then

$$x \leqslant y \quad \text{implies} \quad \bar{x} \leqslant \bar{y},$$

$$x \leqslant \bar{x}, \qquad \bar{\bar{x}} = \bar{x} \,.$$

Moreover, double negation preserves intersection:

$$\overline{x \wedge y} = \bar{x} \wedge \bar{y} \,.$$

Clearly $\overline{x \wedge y} \leqslant \bar{x} \wedge \bar{y}$ (since double-negation preserves order). For the other containment we note first that $x \wedge y = 0$ implies $\bar{x} \wedge y = 0$ (since $x \wedge y = 0$ implies $y \leqslant \neg x \leqslant \neg\neg\neg x$ which implies $\bar{x} \wedge y = 0$). $\bar{x} \wedge \bar{y} \leqslant \overline{x \wedge y}$ is equivalent with $\bar{x} \wedge \bar{y} \wedge \neg\overline{(x \wedge y)}$; but $\neg\overline{(x \wedge y)} = \neg(x \wedge y)$ and, using the last sentence twice, $\bar{x} \wedge \bar{y} \wedge \neg(x \wedge y) = 0$ is a consequence of $x \wedge y \wedge \neg(x \wedge y) = 0$.

1.728. If we adjoin the LAW OF EXCLUDED MIDDLE, $x \vee \neg x = 1$, to the equations, then every element has a complement. Heyting algebras are distributive lattices, hence excluded middle makes them boolean algebras. We may alternatively adjoin the equation $x = \neg\neg x$. Then

$$x \vee \neg x = \neg\neg(x \vee \neg x) = \neg(\neg x \wedge \neg\neg x) = \neg 0 = 1 \,.$$

1.729. Let A and B be Heyting algebras, $f^{\#}: B \longrightarrow A$ an order-preserving map with a left-adjoint $f: A \longrightarrow B$ such that $b \wedge f(a) = f(f^{\#}(b) \wedge a)$. Then $f^{\#}$ preserves the arrow operation:

$$\begin{aligned} a \leqslant f^{\#}(b_1 \rightarrow b_2) \quad &\textit{iff} \quad f(a) \leqslant b_1 \rightarrow b_2 \quad \textit{iff} \quad b_1 \wedge f(a) \leqslant b_2 \\ &\textit{iff} \quad f(f^{\#}(b_1) \wedge a) \leqslant b_2 \\ &\textit{iff} \quad f^{\#}(b_1) \wedge a \leqslant f^{\#}(b_2) \\ &\textit{iff} \quad a \leqslant f^{\#}(b_1) \rightarrow f^{\#}(b_2) \,. \end{aligned}$$

Hence for any logos, $\mathscr{S}ub(-)$ may be viewed as a contravariant functor with values in the category of Heyting algebras [1.721].

If we began by assuming that $f^{\#}$ preserves the arrow operation, then $b \wedge f(a) = f(f^{\#}(b) \wedge a)$ is a consequence: $f(f^{\#}(b) \wedge a) = f(f^{\#}(b) \wedge a)$ implies $f^{\#}(b) \wedge a \leqslant f^{\#}f(f^{\#}(b) \wedge a)$ implies $a \leqslant f^{\#}(b) \rightarrow f^{\#}f(f^{\#}(b) \wedge a)$ implies $a \leqslant f^{\#}(b \rightarrow f(f^{\#}(b) \wedge a))$ implies $f(a) \leqslant b \rightarrow f(f^{\#}(b) \wedge a)$ implies $b \wedge f(a) \leqslant f(f^{\#}(b) \wedge a)$. (The reverse inequality holds for any adjoint pair.)

1.72(10). *Every Heyting algebra can be covered by a Heyting algebra for which the functor* $\Gamma = (1, \_)$ *is a representation of bicartesian categories.*

BECAUSE: The scone of categories in general will be defined in [1.(10)]. The scone of a Heyting algebra **A** may be pictured thusly

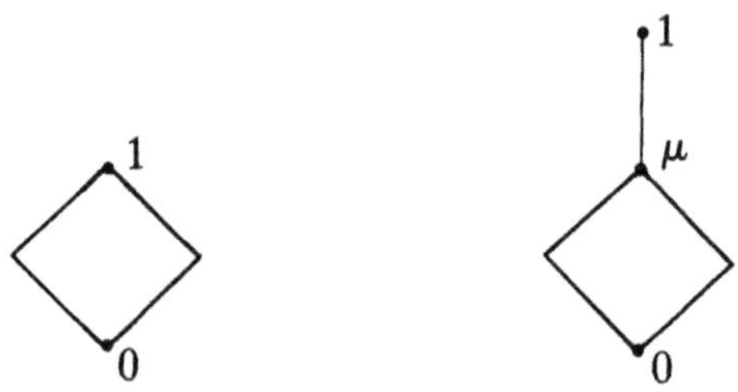

A $\quad\quad$ $\hat{\mathbf{A}}$

The scone $\hat{\mathbf{A}}$ is again a Heyting algebra. Unions and intersections are obvious, and $x \twoheadrightarrow y$ has the maximum possible value. $\hat{\mathbf{A}}/\mu \simeq \mathbf{A}$. 1 in $\hat{\mathbf{A}}$ is clearly not the union of any collection of objects properly below 1. In fact, the functor $\Gamma\colon \hat{\mathbf{A}} \to \mathscr{S}$ is a representation of bicartesian categories, because in a poset, coequalizers exist by default. This functor is always a representation of cartesian categories [1.442].

1.72(11). If $\mathbf{A}$ is a free Heyting algebra, there is a representation of Heyting algebras $\mathbf{A} \to \hat{\mathbf{A}}$, and $(\mathbf{A} \to \hat{\mathbf{A}} \to \hat{\mathbf{A}}/\mu \simeq \mathbf{A}) = 1_{\mathbf{A}}$, i.e., $\mathbf{A}$ is a RETRACT of its scone $\hat{\mathbf{A}}$.

For a free Heyting algebra, Γ is a representation of bicartesian categories. In particular, every finite collection of objects whose join is 1 *contains* 1.

BECAUSE: Their join in $\hat{\mathbf{A}}$ is 1, so at least one of them is 1 [1.72(10)].

1.73. For any representation of logoi $T\colon \mathbf{A} \to \mathbf{B}$ we define $\mathscr{F}il(T) \subset \mathscr{V}al_{\mathbf{A}}$ as the collection of all values U such that $T(U) = 1$. $\mathscr{F}il(T)$ is clearly a filter. Given $A' \subset A$ in $\mathbf{A}$, consider $p_{\mathbf{A}}^{\#\#}(A') \subset 1$, where $p_{\mathbf{A}}$ denotes the unique map from A to 1. Note that $A' \subsetneq A$ iff $p_{\mathbf{A}}^{\#\#}(A') \subsetneq 1$. $T(A') \subsetneq T(A)$ iff $p_{\mathbf{A}}^{\#\#}(A') \not\in \mathscr{F}il(T)$. In particular:

T is a faithful representation of logoi iff $\mathscr{F}il(T)$ is trivial, that is, iff $\mathscr{F}il(T) = \{1\}$.

1.731. *For every filter $\mathscr{F} \subset \mathscr{V}al_{\mathbf{A}}$ there is a logos $\mathbf{A}/\mathscr{F}$ and a representation of logoi $T_{\mathscr{F}}\colon \mathbf{A} \to \mathbf{A}/\mathscr{F}$ such that $\mathscr{F}il(T_{\mathscr{F}}) = \mathscr{F}$. $T_{\mathscr{F}}$ has the universal property: given $T'\colon \mathbf{A} \to \mathbf{B}$ such that $\mathscr{F} \subset \mathscr{F}il(T')$ there exists*

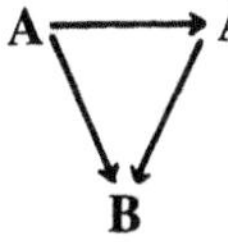

unique up to conjugation. $\mathbf{A}/\mathscr{F} \longrightarrow \mathbf{B}$ *is faithful iff* $\mathscr{F} = \mathscr{F}il(T')$.

BECAUSE: A construction of $\mathbf{A}/\mathscr{F}$ is available from the theory of rational categories [1.48]. Consider $f\colon A \longrightarrow B$ dense in $\mathbf{A}$ iff there exists $U \in \mathscr{F}$ so that $(1 \times f)\colon (U \times A) \longrightarrow (U \times B)$ is an isomorphism. $\mathbf{A}/\mathscr{F}$ is the rational category given by these dense morphisms.

More generally, if $\mathbf{A}$ is cartesian/regular/pre-logos, then so is $\mathbf{A}/\mathscr{F}$.

1.732. The union and equivalence conditions needed for the capitalization lemma are easily verified for logoi. The slice condition: let $\mathbf{A}$ be a logos, A an object therein; $\Sigma\colon \mathbf{A}/A \to \mathbf{A}$ induces an isomorphism between $\mathscr{S}ub_{\mathbf{A}/A}(\_)$ and $\mathscr{S}ub_{\mathbf{A}}(\Sigma(\_))$ and preserves inverse-images; thus double-sharps from **A** yield double-sharps for $\mathbf{A}/A$. $\Delta: \mathbf{A} \to \mathbf{A}/A$ preserves double-sharps because $(\mathbf{A} \xrightarrow{\Delta} \mathbf{A}/A \xrightarrow{\Sigma} \mathbf{A}) \simeq (A \times \_)$ does so, as can be seen by a routine check.

1.733. An object A in a pre-logos is COPRIME if the functor $(A, \_)$ preserves finite unions, i.e. any finite collection of subobjects of A whose union is A must contain A. A is CONNECTED if it has exactly two complemented subobjects. In $\mathscr{SH}(Y)$, the terminator is connected iff Y is a connected space.

A logos is a FOCAL LOGOS if its terminator is a coprime projective, i.e. $\Gamma = (1, \_)$ is a representation of pre-logoi. In $\mathscr{SH}(Y)$, the terminator is projective iff Y is projective in the category $\mathscr{LH}$ of local homeomorphisms, iff Y is empty or every open cover of Y has a refinement which is a partition of Y. This is an apparent contradiction with the connectedness of Y, so how can $\mathscr{SH}(Y)$ ever be focal? Y must be an element of all of its open covers, i.e. the union of all proper open subsets of Y is still a proper open subset, i.e. Y has a maximal proper open subset, i.e. Y has a focal point (to which every sequence converges).

A positive pre-logos is focal iff its terminator is a connected projective.

BECAUSE: If 1 is projective, Γ is a representation of regular categories, so it preserves unions iff it preserves disjoint unions [1.625].

We shall show in 1.(10) that in various free categories, 1 is a coprime projective [1.72(11)]. In particular, the free logos will be shown to be focal.

1.734. A representation of logoi $\mathbf{A} \to \mathbf{F}$ is a FOCAL REPRESENTATION if **F** is focal, i.e. $\mathbf{A} \to \mathbf{F} \xrightarrow{\Gamma} \mathscr{S}$ is a representation of pre-logoi. In [2.217] and [2.343] we will show that every logos may be faithfully represented in a positive logos, hence the word 'positive' will be removable from

Any small (positive) logos has a small, collectively faithful family of focal representations.

BECAUSE: We may concentrate on a small positive capital logos **A** [1.732]. Every representation of pre-logoi $\Gamma_{\bar{\mathscr{F}}}: \mathbf{A} \to \bar{\mathscr{S}}$ [1.635] factors as $\mathbf{A} \to \mathbf{A}/\bar{\mathscr{F}} \xrightarrow{\Gamma} \mathscr{S}$ [1.731]. $\mathbf{A}/\bar{\mathscr{F}}$ is focal, because $\bar{\mathscr{F}}$ is an ultrafilter in $\mathscr{B}$ [1.635].

Moreover,

A logos may be faithfully represented in a single focal logos iff its terminator is coprime.

BECAUSE: Coprimeness of the terminator is easily seen to be reflected by faithful representations. For the converse, let $\mathbf{A}_0$ be a logos with coprime terminator and $\mathbf{A}_0 \to \mathbf{A}$ a capitalization. We seek an ultrafilter $\mathscr{F}$ on the boolean algebra $\mathscr{B}$ of complemented subterminators in $\mathbf{A}$ such that $\mathbf{A}_0 \to \mathbf{A} \to \mathbf{A}/\mathscr{F}$ is faithful. Let $\mathscr{F}_0 \subset \mathscr{B}$ be the filter defined by $U \in \mathscr{F}_0$ iff U is a complemented subterminator in $\mathbf{A}$ for which there exists a proper subterminator $V \subsetneq 1$ in $\mathbf{A}_0$ with $U \cup V = 1$. Let $\mathscr{F}$ be an ultrafilter containing $\mathscr{F}_0$.

1.735. We note that the capitalization does not raise cardinality, unless the original category is finite. The positivization process [2.217] also preserves infinite cardinalities. In particular, every countable logos is faithfully representable in a countable capital positive logos. Therefore

Any countable (positive) logos has a countable, collectively faithful family of focal representations.

Any countable logos with a coprime terminator may be faithfully represented in a countable focal logos.

1.74. There are several representation theorems for logoi. The GEOMETRIC REPRESENTATION THEOREM is proved in this section. Another representation theorem is proved in [1.75–1.77]. A third representation theorem, which combines the first two to a certain extent, will be mentioned in [2.33].

In [2.217] and [2.343], the word 'positive' will be removable from

Every countable (positive) logos may be faithfully represented in a countable power of the logos of sheaves on the real line.

Every countable (positive) logos with a coprime terminator may be faithfully represented in the logos of sheaves on the real line.

1.741. $\Pi\mathscr{S}\!\!\mathscr{h}(\mathbb{R})$ is equivalent to $\mathscr{S}\!\!\mathscr{h}(\Sigma\mathbb{R})$ so we have a faithful representation in a logos of sheaves on an open subset of the real line.

In [1.(10)], we will show that the free logos is focal and thus faithfully representable in $\mathscr{S}\!\!\mathscr{h}(\mathbb{R})$.

1.742. We first show that

For any small (positive) logos $\mathbf{A}$ *there is a small category* $\mathbf{D}$ *and a faithful representation* $\mathbf{A} \to \mathscr{S}^{\mathbf{D}}$.

BECAUSE: We consider a category whose objects are focal representations, and whose morphisms from $(\mathbf{A} \to \mathbf{F})$ to $(\mathbf{A} \to \mathbf{F}')$ are (focal) representations $\mathbf{F} \to \mathbf{F}'$ such that

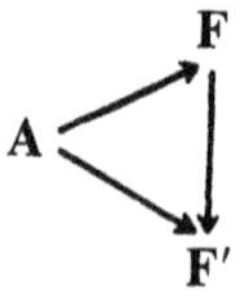

By [1.734], we may consider a small subcategory **D** whose objects give a collectively faithful family of representations of pre-logoi $\mathbf{A} \to \mathbf{F} \xrightarrow{\Gamma} \mathscr{S}$. The functor $T: \mathbf{A} \to \mathscr{S}^{\mathbf{D}}$ given by $T(A)(\mathbf{A} \to \mathbf{F}) = \mathbf{F}(1, A)$, i.e.

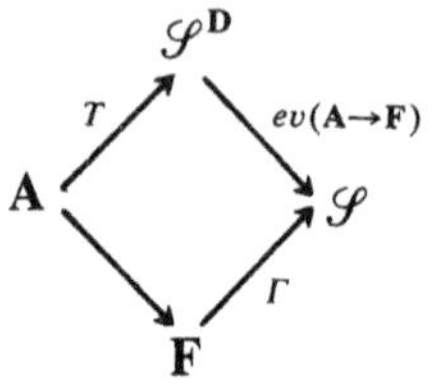

is a faithful representation of pre-logoi. Here and throughout this section, we will be assuming for notational purposes that $|\mathbf{A}|$ is contained in $|\mathbf{F}|$, each **F**.

We shall also require that **D** is large enough so that for any object $\mathbf{A} \to \mathbf{F}$ in **D**, the collection of morphisms in **D** whose source is $\mathbf{A} \to \mathbf{F}$ gives a collectively faithful family of representations of pre-logoi $\mathbf{F} \to \mathbf{F}' \xrightarrow{\Gamma} \mathscr{S}$. This can be obtained by successive applications of [1.734]. (If **A** is itself focal, assume that **D** contains $1_{\mathbf{A}}$.)

The fact that $T: \mathbf{A} \to \mathscr{S}^{\mathbf{D}}$ preserves double-sharps is now a matter of identity checking. However, because of the nature of functors on functors, a notational problem presents itself. We do not consider our notation satisfactory and the readers are invited to finish the argument on their own.

Given $f: A \to B$ and $A' \subset A$ in **A**, we show that $T(f^{\#\#}A') = (Tf)^{\#\#}(TA')$. It suffices to show that $\mathbf{F}(1, f^{\#\#}A') = ((Tf)^{\#\#}(TA'))(\mathbf{A} \to \mathbf{F})$ for every $(\mathbf{A} \to \mathbf{F}) \in |\mathbf{D}|$. By [1.71] and [1.713], the right-hand side consists of all $y \in \mathbf{F}(1, B)$ such that

$$\forall_{(\mathbf{F} \xrightarrow{\varphi} \mathbf{F}') \in \mathbf{D}} \forall_{x \in \mathbf{F}'(1, A)} [f(x) = \varphi(y) \Rightarrow x \in \mathbf{F}'(1, A')] .$$

Clearly, every $y \in \mathbf{F}(1, f^{\#\#}A')$ is such. In the other direction, if $y \in \mathbf{F}(1, B)$ but y is not allowed by $f^{\#\#}A'$ [1.51], then A' does not allow $f^{\#}$ (image of y), so there is a focal representation $\mathbf{F} \xrightarrow{\varphi} \mathbf{F}'$ and $x: 1 \longrightarrow A$ in $\mathbf{F}'$, obtained by the additional requirement on **D**, such that $f(x) = \varphi(y)$, but A' does not allow x.

1.743. $\mathscr{S}^{\mathbf{D}}$ is focal iff $\mathbf{D}$ has a coterminator (e.g. if $\mathbf{A}$ is focal).

1.744. We now replace the category $\mathbf{D}$ [1.742] with a tree.

Any functor $\mathbf{D}' \to \mathbf{D}$ induces a representation of pre-logoi $\mathscr{S}^{\mathbf{D}} \to \mathscr{S}^{\mathbf{D}'}$, given by composition. It is faithful if the functor $\mathbf{D}' \to \mathbf{D}$ is onto on objects.

We say that $\mathbf{D}'$ DOMINATES $\mathbf{D}$ if there is a functor $F\colon \mathbf{D}' \to \mathbf{D}$ which is onto on objects and LEFT-FULL, i.e. for each $D \in |\mathbf{D}'|$ and each $g\colon F(D) \longrightarrow B$ in $\mathbf{D}$, there exists $f\colon D \longrightarrow A$ in $\mathbf{D}'$ so that $F(f) = g$.

Note that such F may not be full, i.e. $\mathbf{D}'(A, A') \to \mathbf{D}(FA, FA')$ may not be a surjection for all $A, A' \in |\mathbf{D}'|$. We shall presently give an example.

If $\mathbf{D}'$ *dominates* $\mathbf{D}$ *then* $\mathscr{S}^{\mathbf{D}}$ *is faithfully representable in* $\mathscr{S}^{\mathbf{D}'}$ *as a logos.*

BECAUSE: The argument is again a matter of identity checking. If $F\colon \mathbf{D}' \longrightarrow \mathbf{D}$ is left-full and onto on objects, the induced representation of pre-logoi $F^{\bullet}\colon \mathscr{S}^{\mathbf{D}} \to \mathscr{S}^{\mathbf{D}'}$ preserves double-sharps. Indeed, recall that double-sharps in a functor category $\mathscr{S}^{\mathbf{C}}$ are given by [1.713]: for $f\colon A \longrightarrow B$ and $A' \subset A$ in $\mathscr{S}^{\mathbf{C}}$, $f^{\#\#}A'$ is the functor such that for any $C \in |\mathbf{C}|$, $(f^{\#\#}A')(C)$ is the set

$$\{b \in B(C) \mid \forall_{(C \xrightarrow{\varphi} C') \in \mathbf{C}} \forall_{a \in A(C')}[f(a) = \varphi(b) \Rightarrow a \in A'(C')]\}\,.$$

Given $f\colon A \longrightarrow B$ and $A' \subset A$ in $\mathscr{S}^{\mathbf{D}}$, we show that for any $D \in |\mathbf{D}'|$, $F^{\bullet}(f^{\#\#}A')(D) = ((F^{\bullet} f)^{\#\#}(F^{\bullet} A'))(D)$, i.e. $(f^{\#\#}A')(FD)$ is

$$\{b \in B(FD) \mid \forall_{(D \xrightarrow{\varphi} D') \in \mathbf{D}'} \forall_{a \in A(FD')}[f(a) = (F\varphi)(b) \Rightarrow a \in A'(FD')]\}\,.$$

But we know that $(f^{\#\#}A')(FD)$ is

$$\{b \in B(FD) \mid \forall_{(FD \xrightarrow{\psi} E) \in \mathbf{D}} \forall_{a \in A(E)}[f(a) = \psi(b) \Rightarrow a \in A'(E)]\}\,,$$

and F is left-full.

1.745. *Every category (with a coterminator) is dominated by a (rooted) tree.*

BECAUSE: Given a category $\mathbf{D}$ (with a coterminator), let $\mathbf{P(D)}$ be the tree of finite nonempty paths (finite paths) of composable morphisms in $\mathbf{D}$, ordered by prolongation on the right. Let $\mathbf{P(D)} \to \mathbf{D}$ be the functor that sends a nonempty path to the target of its last map (and associates a coterminator to the empty path).

1.746. We now make the (rooted) tree $\mathbf{P(D)}$ into a homogeneous (rooted) tree. Let $\mathbf{D}^+$ be the tree of nonempty finite words of morphisms of $\mathbf{D}$ (all finite words, if $\mathbf{D}$ has a coterminator). Consider the greedy functor $\mathbf{D}^+ \to \mathbf{P(D)}$, which for any word $a_1 a_2 \cdots a_m$ searches for the least i, $1 < i \leq m$, so that a_1 is composable with a_i; then for the least j, $i < j \leq m$, so that $a_1 a_i$ is composable with a_j, etc. It associates the empty path to the empty word. $\mathbf{D}^+$ dominates $\mathbf{D}$.

Every countable category with a coterminator is dominated by the binary tree.

BECAUSE: Such a category $\mathbf{D}$ is dominated by a countable homogeneous rooted tree $\mathbf{D}^+$, and therefore by the tree $\mathbb{N}^*$ of finite words of natural numbers, ordered by prolongation on the right. The binary tree $\mathbb{2}^*$ consists of finite words on two symbols a and b, also ordered by prolongation on the right. Let $f\colon \mathbb{N} \longrightarrow \mathbb{N}$ be a function such that $f^{\#}(n)$ is infinite for each n. Let $F\colon \mathbb{2}^* \longrightarrow \mathbb{N}^*$ be given by

$$F(a^{n_0} b a^{n_1} \cdots b a^{n_k}) = f(n_0)\, f(n_1) \cdots f(n_{k-1}).$$

Note that $F(a^{n_0})$ is the empty word. F is left-full, and onto on objects.

1.747. The binary tree may be viewed as a topological space, taking its updeals as the open sets. The category of Lazard sheaves, $\mathscr{LH}(\mathbb{2}^*)$, on this space is isomorphic to the category of right $\mathbb{2}^*$-sets, so it is equivalent to the functor category $\mathscr{S}^{\mathbb{2}^*}$ [1.271, 1.372]. In particular, a right $\mathbb{2}^*$-set X may be topologized by taking its sub-$\mathbb{2}^*$-sets as the open sets. Define $f\colon X \longrightarrow \mathbb{2}^*$ by sending x to $x\Box$, and verify that f is a local homeomorphism. Any morphism of right $\mathbb{2}^*$-sets is continuous, so we obtain a functor (*Right* $\mathbb{2}^*$*-sets*) $\longrightarrow \mathscr{LH}(\mathbb{2}^*)$.

1.748. We wish to replace the space $\mathbb{2}^*$ with the real line.

Any continuous map $g\colon X \longrightarrow Y$ induces a representation of pre-logoi $g^{\#}\colon \mathscr{LH}(Y) \longrightarrow \mathscr{LH}(X)$, given by pulling back along g in the category of continuous maps. It is faithful if g is onto.

If $g\colon X \longrightarrow Y$ *is an open continuous map, then* $g^{\#}\colon \mathscr{LH}(Y) \longrightarrow \mathscr{LH}(X)$ *is a representation of logoi.*

BECAUSE: If g is open continuous, then $g^{\#}$ preserves double-sharps. Indeed, in a category of Lazard sheaves on a space, double-sharps are given as follows [1.713]: for $f: A \longrightarrow B$ and $A' \subset A$, $f^{\#\#}(A')$ is the open subspace of B given by $\bigcup \{B' \subset B \text{ open} \mid f^{\#\#}(B') \subset A'\}$. Given this in $\mathscr{Sh}(Y)$, we have to show $(g^{\#}f)^{\#\#}(g^{\#}A') = g^{\#}(f^{\#\#}A')$, i.e.

$$(1 \times f)^{\#\#}(g^{\#}A') = g^{\#}(f^{\#\#}A'),$$

i.e.

$$\bigcup \{V \subset g^{\#}B \text{ open} \mid (1 \times f)^{\#}V \subset g^{\#}A'\} =$$
$$\bigcup \{g^{\#}B' \subset g^{\#}B \mid B' \subset B \text{ open, and } f^{\#}(B') \subset A'\}.$$

The inclusion of the latter to the former is clear. For the other direction, recall that $g^{\#}B$ is locally homeomorphic to X, and that g is open.

1.749. The space $\mathbb{2}^*$ does not have enough points to have an open, continuous surjection from the real line. We introduce more points, but keep the same lattice of open subsets. Topologize the set $\mathbb{2}^* \cup 2^{\mathbb{N}}$ by letting the basic open sets be the updeals generated by the finite words (Wilson space). The induced topology on $2^{\mathbb{N}}$ is the Cantor topology (the product topology). In $\mathbb{2}^* \cup 2^{\mathbb{N}}$, the empty word is a focal point. The restriction of open sets of $\mathbb{2}^* \cup 2^{\mathbb{N}}$ to $\mathbb{2}^*$ gives an equivalence between $\mathscr{Sh}(\mathbb{2}^* \cup 2^{\mathbb{N}})$ and $\mathscr{Sh}(\mathbb{2}^*)$ [1.373–4]. This construction was obtained by specializing a general process known as sobrification.

1.74(10). We will now exhibit an open, continuous surjection from $[-1/2, 1/2]$ to $\mathbb{2}^* \cup 2^{\mathbb{N}}$. (Therefore, there is such a map from the real line onto $\mathbb{2}^* \cup 2^{\mathbb{N}}$, and the Geometric Representation Theorem follows.) Indeed, for a countable (positive) logos, there is a countable, collectively faithful family of focal representations [1.735]. Each of the focal logoi so obtained may be faithfully represented in $\mathscr{Sh}(\mathbb{2}^*)$ [1.742–7], and thus in $\mathscr{Sh}(\mathbb{2}^* \cup 2^{\mathbb{N}})$ [1.749], and therefore in $\mathscr{Sh}(\mathbb{R})$ [1.748].)

The FREYD CURVE is defined as follows. Any point x in $[-1/2, 1/2]$ can be written (not uniquely) as $x = \Sigma_i a_i/3^i$, where $a_i = -1, 0, +1$. Let $\{a_{n_i}\}_{i=1}^{L}$, $L \leqslant \infty$, be the result of removing all zeros. Consider $\{(-1)^{n_{i-1}} a_{n_i}\}_{i=1}^{L} \in \mathbb{2}^* \cup 2^{\mathbb{N}}$, where n_0 is understood to be 0, and where we think of $\mathbb{2}$ as $\{-1, +1\}$. It is readily checked that this defines an open, continuous surjection from $[-1/2, 1/2]$ to $\mathbb{2}^* \cup 2^{\mathbb{N}}$.

1.75. In this section we prove the STONE REPRESENTATION THEOREM for logoi. We shall use a number of ancient facts from point-set topology, the proofs of which are collected in Appendix A. We shall eventually show [2.217, 2.343] that every logos may be faithfully represented in a positive logos, hence the word 'positive' will be removable from

For every small (positive) logos **A** *there is a space X and a faithful representation* $\mathbf{A} \longrightarrow \mathscr{Sh}(X)$.

X may be chosen to be compact, totally disconnected, and without isolated points.

If the logos is countable, then X may be chosen to be the Cantor space. It may be chosen to be the rational numbers. It may be chosen to be the irrational numbers.

1.751. We call an object an ATOM if 0 is its unique proper subobject. A logos is ATOMICALLY BASED if its atoms form a basis. Atomically based implies boolean: given $B' \subset B$ suppose that excluded middle fails, that is $B' \cup \neg B' \neq B$; let A be an atom and $f: A \longrightarrow B$ a map whose image is not in $B' \cup \neg B'$; then $f^{\#}(B' \cup \neg B') \neq A$ hence $f^{\#}(B' \cup \neg B') = 0$ and $f(A)$ is disjoint from $B' \cup \neg B'$; in particular it is disjoint from B' forcing $f(A) \subset \neg B'$, a contradiction.

We have already obtained the representation theorem for boolean logoi [1.641, 1.71]. (To fit into the Stone Representation Theorem, X in this case may be chosen to be discrete.)

To uniformize matters we go to the other extreme. A logos is ATOMLESS if it has no atoms (note that the definition of atom excludes zero-objects). Given any logos **A** let $\tilde{\Pi}\mathbf{A}$ denote its *periodic power*, that is, the set of periodic functions from $\mathbb{N}$, the natural numbers, to **A**. $\tilde{\Pi}\mathbf{A}$ is a sublogos of $\mathbf{A}^{\mathbb{N}}$ and the diagonal map $\mathbf{A} \longrightarrow \tilde{\Pi}\mathbf{A}$ is a faithful representation. Assuming only that **A** is non-degenerate, $\tilde{\Pi}\mathbf{A}$ is atomless. If **A** is positive and capital, $\tilde{\Pi}\mathbf{A}$ remains so. Thus we concentrate on the case when **A** is a positive capital atomless logos.

1.752. Let $\mathscr{B}$ be the boolean algebra of complemented subterminators of **A**. Since $\mathscr{B}$ is a basis and **A** is atomless, $\mathscr{B}$ is atomless.

Let $\hat{\mathscr{B}}$ be the STONE SPACE of $\mathscr{B}$ [1.389]: its points are the ultra-filters on $\mathscr{B}$; its basic open sets are of the form $\hat{U}$ where $U \in \mathscr{B}$ and $\hat{U} = \{\mathscr{F} \mid U \in \mathscr{F}\}$. The Stone space of any boolean algebra is compact and totally disconnected. The given boolean algebra is isomorphic to its lattice of CLOPENS, that is, the sets that are both open and closed. Hence the atoms of the boolean algebra correspond to the isolated points of the Stone space. In the case at hand, $\hat{\mathscr{B}}$ is without isolated points.

For any subspace $X \subset \hat{\mathscr{B}}$ we shall construct a functor $T: \mathbf{A} \rightarrow \mathscr{Sh}(X)$ with the following properties:

(1) For $\mathscr{F} \in X$, $(\mathbf{A} \xrightarrow{T} \mathscr{Sh}(X) \xrightarrow{(-)_{\mathscr{F}}} \mathscr{S}) = \Gamma_{\mathscr{F}}$ where $(-)_{\mathscr{F}}$ denotes the stalk-functor and $\Gamma_{\mathscr{F}}$ is as defined in [1.634].

From this first property we may conclude that T is a representation of pre-logoi: the stalk-functors from $\mathscr{Sh}(X)$ to $\mathscr{S}$ always form a collectively faithful family of pre-logos representations; each $\Gamma_{\mathscr{F}}$ is a pre-logos representation; hence T is such.

In particular, T preserves subobjects and we obtain for each object A a map $\mathcal{Sub}(A) \longrightarrow \mathcal{Sub}(TA)$. The second critical property on T is:

(2) The image of $\mathcal{Sub}(A) \longrightarrow \mathcal{Sub}(TA)$ is a basis, that is: every subobject $Y \subset TA$ is a union of subobjects of the form TA' where $A' \subset A$.

1.753. The two conditions above characterize T, indeed, as we shall see below, force its construction. First, however, we show how the Stone Representation Theorem is a consequence using the notation of the previous section.

If $X \subset \hat{\mathcal{B}}$ is such that $T: \mathbf{A} \longrightarrow \mathcal{Sh}(X)$ is faithful, then T is a faithful representation of logoi.

BECAUSE: Since T is known to be a representation of pre-logoi we need only show that it preserves double-sharps: given $f: A \longrightarrow B$ and $A' \subset A$ we wish to show that $T(f^{\#\#}A') = (Tf)^{\#\#}(TA')$. It suffices to show for arbitrary $Y \subset TB$ that $Y \subset T(f^{\#\#}A')$ iff $(Tf)^{\#}(Y) \subset T(A')$. By hypothesis, there exists a family $\{A_i\}$ of subobjects of A such that $Y = \bigcup TA_i$. Hence $Y \subset T(f^{\#\#}A')$ iff $TA_i \subset T(f^{\#\#}A')$ (for all i). By definition of double-sharps, such is equivalent to $f^{\#}A_i \subset A'$ (for all i). Since T is faithful and preserves inverse-images, such is equivalent to $(Tf)^{\#}(TA_i) \subset TA'$ (for all i), which is equivalent, in turn, to $\bigcup (Tf)^{\#}(TA_i) \subset TA'$. Inverse images in logoi preserve arbitrary unions [1.711], hence we finally obtain $(Tf)^{\#} \bigcup (TA_i) \subset TA'$, that is $(Tf)^{\#}(Y) \subset TA'$.

1.754. If $X = \hat{\mathcal{B}}$ then T is faithful [1.635] and $\mathbf{A} \longrightarrow \mathcal{Sh}(\hat{\mathcal{B}})$ is a representation of logoi.

For the other statements in the Stone Representation Theorem, we recall [1.735] that every countable logos is faithfully representable in a countable capital positive atomless logos. The boolean algebra of complemented subterminators therein is countable and atomless. Up to isomorphism there is only one such boolean algebra and its Stone space is the Cantor space.

For $X \subset \hat{\mathcal{B}}$, $T: \mathbf{A} \longrightarrow \mathcal{Sh}(X)$ is faithful iff for each $V \subsetneq 1$ there exists $\mathcal{F} \in X$ such that $\Gamma_{\mathcal{F}}(V) \neq 1$. We may, therefore, choose a subspace $X \subset \hat{\mathcal{B}}$ of the same cardinality as $\mathbf{A}$. Necessarily such X meets each non-empty clopen, hence is dense in $\hat{\mathcal{B}}$. If $\mathbf{A}$ is countable, X is a countable dense subset of the Cantor space. Any such subspace is homeomorphic to the rationals.

Finally, given such countable X choose another countable dense subset Y disjoint from X. Then $\mathbf{A} \longrightarrow \mathcal{Sh}(\hat{\mathcal{B}} - Y)$ is faithful. The complement of a countable dense subset of the Cantor space is homeomorphic to the irrationals.

1.755. We finish the proof of the Stone Representation Theorem, therefore, by constructing $T: \mathbf{A} \longrightarrow \mathcal{Sh}(X)$ as required by [1.752].

For $X \subset \hat{\mathcal{B}}$, let $|X|$ denote the underlying set.

$$\mathbf{A} \xrightarrow{T} \mathcal{Sh}(X) \longrightarrow \mathcal{S}/|X| \simeq \mathcal{S}^{|X|}$$

is forced by condition (1): For $\mathcal{F} \in X$ the corresponding coordinate functor is $\Gamma_{\mathcal{F}}$. For $A \in \mathbf{A}$, TA will be a space together with a local homeomorphism to X. We know from the above that the underlying set of the space in question is $\bigcup_{|X|} \Gamma_{\mathcal{F}}(A)$. The function to X is the obvious function.

Condition (2) forces the topology. For any $A' \subset A$, $\bigcup_{|X|} \Gamma_{\mathscr{F}}(A')$ must be an open subset of $\bigcup_{|X|} \Gamma_{\mathscr{F}}(A)$, and every open subset is a union of such; that is, the subsets of the form $\bigcup_{|X|} \Gamma_{\mathscr{F}}(A')$ are taken as a basis for the topology.

Given $f: A \longrightarrow B$ it is now routine that the induced function from $\bigcup_{|X|} \Gamma_{\mathscr{F}}(A)$ to $\bigcup_{|X|} \Gamma_{\mathscr{F}}(B)$ is continuous. Moreover, if f is monic then the induced function is a topological embedding onto an open subset.

Only one thing more need be shown, namely that the obvious function $\bigcup_{|X|} \Gamma_{\mathscr{F}}(A) \longrightarrow X$ is a local homeomorphism. We may verify first the special case $A = 1$, in which case the obvious function is a global homeomorphism. Next, for any A and $x \in \bigcup_{|X|} \Gamma_{\mathscr{F}}(A)$ we seek an open neighborhood on which the obvious function is a homeomorphism. Let $\mathscr{F} \in X$ be such that $x \in \Gamma_{\mathscr{F}}(A)$. Recalling the definition of $\Gamma_{\mathscr{F}}$ let $g: U \longrightarrow A$, $U \in \mathscr{F}$ name the element x. $T(U) \longrightarrow T(A)$ is an open embedding whose image contains x. When followed by the obvious function we obtain an open embedding $T(U) \longrightarrow X$.

1.76. Given a point $x \in X$ let $\mathscr{F}$ be the filter of open sets that contain x. $\mathscr{S}\!\mathscr{h}(X)/\mathscr{F}$ [1.731] is the category of MICRO-SHEAVES over x.

Micro-sheaves seldom yield locally complete categories. Let X be the unit interval and x its mid-point. Let $\mathbb{N}$ be the discrete space of natural numbers. $\mathbb{N} \times X \longrightarrow X$, as an object in $\mathscr{S}\!\mathscr{h}(X)$, is a countable copower of the terminator. Its only points are of the form $\{n\} \times X \subset \mathbb{N} \times X$ and it is the union of its points.

When we view $\mathbb{N} \times X$ as an object in the category of micro-sheaves it is no longer a copower, nor is it the union of its points. Note first, that it has gained no points: $A \subset \mathbb{N} \times X$ is isomorphic to the terminator in $\mathscr{S}\!\mathscr{h}(X)/\mathscr{F}$ iff it represents the same subobject as $\{n\} \times X$, for some $n \in X$. $B \subset \mathbb{N} \times X$ contains (in $\mathscr{S}\!\mathscr{h}(X)/\mathscr{F}$) all points iff for each $n \in \mathbb{N}$ there exists an open interval $x \in U_n$ such that $\{n\} \times U_n \subset B$. Let U_n be the largest such open interval. Define U'_n as the result of shrinking U_n around x by a factor of n and define B' as $\bigcup \{n\} \times U'_n$. B' still contains all points. But for no $V \in \mathscr{F}$ does $V \times B' \longrightarrow V \times B$ become an isomorphism. Hence there is no smallest subobject that contains all the points of $\mathbb{N} \times X$.

If there were a countable copower of 1 then we could construct the union of points in $\mathbb{N} \times X$ as the image of a map $\Sigma 1 \longrightarrow \mathbb{N} \times X$. Hence this category of micro-sheaves is without countable copowers.

1.761. If $\mathscr{F} \subset \mathscr{V}\!\mathscr{a}\!\mathscr{l}_{\mathbf{A}}$ is closed under countable intersection then $\mathbf{A}/\mathscr{F}$ inherits any countable unions and coproducts possessed by $\mathbf{A}$. $\mathbf{A} \longrightarrow \mathbf{A}/\mathscr{F}$, in other words, will preserve countable unions and coproducts. If $\mathscr{F}$ fails to be closed under some size of intersections we may modify the argument in the last section to show that $\mathbf{A} \longrightarrow \mathbf{A}/\mathscr{F}$ fails to preserve that size union. Note that if $\mathscr{F}$ is closed under arbitrary intersection, then $U \in \mathscr{F}$ iff $(\bigcap \mathscr{F}) \subset U$ and $\mathbf{A}/\mathscr{F} = \mathbf{A}/(\bigcap \mathscr{F})$.

The only time that $x \in X$ yields a locally complete category of micro-sheaves is if there is a minimal open set U containing x, in which case the category of micro-sheaves over x is equivalent to the category of sheaves over U. If X is a T_1 space, then x is necessarily an isolated point and the category of micro-sheaves over x is equivalent to the category of sets.

1.77. Given a binary endo-relation R on an object in a regular category we define its TRANSITIVE CLOSURE, $R^{\dagger}$, if such exists, as the minimal transitive relation containing R. Its TRANSITIVE-REFLEXIVE CLOSURE, R^*, if such exists, is the minimal transitive-reflective relation containing R. In a pre-logos, if either $R^{\dagger}$ or R^* exists then so does the other: $R^* = 1 \cup R^{\dagger}$ and $R^{\dagger} = RR^*$. (To see the latter, suppose $R \subset T$ and $T^2 \subset T$; then $1 \cup T$ is transitive-reflexive, hence $R^* \subset 1 \cup T$; from which we obtain $RR^* \subset R(1 \cup T) \subset R \cup RT \subset T \cup T^2 \subset T$.)

A TRANSITIVE (PRE-) LOGOS is a (pre-) logos in which each endo-relation has a transitive closure.

1.771. Suppose $\mathbf{A}$ is a pre-logos in which each lattice of subobjects has countable unions and suppose that inverse images preserved countable unions (as is necessary in a logos). We may then construct $R^{\dagger}$ as $R \cup R^2 \cup \cdots \cup R^n \cup \cdots$. (Its transitivity is a consequence of the fact that composition distributes with countable union whenever inverse images preserve countable unions.)

Our prime examples, $\mathscr{S}^{\mathbf{A}}$ and $\mathscr{Sh}(X)$, are thus easily seen to be transitive logoi.

1.772. A σ-TRANSITIVE LOGOS is a transitive logos such that for all endo-relations R, $R^{\dagger}$ is the least upper bound of the finite powers of R. Note that if $\bigcup_{n=1}^{\infty} R^n$ exists, then necessarily it is $R^{\dagger}$.

A σ-TRANSITIVE PRE-LOGOS is a transitive pre-logos such that for all endo-relations R, $R^{\dagger}$ is the *stable* least upper bound of the finite powers of R, that is, $R^{\dagger} = \bigcup_{n=1}^{\infty} R^n$ and the countable union in question is preserved under inverse images.

σ-transitivity is an infinitary Horn sentence: for transitive logoi it may be expressed as,

$$(R \subset T) \wedge (R^2 \subset T) \wedge \cdots \wedge (R^n \subset T) \wedge \cdots \quad \textit{implies} \quad R^{\dagger} \subset T\,.$$

For transitive pre-logoi a more complicated expression is needed. Given $R \subset B \times B$, $f\colon A \longrightarrow B \times B$ and $A' \subset A$ then

$$(f^{\#}(R) \subset A') \wedge (f^{\#}(R^2) \subset A') \wedge \cdots \wedge (f^{\#}(R^n) \subset A') \\ \wedge \cdots \quad \textit{implies} \quad f^{\#}(R^{\dagger}) \subset A'$$

Structures satisfying an infinitary Horn sentence are, just as with ordinary Horn sentences, closed under arbitrary products and substructures. Since the category of sets is σ-transitive, any attempt to represent transitive pre-logoi in a power of $\mathscr{S}$ will require σ-transitivity.

1.773. Any locally complete regular category is transitive: $R^{\dagger}$ is constructible as the intersection of all transitive relations that contain R.

It need not be σ-transitive: Define R_α, α an ordinal number, as follows: $R_1 = R$, $R_{\alpha+1} = R_\alpha \cup (R_\alpha)^2$ and for limit ordinals β, $R_\beta = \bigcup_{\beta<\alpha} R_\beta$.

The category of compact Hausdorff spaces is a locally complete pre-topos. It is transitive but not σ-transitive. For any ordinal γ consider the relation R on the space of ordinals $\{\beta \mid \beta \leq \gamma\}$, defined by $\alpha R \beta$ iff $\alpha = \beta$ or β is the immediate successor of α. $R^\dagger$ is easily verified to be the ordering relation. If γ is taken as $\omega + 1$ then $R_\omega \neq R^\dagger$ (but $R_{\omega+1} = R^\dagger$). For arbitrarily large ordinals α we can, in this manner, find examples of $R_\alpha \neq R^\dagger$. (For limit ordinals α, $RR_\alpha \neq R_\alpha R$ for sufficiently large γ.)

1.774. The pre-logoi obtainable from either recursive or primitive recursive functions [1.572, 1.573] are transitive, but only in the first case do we obtain σ-transitivity.

Given an object A in either category, there are well-known constructions of the object W of finite words of A-elements. Given a relation on A tabulated by $\langle l, r\rangle: R \longrightarrow A \times A$, we may construct $R^\dagger$ as the relation tabulated by the image of $\langle l', r'\rangle: W' \longrightarrow A \times A$ where $W' \subset W$ is defined by $\langle a_1, a_2, \ldots, a_n\rangle \in W'$ iff $(ra_1 = la_2) \wedge (ra_2 = la_3) \wedge \cdots \wedge (ra_{n-1} = la_n)$. Define $l'\langle a_1, \ldots, a_n\rangle = a_1$ $r'\langle a_1, \ldots, a_n\rangle = a_n$.

The forgetful functor from the category of recursive functions to the category of sets is faithful and preserves transitive closures, guaranteeing σ-transitivity. For primitive recursive functions, consider the relation on the natural numbers, $\mathbb{N}$, defined by nRm iff $n = m$ or $n + 1 = m$. $R^\dagger$ is easily verified to be the ordering relation, but $R^\dagger$ is not the union of the finite powers of R (not even an unstable union). Let $f: A \longrightarrow \mathbb{N}$ be a one-to-one onto primitive recursive function such that f^{-1} is not primitive recursive. Let $B \subset A \times \mathbb{N}$ be defined by $\langle a, n\rangle \in B$ iff $f(a) \leq n$ and define $g: B \longrightarrow R^\dagger \subset \mathbb{N} \times \mathbb{N}$ by $g\langle a, n\rangle = \langle n - f(a), n\rangle$. g is one-to-one and onto but g^{-1} is not primitive recursive, hence it represents a proper subobject of $R^\dagger$. For any finite m, however, the restriction of g^{-1} to R^m is primitive recursive. Thus the subobject represented by g contains all finite powers of R.

(Viewing g as a relation on $\mathbb{N} \times \mathbb{N}$, we note that the forgetful functor into the category of recursive functions carries it to a transitive relation. If it were transitive in the category of primitive recursive functions, then of course it would have to be the same as the ordering relation. The 'transitivity map' is recursive, but not primitive recursive.)

1.775. Given an endo-relation R in a regular category, its EQUIVALENCE CLOSURE, $R^{\equiv}$, if such exists, is the minimal equivalence relation that contains R. In a transitive pre-logos, $R^{\equiv}$ is constructible as $(R \cup R^\circ)^*$. Every endo-relation in a given effective regular category has an equivalence closure iff the category has coequalizers.

We say that a pre-logos is E-STANDARD if every endo-relation R has an equivalence closure, $R^{\equiv}$, and if $R^{\equiv}$ is always the stable union of finite powers of $R^\circ \cup 1 \cup R$.

Clearly, a σ-transitive pre-logos is E-standard, and as with σ-transitivity, E-standardness is a necessary condition for the existence of a faithful family of set-valued representations that preserve equivalence closures.

1.776. The word 'positive' will eventually be removable from the statements below [2.217, 2.343]. The word 'countable' is not removable. The proof of the statements below is in section [1.777].

Every countable positive σ-transitive pre-logos is faithfully representable in a power of $\mathscr{S}$.

Every countable positive σ-transitive logos is faithfully representable in $\mathscr{SH}(X)$, where X may be taken as either the rationals or the irrationals.

Every countable E-standard bicartesian pre-topos is faithfully representable in a power of $\mathscr{S}$.

The last sentence characterizes the countable bicartesian categories representable in a power of $\mathscr{S}$: the statement that a bicartesian category is a pre-topos is equivalent to a set of Horn-sentences in the bicartesian predicates.

1.777. The category of (pre-) logoi and representations that preserve (stable) unions satisfies the three conditions needed for the capitalization lemma [1.54]. That is, the inclusion of a (pre-) logos into its capitalization preserves all stable unions. The atom-splitting process also preserves stable unions. Each of the sentences above thus becomes a consequence of:

Given a countable atomless positive capital (pre-) logos $\mathbf{A}$ *and a countable family of stable unions therein, there exists a dense G_δ, Z, in the Stone-space $\hat{\mathscr{B}}$ ($\mathscr{B}$ the boolean algebra of complemented subterminators) such that $\mathbf{A} \longrightarrow \mathscr{SH}(Z)$ is faithful and preserves each of the stable unions in the given countable family.*

The construction of $Z \subset \hat{\mathscr{B}}$ proceeds as follows. We first replace the given family of stable unions with a family of unions of 1. Given an arbitrary stable union $A = \bigcup A_\iota$ and any $U \in \mathscr{B}$, $f: U \longrightarrow A$, then U is a stable union of $\{f^{\#}A_\iota\}$. Conversely, given an ultra-filter $\mathscr{F}$ such that for all $U \in \mathscr{B}$ and $f: U \longrightarrow A$ it is the case that $\Gamma_{\mathscr{F}}(U) = \bigcup \Gamma_{\mathscr{F}}(f^{\#}(A_\iota))$, then $\Gamma_{\mathscr{F}}(A) = \bigcup \Gamma_{\mathscr{F}}(A_\iota)$. Hence each stable union in $\mathbf{A}$ may be replaced by a countable family of stable unions of the form $U = \bigcup V_\iota$, $U \in \mathscr{B}$. If U' denotes the complement of U, then 1 is a union of $\{(V_\iota \cup U')\}$. Conversely if $1 = \bigcup \Gamma_{\mathscr{F}}(V_\iota \cup U')$ then $\Gamma_{\mathscr{F}}(U) = \bigcup \Gamma_{\mathscr{F}}(V_\iota)$. Thus the preservation of a countable family of stable unions can be reduced to the preservation of a countable family of unions of 1.

We momentarily concentrate on a single union $1 = \bigcup V_\iota$. It is preserved by $\Gamma_{\mathscr{F}}$ iff $V_i \in \mathscr{F}$ for some i. Let $X \subset \hat{\mathscr{B}}$ denote the set of such ultra-filters, that is, $X = \bigcup \hat{V}_\iota$. X is open. For any $W \subset 1$ in $\mathbf{A}$ such that $X \subset \hat{W}$, it is the case that $W = 1$ (because $1 = \bigcup V_\iota$). X is not only dense in $\hat{\mathscr{B}}$, it is dense in all of the closed sets of the form $\hat{\mathscr{B}} - \hat{W}$, $W \subset 1$.

Thus for each union $1 = \bigcup V_i$ there is an open $X \subset \hat{\mathscr{B}}$ which is dense in $\hat{\mathscr{B}} - \hat{W}$ for all $W \subset 1$, such that $\bigcup \Gamma_{\mathscr{F}}(V_i) = 1$ for all $\mathscr{F} \in X$. Given a countable family of unions of 1, therefore, we may intersect the X's to obtain a G_δ, Z, which meets $\hat{\mathscr{B}} - \hat{W}$ for all $W \subsetneqq 1$ and such that $\Gamma_{\mathscr{F}}$ preserves each of the given unions for all $\mathscr{F} \in Z$. Since for each $W \subsetneqq 1$ there exists $\mathscr{F} \in Z$ such that $W \notin \mathscr{F}$, it is the case that $\{\Gamma_{\mathscr{F}}\}_{\mathscr{F} \in Z}$ is faithful.

When $\mathbf{A}$ is a logos, we may as in [1.754] choose a countable dense subset $X \subset Z$ so that $\mathbf{A} \longrightarrow \mathscr{Sh}(X)$ is faithful. X is homeomorphic to the rationals. Alternatively we may choose a countable dense $Y \subset \hat{\mathscr{B}} - X$. $\mathbf{A} \longrightarrow \mathscr{Sh}(Z - Y)$ is still a faithful representation of logoi, it still preserves the given stable unions, $Z - Y$ is still a G_δ but now with a dense complement in Cantor space. Any such subspace is homeomorphic to the irrationals.

1.78. In a regular category let R, S be relations with a common target. R/S denotes the maximum relation T, if it exists, such that $TS \subset R$. That is, $T \subset R/S$ iff $TS \subset R$.

The source-target information is pictured by

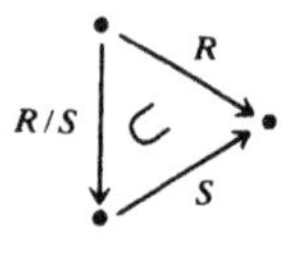

Note that the diagram only *semi-commutes* as indicated by the containment sign.

1.781. In $\mathscr{S}$, $x(R/S)y$ iff $\forall_z (ySz) \Rightarrow (xRz)$.

1.782. *In a regular category, let R be a relation, f a map, R and f with the same target. R/f exists and is equal to Rf°.*

BECAUSE: As may be readily verified in $\mathscr{S}$, $Tf \subset R$ iff $T \subset Rf^\circ$.

1.783. $(R/S_1)/S_2 \simeq R/(S_2S_1)$. *That is, if $(R/S_1)S_2$ exists then so does $R/(S_2S_1)$ and they are equal.*

BECAUSE: $T \subset (R/S_1)/S_2$ iff $TS_2 \subset R/S_1$, iff $TS_2S_1 \subset R$.

1.784. *In a logos, R/S exists for every pair of relations with a common target.*

BECAUSE: Since every relation S is of the form $l^\circ r$ we see from the above that R/S exists iff $(Rr^\circ)/l^\circ$ exists. We may specialize to the case R/f°.

Suppose $R \subset A \times B$ and $f\colon C \longrightarrow B$. Given $T \subset A \times C$ then $Tf^\circ = (1 \times f)^{\#}(T)$. We may construct R/f°, therefore, as $(1 \times f)^{\#\#}(R)$.

1.785. The converse: given $A' \subset A$ and $f: A \longrightarrow B$ view $A' \mapsto A \xrightarrow{\langle 1,1 \rangle} A \times A$ as a relation on A. If A'/f° exists then $f^{\#\#}(A')$ is $1_B \cap (A'/f^\circ)^\circ(A'/f^\circ)$.

1.786. Given R, S, T with a common target, it is the case that $(R/S)(S/T) \subset R/T$ (since $(R/S)(S/T)T \subset (R/S)S \subset R$).

For any relation R, R/R is transitive and reflexive ($1 \subset R/R$ since $1R \subset R$).

1.787. For an endo-relation R in a logos, let $\bar{R}$ denote the minimum reflexive relation, if it exists, such that $R\bar{R} \subset \bar{R}$. Recall that R^* denotes the minimum reflexive transitive relation if it exists, such that $R \subset R^*$.

In a logos, $\bar{R} = R^*$. That is, if either $\bar{R}$ or R^* exists in a logos then they both exist and are equal.

BECAUSE: If R^* exists then clearly $RR^* \subset R^*R^* \subset R^*$. Given any reflexive S much that $RS \subset S$ then $R \subset S/S$. Since S/S is reflexive transitive, $R^* \subset S/S$. Hence $R^*S \subset S$ and $R^* \subset R^*1 \subset R^*S \subset S$. Thus R^*, if it exists, is $\bar{R}$.

If $\bar{R}$ exists then it is transitive. Indeed, $\bar{R}/\bar{R}$ is reflexive and $R(\bar{R}/\bar{R})\bar{R} \subset R\bar{R} \subset \bar{R}$, so $R(\bar{R}/\bar{R}) \subset \bar{R}/\bar{R}$. Thus $\bar{R} \subset \bar{R}/\bar{R}$ and so $\bar{R}\bar{R} \subset \bar{R}$. On the other hand, if S is reflexive transitive and $R \subset S$, then $RS \subset SS \subset S$, and so $\bar{R} \subset S$.

1.8. ADJOINT FUNCTORS, GROTHENDIECK TOPOI, AND EXPONENTIAL CATEGORIES

1.81. $F: \mathbf{A} \to \mathbf{B}$ and $G: \mathbf{B} \to \mathbf{A}$ comprise an ADJOINT PAIR OF FUNCTORS if there exists a natural equivalence $(FA, B) \simeq (A, GB)$ (natural in both variables, i.e. as functors from $\mathbf{A}^\circ \times \mathbf{B}$ to the category of sets). F is called a LEFT-ADJOINT of G and G is called a RIGHT-ADJOINT of F.

1.811. We have already seen several adjoint pairs between posets. Since hom-sets have at most one element, adjointness reduces to $(FA, B) \neq \emptyset$ iff $(A, GB) \neq \emptyset$, which translates to $FA \leqslant B$ iff $A \leqslant GB$.

(Direct images, for example, are left-adjoints of inverse images.)

1.812. When 'forgetful functors' have left-adjoints their values are usually called 'free objects'. Thus if $\mathscr{G}$ is the category of groups and $F: \mathscr{S} \to \mathscr{G}$ sends a set B to a group freely generated by B, then F is a left-adjoint of the forgetful functor.

1.813. If the inclusion functor of a subcategory $\mathbf{A}' \subset \mathbf{A}$ has a left-adjoint, $\mathbf{A}'$ is said to be a REFLECTIVE SUBCATEGORY and the values of the left-adjoint are called REFLECTIONS.

(The strange habit of writing 'connection' as 'connexion' has led some to write 'reflection' as 'reflexion' which in turn leads to the abomination 'reflexive' category for 'reflective'.)

The reflectiveness of a given subcategory $\mathbf{A}' \subset \mathbf{A}$ is an elementary condition: it says that for every $A \in \mathbf{A}$ there exists $A \to \bar{A}$, $\bar{A} \in \mathbf{A}'$ such that for every $A \to B$, $B \in \mathbf{A}'$ there is a unique $(\bar{A} \to B) \in \mathbf{A}'$ such that

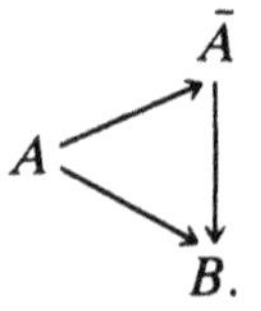

With the axiom of choice we can choose for each $A \in \mathbf{A}$ such $A \to \bar{A}$. Given $x: A_1 \to A_2$ in $\mathbf{A}$, define $\bar{x}: \bar{A}_1 \to \bar{A}_2$ as the unique map such that

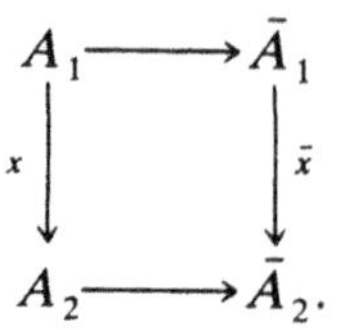

1.814. Many reflection functors were used before the general concept and the names for them vary. For example: the abelian reflection of a group is called its 'abelianization'; the compact reflection of a Hausdorff space is called its 'Stone–Čech compactification'; the reflection of an integral domain in the subcategory of fields is called its 'field of quotients'.

Most reflective subcategories that arise are full subcategories. Fullness is the necessary and sufficient condition for the reflection functor to be idempotent.

For an example of a non-full reflective subcategory, let **A** be the category of lattices and lattice homomorphisms and **A**′ the subcategory of complete lattices and maps which preserve finite $\wedge$ and arbitrary $\bigvee$ (sometimes called 'andOR maps'). The reflection of A is the lattice of ideals of A.

1.815. Given a full reflective subcategory P' of a poset P, its reflection functor is called a CLOSURE OPERATION, that is, an order-preserving, idempotent and inflationary unary operation:

$$x \leqslant y \quad \textit{implies} \quad \bar{x} \leqslant \bar{y}\,,$$

$$\bar{\bar{x}} = \bar{x}\,,$$

$$x \leqslant \bar{x}\,.$$

Given a closure operation, the set of its values (with the induced ordering) is a full reflective subcategory.

1.816. If an inclusion functor $\mathbf{A}' \subset \mathbf{A}$ has a right-adjoint, **A**′ is said to be COREFLECTIVE. In the category of abelian groups, the full subcategory of torsion groups is coreflective; in the category of pointed CW-complexes the full subcategory of simply connected spaces is coreflective ('universal covering spaces').

1.817. If $G: \mathbf{B} \to \mathbf{A}$ has a left-adjoint $F: \mathbf{A} \to \mathbf{B}$ then for every $A \in \mathbf{A}$, $(A, G(-))$ is representable, namely by FA. Conversely, given $G: \mathbf{B} \to \mathbf{A}$ such that $(A, G(-))$ is representable for all A, then we may construct a left-adjoint for G: for each $A \in \mathbf{A}$ choose $FA \in \mathbf{B}$ and an equivalence $(A, G(-)) \simeq (FA, -)$; given $x: A_1 \to A_2$ define Fx as the unique map such that

$$\begin{array}{ccc} (A_2, G(-)) & \simeq & (FA_2, -) \\ {\scriptstyle (x, G(-))}\downarrow & & \downarrow{\scriptstyle (Fx, -)} \\ (A_1, G(-)) & \simeq & (FA_1, -). \end{array}$$

1.818. A pair of contravariant functors $F\colon \mathbf{A}\to\mathbf{B}$, $G\colon\mathbf{B}\to\mathbf{A}$ are said to be ADJOINT ON THE RIGHT if (B, FA) and (A, GB) are naturally equivalent. They are ADJOINT ON THE LEFT if (FA, B) and (GB, A) are naturally equivalent.

A pair of contravariant functors between posets which are adjoint on the right is a *Galois connection.*

F, G are adjoint on the right iff $\mathbf{A}\xrightarrow{F}\mathbf{B}\to\mathbf{B}^\circ$ is the left-adjoint of $\mathbf{B}^\circ\to\mathbf{B}\xrightarrow{G}\mathbf{A}$. They are adjoint on the left iff $\mathbf{A}^\circ\to\mathbf{A}\xrightarrow{F}\mathbf{B}$ is the left-adjoint of $\mathbf{B}\xrightarrow{G}\mathbf{A}\to\mathbf{A}^\circ$. Theorems about covariant adjoint pairs thus automatically yield theorems about contravariant adjoint pairs.

1.82. For categories $\mathbf{D}$ and $\mathbf{B}$, $\mathbf{B}^{\mathbf{D}}$ denotes the category whose objects are functors from $\mathbf{D}$ to $\mathbf{B}$ and whose maps are natural transformations. In most contexts $\mathbf{D}$ is a small category.

The DIAGONAL FUNCTOR $\Delta\colon\mathbf{B}\to\mathbf{B}^{\mathbf{D}}$ sends an object B to the functor with a constant value, B, and sends a map $x\colon B_1\to B_2$ to the transformation with a constant value, x. If $\mathbf{D}\neq\emptyset$ then Δ is an embedding and we may view $\mathbf{B}$ as the subcategory of constant functors and transformations. If $\mathbf{D}=\emptyset$ then $\mathbf{B}^{\mathbf{D}}$ is the terminator among categories and Δ is the unique functor from $\mathbf{B}$. In this case Δ has a right-adjoint iff $\mathbf{B}$ has a terminator, and it has a left-adjoint iff $\mathbf{B}$ has a coterminator.

1.821. The objects of $\mathbf{B}^{\mathbf{D}}$ can be thought of as *diagrams in* $\mathbf{B}$ *modelled on* $\mathbf{D}$. In this context it is useful to denote the objects of $\mathbf{D}$ with lower-case letters i, j, k (they can be thought of as 'vertices'). Given $D\colon\mathbf{D}\to\mathbf{B}$ a map from $\Delta(B)$ to D can be thought of as a *lower-bound* of D. It is described by a collection $\{B\to D_i\}_{i\in|\mathbf{D}|}$ which satisfies the *compatibility condition* that for every x in $\mathbf{D}$ it is the case that

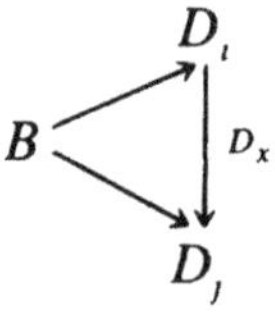

D has a coreflection in Δ iff it has a *greatest lower-bound*, that is, a lower-bound $\{L\to D_i\}_{|D|}$ such that for every lower-bound $\{B\to D_i\}_{i\in\mathbf{D}}$ there exist *unique* $B\to L$ such that

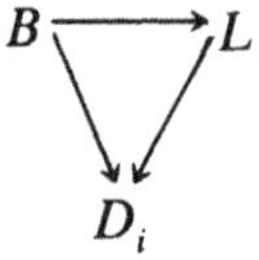

for all i.

If **D** is discrete then the compatibility condition evaporates and we see that $\{P \to D_i\}_{|\mathbf{D}|}$ is a greatest lower-bound of D iff it is a product diagram.

1.822. For arbitrary **D** the values of a right-adjoint of $\Delta: \mathbf{B} \to \mathbf{B}^{\mathbf{D}}$ are called LIMITS. $\varprojlim D$ denotes, if it exists, a greatest lower bound of D. The values of a left-adjoint of D are called COLIMITS and are denoted $\varinjlim D$. (Colimits may be thought of as least upper bounds.)

As already noted, $\varprojlim \emptyset$ is a terminator, $\varinjlim \emptyset$ a coterminator.

Let **D** be the category finitely presented by $\bullet \rightrightarrows \bullet$ (It has just two objects, two identity maps, and two non-identity maps.) A right-adjoint of $\Delta: \mathbf{B} \to \mathbf{B}^{\mathbf{D}}$ delivers equalizers, a left-adjoint delivers coequalizers.

1.823. A category is said to be COMPLETE if every small diagram has a limit and is said to be COCOMPLETE if every small diagram has a colimit.

1.824. A collection of monic maps $\{A_i \rightarrowtail A\}_I$ may be viewed as a diagram (modelled on I' where I' is the poset obtained by adjoining a terminator to the discrete poset I). Its limit, if it exists, determines a subobject of A which is clearly an intersection in $\mathcal{S}ub(A)$ of the subobjects determined by the A_i's. It is more than just a greatest lower bound in $\mathcal{S}ub(A)$. $B \to A$, monic or not, factors through $\varprojlim A_i$ iff it factors through each of the A_i's.

1.825. *A category is complete iff it has equalizers and arbitrary products.*

BECAUSE: First, given a collection of monic maps $\{A_i \rightarrowtail A\}_I$ each of which is an equalizer, say of f_i, g_i: $A \to B_i$, we may construct $\varprojlim A_i$ as the equalizer of f, g: $A \to \Pi B_i$ where the i-th coordinate of f is f_i and the i-th coordinate of g is g_i.

Given any small **D** and diagram D modelled on **D** let $\{P \to D_i\}_{|\mathbf{D}|}$ be a product. For each $x: i \to j$ in **D** let $E_x \rightarrowtail P$ be an equalizer of $P \to D_j$ and $P \to D_i \xrightarrow{x} D_j$. Then

$$\varprojlim D = \varprojlim \{E_x \rightarrowtail P\}_{x \in \mathbf{D}} .$$

(The axiom of choice can be avoided by putting the two halves together. $\varprojlim D$ is an equalizer of a single pair of maps between $\Pi_{|\mathbf{D}|} D_i$ and $\Pi_{\mathbf{D}} D_{(x\Box)}$.)

The last section dualizes to

A category is cocomplete iff it has coequalizers and arbitrary coproducts.

The proof works as well for

A category is (co-) cartesian iff every finite diagram has a (co-) limit.

1.827. A functor is CONTINUOUS if it preserves all small limits. When its source is complete, a functor is continuous iff it preserves equalizers and arbitrary products. Dually, a functor is COCONTINUOUS if it preserves all small colimits. If its source is cocomplete then it is continuous iff it preserves coequalizers and arbitrary coproducts.

A contravariant functor is called CONTINUOUS if it carries colimits to limits.

1.828. The prefix 'weak-' is used to remove uniqueness conditions from definitions. A WEAK-LIMIT of a diagram D is a greatest lower bound without the uniqueness: it is a compatible family $\{W \to D_i\}_{|\mathbf{D}|}$ such that for any compatible family $\{B \to D_i\}_{|\mathbf{D}|}$ there exists at least one $B \to W$ such that

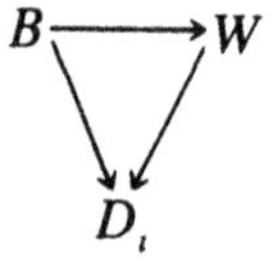

for all i.

A category is WEAKLY-COMPLETE if every small diagram has a weak-limit. There are several important examples such as the homotopy category of pointed CW-complexes and the category whose objects are small categories and whose maps are natural equivalence classes of functors (sometimes called COSCANECOF).

If a functor preserves any weak-limit of a given diagram then it may easily be seen to preserve them all. In particular, a continuous functor from a complete category preserves all weak-limits.

1.829. The phrase 'weakly continuous' should be avoided:

A functor that preserves weak-limits preserves limits.

BECAUSE: Suppose that T preserves weak-limits. If $\{L \to D_i\}_{|\mathbf{D}|}$ is a limit of D then $\{TL \to TD_i\}$ is certainly a weak-limit. It is a plain limit iff it is a collectively monic family of maps. We need only show that the preservation of weak-limits implies the preservation of monic families.

Given $\{A \to A_i\}$ let J be the poset obtained by adjoining to the discrete poset I two new elements l, r, ordered by $l < i$, $r < i$ for all

$i \in I$. Let D be the diagram modelled on J that sends i to A_i and both l and r to A. D sends $l \to i$ and $r \to i$ to the given map $A \to A_i$. Consider the lower bound $\{A \to D_j\}$ where $A \to D_l$ and $A \to D_r$ are identity maps and $\{A \to D_i\}$ is the original family. This lower bound is a weak limit iff the original family is monic.

1.82(10). A PRE-LIMIT for a diagram $D: \mathbf{D} \to \mathbf{B}$ is a *set* of lower-bounds which is cofinal in the class of all lower-bounds; that is, a set of compatible families $\{\{P_j \to D_i\}_{|\mathbf{D}|}\}_J$ such that for any lower-bound $\{B \to D_i\}$ there exists $j \in J$, $B \to P_j$ such that

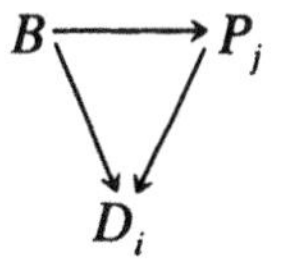

for all $i \in |\mathbf{D}|$.

A category is PRE-COMPLETE if every small diagram has a pre-limit. It is routine that

A functor that preserves weak-limits preserves limits.

1.83. Given $G: \mathbf{B} \to \mathbf{A}$ and an object $A \in \mathbf{A}$ we say that a set $\{A \to G(B_i)\}_I$ is a PRE-ADJOINT for A if for every $A \to G(B)$ there exists $i \in I$, $x: B_i \to B$ such that

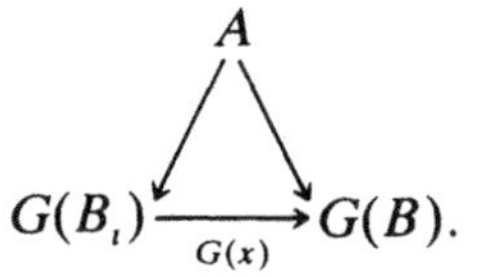

If, as is the case in all our examples, $\mathbf{A}$ is locally small, A has a pre-adjoint iff there is a set of objects $\{B_i\}_I$ in $\mathbf{B}$ such that for every $A \to G(B)$ there exists $i \in I$, $x: B_i \to B$, $A \to G(B_i)$ such that

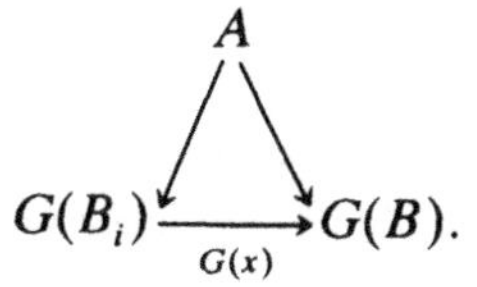

Note that if $\mathbf{B}$ is a full subcategory, G its inclusion, then a pre-adjoint, or if one wishes, a PRE-REFLECTION of A exists if there is a set $\{B_i\}$ in $\mathbf{B}$

such that every map from A to a $\mathbf{B}$-object factors through at least one of the B_i's.

$G\colon \mathbf{B} \to \mathbf{A}$ is called a PRE-ADJOINT FUNCTOR if every $A \in \mathbf{A}$ has a pre-adjoint.

THE GENERAL ADJOINT FUNCTOR THEOREM. *If* $\mathbf{B}$ *is locally small and complete then* $G\colon \mathbf{B} \to \mathbf{A}$ *has a left-adjoint iff it is continuous and pre-adjoint.*

We will obtain this as a corollary of the More General Adjoint Functor Theorem below.

1.831. A functor $G\colon \mathbf{B} \to \mathbf{A}$ is UNIFORMLY CONTINUOUS if for every small diagram $D\colon \mathbf{D} \to \mathbf{B}$, G sends the lower bounds of D to a cofinal family of lower-bounds of $\mathbf{D} \xrightarrow{D} \mathbf{B} \xrightarrow{G} \mathbf{A}$; that is, given any lower-bound $\{A \to G(D_i)\}$ there exists a lower-bound $\{B \to D_i\}$ and a map $A \to G(B)$ such that

$$\begin{array}{ccc} A & \longrightarrow & G(B) \\ & \searrow \swarrow & \\ & G(D_i) & \end{array} \qquad \text{for all } i\,.$$

Clearly a uniformly continuous functor preserves pre-limits, hence weak-limits, hence [1.829] limits. If $\mathbf{B}$ is pre-complete then G is uniformly continuous iff it preserves pre-limits. Consequently if $\mathbf{B}$ is (weakly) complete then G is uniformly continuous iff it preserves (weak) limits.

THE MORE GENERAL ADJOINT FUNCTOR THEOREM. If $\mathbf{B}$ *is locally small and if idempotents split in* $\mathbf{B}$ *then* $G\colon \mathbf{B} \to \mathbf{A}$ *has a left-adjoint iff it is uniformly continuous and pre-adjoint.*

BECAUSE: [1.831–1.835], G has a left-adjoint iff $(A, G(-))$ is representable for each $A \in \mathbf{A}$, [1.817]. We thus convert the problem to the characterization of representable functors (which, unfortunately, is not a special case: representables have left-adjoints only when arbitrary co-powers exist).

The uniform-continuity and the pre-adjointness of G lead to two properties on $(A, G(-))$ which we examine separately:

1.832. A set-valued functor $T\colon \mathbf{B} \to \mathscr{S}$ is POINTWISE CONTINUOUS if for every small diagram $D\colon \mathbf{D} \to \mathbf{B}$ and lower-bound $\{1 \to T(D_i)\}$ in $\mathscr{S}$

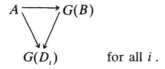

there exists a lower-bound $\{B \to D_i\}$ in **B** and a point $1 \to T(B)$ such that

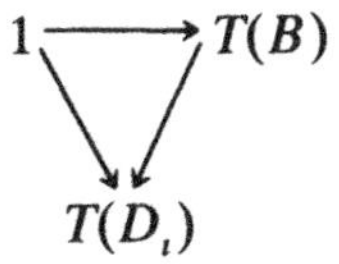

for all i.

Clearly, $G: \mathbf{B} \to \mathbf{A}$ is uniformly continuous iff $(A, G(-))$ is pointwise continuous for each $A \in \mathbf{A}$.

1.833. Given a set-valued functor $T: \mathbf{B} \to \mathscr{S}$ and a collection of elements $\{x \in T(B_i)\}_I$ we may construct the smallest subfunctor $T' \subset T$ that contains all the elements: $y \in T'(B)$ iff there exists $i \in I, f: B_i \to B$ such that $(Tf)(x) = y$. If $T' = T$ we say that *T is generated by the* x_i*'s*. T is a PETTY-FUNCTOR if it is generated by some *set* of elements.

Given $G: \mathbf{B} \to \mathbf{A}$ then a pre-adjoint for $A \in \mathbf{A}$ yields a generating set for $(A, G(-))$ and conversely. Hence, G is pre-adjoint iff $(A, G(-))$ is a petty-functor for each $A \in \mathbf{A}$.

The More General Adjoint Functor Theorem thus converts to:

1.834. THE GENERAL REPRESENTABILITY THEOREM. *If* **B** *is locally small and if idempotents split in* **B** *then a set-valued functor* $T: \mathbf{B} \to \mathscr{S}$ *is representable iff it is a pointwise continuous petty-functor.*

BECAUSE: We make one more conversion of the problem. For a set-valued functor $T: \mathbf{B} \to \mathscr{S}$ we define its *category of elements*, $\mathscr{E}\ell(T)$, as the category founded on **B** with objects as the class of pairs $\{\langle x, B\rangle \mid x \in TB\}$ and with the source-target predicate given by $\langle x, B\rangle \xrightarrow{f} \{x', B'\rangle$ iff $(Tf)(x) = x'$.

It is easily seen that *T is representable iff* $\mathscr{E}\ell(T)$ *has a coterminator.*

T is pointwise continuous iff every small diagram in $\mathscr{E}\ell(T)$ has at least one lower-bound (the nearest trace of completeness.)

T is petty iff there is a pre-coterminator in $\mathscr{E}\ell(T)$, that is a set of objects $\{C_i\}$ such that for every object X there exists i and a map $C_i \to X$.

If idempotents split in **B** then they easily split in $\mathscr{E}\ell(T)$. Hence we finish the More General Adjoint Functor Theorem and the General Representability Theorem with

1.835. *A locally small category in which idempotents split has a coterminator iff it has a pre-coterminator and if all small diagrams have lower-bounds.*

BECAUSE: Clearly a coterminator yields a pre-coterminator and is a lower-bound for every small diagram (indeed, for every diagram, small or not).

For the other direction, we start by viewing the pre-coterminator as a discrete diagram and let W be a lower-bound. W is a weak-coterminator. Next, regard W with all its endomorphisms as a one-object diagram and let $f: W' \to W$ be a lower-bound. For all $g: W \to W$ it is the case that $fg = f \cdot 1 = f$. Since W is a weak-coterminator there exists $h: W \to W'$. Define e as hf. For all $g: W \to W$ it is the case that $eg = e$. In particular, e is an idempotent. Let $p: W \to C$, $i: C \to W$ split e; that is $pi = e$ and $ip = 1$. All endomorphisms on C are one: given $x: C \to C$ it is the case that

$$x = ipipxip = ie(pxi)p = iep = ipip = 1 .$$

Since C has a map to W, it is a weak-coterminator. In fact, it is a plain coterminator. Given $f, g: C \to X$ let $y: E \to C$ be a lower-bound of f, g; that is, $yf = yg$. C is a weak coterminator and there exists $x: C \to E$. But $xy = 1$ and it is the case that $f = xyf = xyg = g$.

1.836. Suppose idempotents do not split in $\mathbf{B}$. Let $\hat{\mathbf{B}} = \mathscr{S}\!plit\,(\mathscr{I}\!d\,)$ be the result of formally splitting all idempotents. The canonical embedding $\mathbf{B} \to \hat{\mathbf{B}}$ is full, uniformly continuous and pre-adjoint. If it had a left-adjoint, then as any full reflective subcategory, $\mathbf{B}$ would be a retract of $\hat{\mathbf{B}}$. But any retract of a category in which idempotents split is one in which idempotents split.

Suppose $e: B \to B$ is an idempotent that does not split. Let $T: \mathbf{B} \to \mathscr{S}$ be the image of the transformation $(e, -): (B, -) \to (B, -)$. T is a point-wise continuous petty functor. If it were representable then e would split. In the category $\mathscr{E}\ell(T)$ the object $\langle e, B \rangle$ is a weak-coterminator that appears as a lower-bound of every diagram.

Given any pointwise continuous petty-functor $T: \mathbf{B} \to \mathscr{S}$, the extension to $\hat{\mathbf{B}}$ remains pointwise continuous and a petty-functor, from which one may conclude that any set-valued functor from a locally small category is a retract of a representable iff it is a pointwise continuous petty-functor.

1.837. If $\mathbf{B}$ is complete then for any $\mathbf{D}$, $\Delta: \mathbf{B} \to \mathbf{B}^{\mathbf{D}}$ is continuous. It is pre-adjoint iff all diagrams modelled on $\mathbf{D}$ have pre-colimits. Hence.

A complete locally small category is cocomplete iff it is pre-cocomplete.

1.838. Suppose that $\mathbf{B}$ is complete and WELL-POWERED, that is, $\mathscr{S}\!ub\,(B)$ is a *set* for all $B \in \mathbf{B}$. (Different maps have different graphs, hence well-powered implies local smallness.) Given continuous $T: \mathbf{B} \to \mathscr{S}$, $\mathscr{E}\ell(T)$ is again complete and well-powered.

Each object in a complete well-powered category has a unique minimal subobject.

By a *minimal object* is meant an object with no proper subobjects. (Atomic objects are not minimal: an atom has a unique proper subobject.) The minimal objects in a complete well-powered category, therefore, form an initial class, $\mathscr{M}$. The category has an initial *set* iff $\mathscr{M}$, regarded as a full subcategory, has an initial set.

$\mathscr{M}$ is a pre-ordered class: for any minimal object M the equalizer of $f, g: M \to A$ cannot be proper, hence $f = g$. Every subset of $\mathscr{M}$ has a lower bound: a minimal subobject of $\varprojlim \mathscr{M}'$ is a lower bound of $\mathscr{M}'$. Any map between minimal objects is a cover, hence any sort of well-co-powered condition says that the existence of a coterminator implies that there are only a set of isomorphism types of minimal objects.

Given continuous $T: \mathbf{B} \to \mathscr{S}$, where $\mathbf{B}$ is complete and well-powered, we are thus advised to classify the minimal objects of $\mathscr{E}\ell(T)$. Consider the particular case where $\mathbf{B}$ is the category of rings, T the forgetful functor. An object of $\mathscr{E}\ell(T)$ is a ring with a distinguished element. It is minimal iff no proper subring contains the distinguished element, that is, if the ring is generated by the element. Any minimal object, therefore, is of the form $\mathbb{Z}[X]/P$, X being the distinguished element. $\mathbb{Z}[X]$ is the coterminator for $\mathscr{E}\ell(T)$. The forgetful functor is represented by $\mathbb{Z}[X]$.

Now change T to the functor that sends a ring to its set of units. T is still continuous. An object in $\mathscr{E}\ell(T)$ is a ring with a distinguished *unit*. It is minimal if no proper subring contains the distinguished unit *as a unit*, that is, no proper subring contains the distinguished unit *and its inverse*. Hence a minimal subobject is a ring generated by a unit and its inverse. Any such ring is the group-ring of a cyclic group. The coterminator is $\mathbb{Z}[\mathbb{Z}]$, the group-ring of the integers.

1.839. In finding pre-adjoints one usually invokes a *cardinality function*, that is, a function K from the objects of the source category to the cardinal numbers with the property that for any cardinal α, $K^{\#}(\alpha)$ is a *set* of isomorphism types. For a concrete category, the ordinary cardinality of the underlying set of an object is the first candidate for a cardinality function.

Given continuous $G: \mathbf{B} \to \mathbf{A}$ where $\mathbf{B}$ is complete and well-powered, a minimal object of $\mathscr{E}\ell(A, G(-))$ is a map $A \to G(B)$ which cannot be factored through any $G(B') \subset G(B)$ where $B' \subsetneqq B$. We say, in this case, that B is *generated by A through G*.

As an example, let $\mathbf{B}$ be the category of compact Hausdorff groups, $\mathbf{A}$ the category of all topological groups, G the inclusion functor. $f: A \to B$ generates iff its image is dense in B, in which case every element of B is the limit of a filter on A and $K(B) \leq 2^{2^{K(A)}}$ where K denotes the usual cardinality.

Now let $\mathbf{A}$ be the category of topological spaces (no group structure), G the forgetful functor. $f: A \to G(B)$ generates iff the subgroup generated by $\operatorname{Im}(f)$ is dense in $G(B)$, in which case $K(B) \leq 2^{2^{\aleph_0 + K(A)}}$.

1.83(10). A set of objects $\{C_i\}_I$ is a COGENERATING SET if $\{(-, C_i)\}_I$ is collectively an embedding. That is, given $f, g: A \to B$, $f \neq g$ there exists $i \in I$, $h: B \to C$ such that $fh \neq gh$. In a complete category, $\{C_i\}_I$ is a cogenerating set iff every object is embeddable in a product of the C_i's.

THE SPECIAL ADJOINT FUNCTOR THEOREM. *If* **B** *is complete, well-powered and has a cogenerating set, and if* **A** *is locally small, then every continuous* $G: \mathbf{B} \to \mathbf{A}$ *has a left-adjoint.*

BECAUSE: Using the last section, we can define $K(B)$ as the sum of the cardinalities of (B, C_i) where $\{C_i\}_I$ is a cogenerating set. K is a cardinality function: if $K(B) = \alpha$ then B can be embedded in a α-fold product of the C_i's. If $f: A \to G(B)$ generates, then $K(B)$ is bounded by the sum of the cardinalities of $(A, G(C_i))$ because $(B, C_i) \to (G(B), G(C_i)) \to (A, G(C_i))$ is monic, each i.

A direct construction of the left-adjoint can be given as follows: we reduce to the case of the continuous set valued functors $(A, G(-))$. That is, we may specialize to the case that $\mathbf{A} \simeq \mathcal{S}$.

If $T: \mathbf{B} \to \mathcal{S}$ is continuous, **B** complete and well-powered then $\mathscr{E}\ell(T)$ is also complete and well-powered. If $\{C_i\}$ is a cogenerating set for **B** then $\{\langle x, C_i \rangle \mid x \in T(C_i), i \in I\}$ is a cogenerating set for $\mathscr{E}\ell(T)$.

We thus finish with:

A complete well-powered category with a cogenerating set has a coterminator.

If $\{C_i\}_I$ is a cogenerating set then the minimal subobject of $\Pi_I C_i$ is a coterminator: given an object B, then the minimal subobject, M, of $B \times \Pi C_i$ has a map to B. It suffices to show that M is isomorphic to the minimal subobject of ΠC_i. M has a map to each C_i. As any minimal object, it can have no more than one map to each C_i. Hence $M \to \Pi_I C_i$ is monic and represents the minimal subobject of ΠC_i.

1.83(11). We record two of the dualizations of the above:

If **A** *is co-complete, well-co-powered and with a generating set, then every cocontinuous functor from* **A** *to a locally small category has a right adjoint and every contravariant continuous functor (carrying colimits to limits) to a locally small category has an adjoint on the right.*

1.84. Grothendieck topoi

We shall eventually give a total of four definitions of Grothendieck topoi. Our first (and history's second) is the GIRAUD DEFINITION: a locally small cocomplete effective regular category in which pullbacks preserve arbitrary unions, in which disjoint unions of objects exist, and finally, in which there is a generating set. (Note that such implies pre-topos and logos.)

1.841. Our prime examples satisfy the Giraud definition: in $\mathscr{S}^{\mathbf{A}}$ the representables generate [1.632, 1.442]; in $\mathscr{S}\!\mathscr{h}(Y)$ the subterminators generate.

1.842. In the next chapter we will give a direct proof of the following theorem that does not depend on the special adjoint functor theorem. The theorem was known from the beginning by the Grothendieck school. Its importance, however, was not appreciated until Lawvere made it the basis of the theory of elementary topoi as will be exposed in [1.9].

If $\mathbf{E}$ *is a Grothendieck topos (Giraud definition) then the graphing functor* $\mathbf{E} \to \mathscr{R}\!\mathit{el}(\mathbf{E})$ *has a right adjoint.*

We separate the argument:

1.843. *A Grothendieck topos is well-powered.*

BECAUSE: If $\{G_\iota\}_I$ is a generating set then it is also a basis: as in any pre-topos, every subobject appears as an equalizer, hence $B' \subset B$ is distinguished by the family of subsets $\{(G_\iota, B') \subset (G_i, B)\}_I$.

Consequently,

$\mathscr{R}\!\mathit{el}(\mathbf{E})$ *is locally small.*

Moreover, as in any pre-topos, comonics coincide with covers, and the isomorphism-types of covers with A as source are in one-to-one correspondence with the equivalence relations on A. Hence

A Grothendieck topos is well-copowered.

1.844. *A Grothendieck topos is locally complete.*

BECAUSE: Given a family of subobjects, $\{A_i \rightarrowtail A\}$, the image of the induced map $\Sigma A_i \to A$ is their union, where ΣA_ι denotes a disjoint union.

Since inverse images preserve arbitrary unions, we further obtain:

The composition of relations distributes with arbitrary unions.

1.845. If $\{u_\iota\colon A_\iota \to S\}$ is a coproduct, then $u_i u_\iota^\circ = 1$ for each $i \in I$. Using only that pairs of objects have disjoint unions, indeed only that $\ell, \imath\colon 1 \to 1 + 1$ are disjoint, we can verify that arbitrary coproducts are disjoint unions: given any $j \in I$ let $f\colon S \to 1 + 1$ be the map such that $u_j f = A_j \to 1 \xrightarrow{\ell} 1 + 1$ and $u_\iota f = A_i \to 1 \xrightarrow{\imath} 1 + 1$, $i \neq j$. Then $f^\#(\ell) = A_j$ and $f^\#(\imath)$ contains all the other A_i's. Consequently $u_i u_j^\circ = 0$ for $i \neq j$.

$\bigcup u_i^\circ u_i$ must be the identity function on S. It is clearly contained in the identity; if it were tabulated by $S' \rightarrowtail S \xrightarrow{\langle 1,1 \rangle} S \times S$ then each $A_i \subset S'$ and hence $S' = S$ since the u_i's are collectively a cover.

All of which yields,

A coproduct in **E** *remains a coproduct in* $\mathscr{R}el(\mathbf{E})$.

BECAUSE: Given $\{R_i: A \to B\}$ in $\mathscr{R}el(\mathbf{E})$ let $R = \bigcup u_i^\circ R_i$. It is routine computation that R is the unique relation such that $u_i R = R_i$, for all $i \in I$.

1.846. We now have that a Grothendieck topos **E** is a cocomplete well-copowered category with a generating set, that $\mathscr{R}el(\mathbf{E})$ is locally small and that $\mathbf{E} \to \mathscr{R}el(\mathbf{E})$ preserves coproducts. By the special adjoint functor theorem only one thing needs yet to be verified:

A coequalizer in **E** *remains a coequalizer in* $\mathscr{R}el(\mathbf{E})$.

BECAUSE: Suppose that $h: B \twoheadrightarrow C$ is a coequalizer of $f, g: A \to B$. We recall that $h^\circ h = 1$ and as in any effective regular category, that hh° is the smallest equivalence relation that contains $f^\circ g$. A Grothendieck topos is E-standard, that is, the smallest equivalence relation that contains R is $\bigcup_{n=-\infty}^{\infty} R^n$ where $R^0 = I$, $R^{-n} = (R^\circ)^n$.

We wish to show that if R is a relation such that $fR = gR$ then there exists a unique R' such that $hR' = R$. The uniqueness is easy: h has a left-inverse in $\mathscr{R}el(\mathbf{E})$.

As is typical when working with relations, existential assertions are replaced with formulas. We take R' to be $h^\circ R$. We must show that $fR = gR$ implies $hh^\circ R = R$.

(At this point we could invoke the metatheorem obtainable from [1.563].)

Since $1 \subset hh^\circ$ we easily have $R \subset hh^\circ R$. For the other containment we note first that $f^\circ gR \subset f^\circ fR \subset R$ and that $g^\circ fR \subset g^\circ gR \subset R$ hence that $SR = R$ (where $S = g^\circ f \cup 1 \cup f^\circ g$). By induction $S^{n+1}R \subset S^n SR \subset S^n R \subset R$, and finally by the distributivity with arbitrary unions $(\bigcup_{n=1}^{\infty} S^n)R \subset R$.

1.847. This application of the special adjoint functor theorem illustrates the irrelevancy of completeness conditions on the target category. As we saw, disjoint unions provide coproducts for $\mathscr{R}el(\mathbf{E})$ (and since $\mathscr{R}el(\mathbf{E})$ is self-dual via the reciprocation operation, they also provide products). $\mathscr{R}el(\mathbf{E})$ is not, however, cocomplete. Indeed, idempotents do not always split: in $\mathscr{R}el(\mathscr{S})$ the ordering relation on the two-element ordered set is idempotent but cannot be split. (As will be seen in chapter 2, $\mathscr{R}el(\mathbf{E})$ is weakly bicomplete.)

1.85. Exponential categories

A category **A** with binary products is EXPONENTIAL (sometimes *cartesian closed*) if for every object A the functor $A \times -: \mathbf{A} \to \mathbf{A}$ has a right adjoint; equivalently, if for every A, B the set-valued functor $(A \times -, B)$ is representable by an object, denoted B^A. We obtain an elementary definition as follows: if $\eta:(-, B^A) \to (A \times -, B)$ is an equivalence then η sends 1 in (B^A, B^A) to a map $e\colon A \times B^A \to B$ which will be called an EVALUATION MAP; for any $f\colon A \times X \to B$ there exists a unique map from X to B^A denoted λf, such that

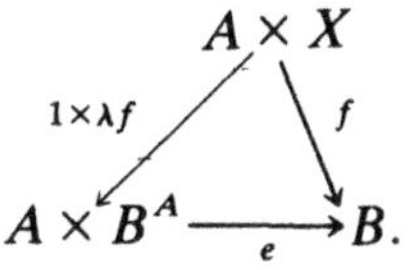

In the diagrammatic notation, we may define an exponential category as a category with binary products such that:

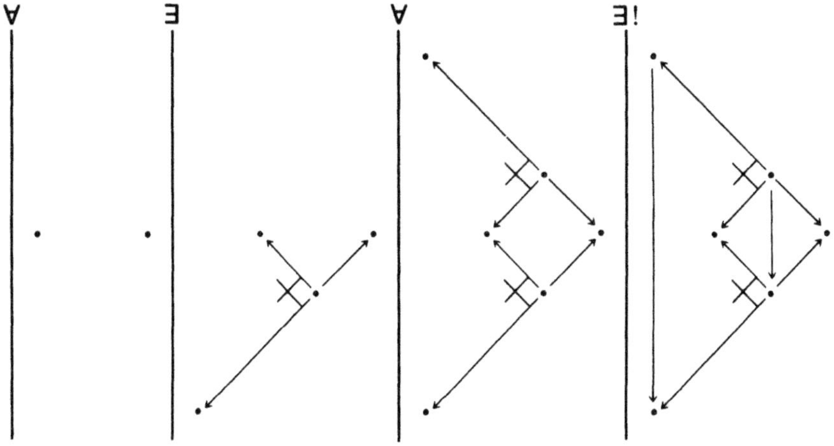

1.851. We will show in [1.9] that Grothendieck topoi, hence our prime examples $\mathscr{S}^{\mathbf{A}}$ and $\mathscr{Sh}(Y)$, are exponential. For the category of sets, of course, B^A is constructible as the set of functions from A to B. In the category of posets (which is not regular, hence not a topos) B^A is constructible as the set of order-preserving maps from A to B with the usual ordering thereon: $f \leq g$ if $f(a) \leq g(a)$ for all $a \in A$. The category of posets is a full subcategory of the category of small categories. $\mathbf{B}^{\mathbf{A}}$ is as already described: its objects are functors from $\mathbf{A}$ to $\mathbf{B}$ and its maps are natural transformations.

1.852. A poset, when viewed as a category, is exponential iff it has binary intersections and if for every a, b there exists an element b^a such that

$$x \leq b^a \quad \textit{iff} \quad a \wedge x \leq b\,.$$

We have already encountered it: the usual notation in a poset for b^a is $a \to b$, the Heyting arrow operation [1.72].

1.853. In an exponential category, B^A may be viewed as a bifunctor, covariant in the lower variable, contravariant in the upper. Given $f: B_1 \to B_2$ then $f^A: B_1^A \to B_2^A$ is the unique map such that

$$\begin{array}{ccc}
(-, B_1^A) & \xrightarrow{(-,f^A)} & (-, B_2^A) \\
\downarrow \wr & & \downarrow \wr \\
(A \times -, B_1) & \xrightarrow[(1\times -,f)]{} & (A \times -, B_2)
\end{array}$$

Given $g: A_1 \to A_2$ then $B^g: B^{A_2} \to B^{A_1}$ is the unique map such that

$$\begin{array}{ccc}
(-, B^{A_2}) & \xrightarrow{(-,B^g)} & (-, B^{A_1}) \\
\downarrow \wr & & \downarrow \wr \\
(A_2 \times -, B) & \xrightarrow[(g\times -,1)]{} & (A_1 \times -, B)
\end{array}$$

(Elementary constructions are given by

$$f^A = \lambda(A \times B_1{}^A \xrightarrow{e} B_1 \xrightarrow{f} B_2) \quad \text{and}$$

$$B^g = \lambda(A_1 \times B^{A_2} \xrightarrow{g\times 1} A_2 \times B^{A_2} \xrightarrow{e} B.)$$

The word *bifunctor* means, in this case:

$$\begin{array}{ccc}
B_1{}^{A_2} & \xrightarrow{f^{A_2}} & B_2{}^{A_2} \\
{\scriptstyle B_1{}^g}\downarrow & & \downarrow{\scriptstyle B_2{}^g} \\
B_1{}^{A_1} & \xrightarrow[f^{A_1}]{} & B_2{}^{A_1}.
\end{array}$$

That is, B^A may be regarded as a functor from $\mathbf{A}^\circ \times \mathbf{A}$ to $\mathbf{A}$.

$(-)^A$, of course, has a left-adjoint, $A \times -$, hence $(-)^A$ preserves limits.

The commutativity of products implies that $B^{(-)}$ is a contravariant functor with an adjoint on the right: itself. That is

$$(A_1, B^{A_2}) \simeq (A_2 \times A_1, B) \simeq (A_1 \times A_2, B) \simeq (A_2, B^{A_1}).$$

$B^{(-)}$ therefore is a continuous contravariant functor: it carries colimits to limits.

In a bicartesian exponential category we therefore obtain:

$$\begin{aligned} &1^A \simeq 1, && (B_1 \times B_2)^A \simeq B_1^{\,A} \times B_2^{\,A}, \\ &B^0 \simeq 1, && B^{A_1+A_2} \simeq B^{A_1} \times B^{A_2}. \end{aligned}$$

The associativity of products yields

$$(B^{A_1})^{A_2} \simeq B^{A_1 \times A_2}$$

(since $(-, (B^{A_1})^{A_2}) \simeq (A_2 \times -, B^{A_1}) \simeq (A_1 \times (A_2 \times -), B) \simeq ((A_1 \times A_2) \times -, B) \simeq (-, B^{A_1 \times A_2})$).

1 is a unit for products, hence

$$B^1 \simeq B$$

(since $(-, B^1) \simeq (1 \times -, B) \simeq (-, B)$).

$A \times -$ has a right-adjoint, therefore it must preserve colimits. Hence

$$A \times (B_1 + B_2) \simeq (A \times B_1) + (A \times B_2),$$

$$A \times 0 \simeq 0.$$

The last isomorphism is equivalent to the strictness of the coterminator. Given $A \to 0$ then A is a retract of $A \times 0$. But a coterminator has no proper retracts and $A \to 0$ is an isomorphism.

We previously encountered these equivalences (therein called identities) for the special case of Heyting algebras [1.72].

Caution: 0^A is not 0. Indeed $0^0 \simeq 1$. Since $(-)^A$ preserves limits, hence monics, 0^A is a subterminator. It is the largest subterminator U such that $A \times U = 0$.

1.854. In any category with binary products $\Sigma: \mathbf{A}/B \to \mathbf{A}$ is a left-adjoint of $\Delta: \mathbf{A} \to \mathbf{A}/B$. That is,

$$\left(\Sigma\left(\begin{array}{c} A \\ \downarrow{\scriptstyle f} \\ B \end{array}\right), C\right) \simeq \left(\left(\begin{array}{c} A \\ \downarrow{\scriptstyle f} \\ B \end{array}\right), \left(\begin{array}{c} B \times C \\ \downarrow{\scriptstyle l} \\ B \end{array}\right)\right).$$

When does Δ have a right-adjoint?

In the case $\mathbf{A} = \mathscr{S}$, $\mathscr{S} \xrightarrow{\Delta} \mathscr{S}/B \simeq \mathscr{S}^B$ is the diagonal functor in the category of diagrams modelled on the discrete category B. The right-adjoint delivers products. When $\Delta: A \to A/B$ has a right-adjoint we will denote it as $\Pi: A/B \to A$. It may be viewed as delivering products of 'families in $\mathbf{A}$ indexed on the object B'.

If $\Pi: A/B \to A$ exists, then for every A, A^B exists.

BECAUSE: $(B \times -, A) \simeq (\Sigma\Delta(-), A) \simeq (\Delta(-), \Delta A) \simeq (-, \Pi\Delta(A))$.

If $\mathbf{A}$ is cartesian and exponential, then $\Delta: \mathbf{A} \to \mathbf{A}/B$ has a right-adjoint $\Pi: \mathbf{A}/B \to \mathbf{A}$ for every B.

BECAUSE: For an object $f: A \to B$ in A/B we may construct $\Pi(f)$ via the pullback

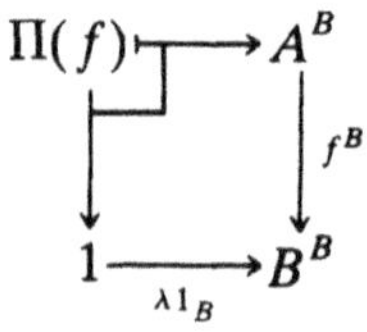

For any $C \in \mathbf{A}$ consider

$$
\begin{array}{ccccc}
(C, \Pi(f)) & \longmapsto & (C, A^B) & \xrightarrow{\sim} & (B \times C, A) \\
\downarrow & & \downarrow & & \downarrow {\scriptstyle (1,f)} \\
(C, 1) & \longrightarrow & (C, B^B) & \xrightarrow{\sim} & (B \times C, B)
\end{array}
$$

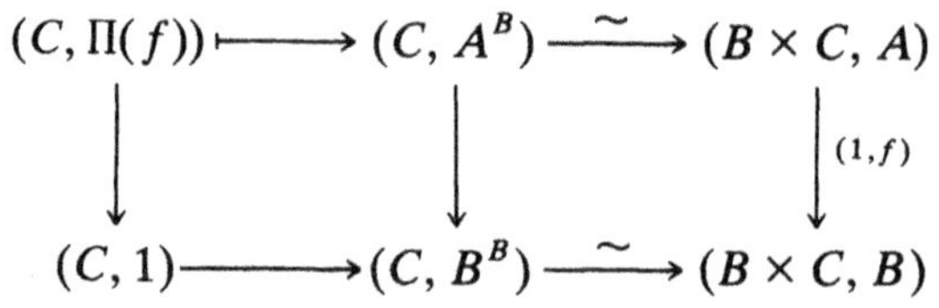

The function from $(C, 1)$ to $(B \times C, B)$ sends the unique element of $(C, 1)$ to the projection map $\ell: B \times C \to B$. Hence $(C, \Pi(f))$, viewed as a subset of $(B \times C, A)$, is the set of maps $g: B \times C \to A$ such that

$$
\begin{array}{ccc}
B \times C & \xrightarrow{g} & A \\
& {\scriptstyle \ell} \searrow \downarrow {\scriptstyle f} & \\
& B &
\end{array}
$$

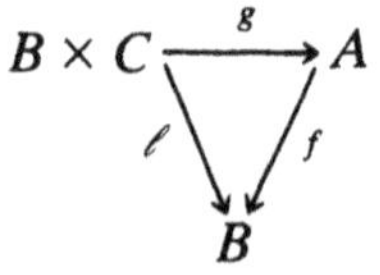

which precisely describes $(\Delta(C), (f))$.

1.855. Given a B-indexed family of sets $\{A_b\}$ in $\mathscr{S}$ the classical construction of $\Pi_B A_b$ is as the set of functions from B such that $g(b) \in A_b$ for all $b \in B$. If the A_b's are disjoint, $\Pi_B A_b$ is the set of functions from B to A such that $B \xrightarrow{g} A \xrightarrow{f} B = 1_B$ where $A = \cup_B A_b$ and $A \xrightarrow{f} B$ is the obvious. This is precisely the construction of $\Pi_B(f)$ just given.

1.856. The slice lemma does not hold for exponential categories. Let **P** denote the category of posets, 3 the three-element linearly ordered set. If **P**/3 were exponential, then $A \times -$ in **P**/3 would preserve colimits, in particular, coequalizers (it does preserve coproducts). Let B denote the **P**/3-object, $f: 2 + 2 \twoheadrightarrow 3$, where $2 + 2$ is the disjoint union of two linearly ordered sets of two elements each, and f is onto and sends the top element of one linearly ordered half and the bottom element of the other half to the middle element of 3. Let C denote $g: 1 \to 3$ where the image of g is the middle element of 3. There are two maps from C to B. Their coequalizer is the terminator. Let A denote the object $h: 2 \to 3$ where 2 is the two-element linearly ordered set, h the inclusion into 3 whose image excludes the middle element. $A \times B$ is the two-element discrete set. $A \times B \to A \times 1$ is not the coequalizer of any pair of maps to $A \times B$.

(Curiously, **P**/2 is exponential. Indeed, **P**/A is exponential iff A contains no three-element chain.)

1.857. *If* **A**′ *is a full coreflective subcategory of an exponential category*, **A**, *closed under binary products, then it is exponential.*

BECAUSE: Suppose G denotes the coreflection functor. For $A, B, C \in |\mathbf{A}'|$ we have $(A \times C, B) \simeq (C, B^A) \simeq (C, G(B^A))$.

If **A** is an exponential category we say that a full subcategory, **A**′, is an EXPONENTIAL IDEAL if for $A \in |\mathbf{A}|, \mathrm{B} \in |\mathbf{A}'|$ it is the case that $B^A \in |\mathbf{A}'|$. A REPLETE SUBCATEGORY is a subcategory closed under isomorphism type.

A full replete reflective subcategory of an exponential category is an exponential ideal iff reflections preserve products.

BECAUSE: If $\mathbf{A}' \subset \mathbf{A}$ is a reflective exponential ideal we wish to show that $\overline{A_1 \times A_2} \simeq \bar{A}_1 \times \bar{A}_2$ for all A_1, A_2 where the over-score denotes reflections. For all $B \in |\mathbf{A}'|$,

$$(\overline{A_1 \times A_2}, B) \simeq (A_1 \times A_2, B) \simeq (A_2, B^{A_1}) \simeq (\bar{A}_2, B^{A_1})$$
$$\simeq (A_1, B^{\bar{A}_2}) \simeq (\bar{A}_1, B^{\bar{A}_2}) \simeq (\bar{A}_1 \times \bar{A}_2, B)\,.$$

If **A**′ is a full replete reflective subcategory for which reflections preserve products, then $A \times C$ and $A \times \bar{C}$ have the same reflection. Hence for $B \in |\mathbf{A}'|$ we obtain $(C, B^A) \simeq (A \times C, B) \simeq (A \times \bar{C}, B) \simeq (\bar{C}, B^A)$. Given any $B \in |\mathbf{A}'|$, therefore, and $C \to B^A$ there exists unique

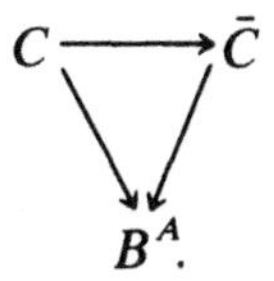

In the case $C = B^A$, such forces $C \to \bar{C}$ to be an isomorphism.

1.858. A KURATOWSKI INTERIOR OPERATION on a lattice is a deflationary, idempotent, intersection-preserving operation:

$$x^{\bullet} \leqslant x\,, \qquad x^{\bullet} = (x^{\bullet})^{\bullet}\,, \qquad (x \wedge y)^{\bullet} = x^{\bullet} \wedge y^{\bullet}\,.$$

The set of its values (equivalently, its fixed points) is called the set of *open elements*. The first lemma in the last section says that the sublattice of open elements for a Kuratowski interior operation is a Heyting algebra if the ambient lattice is a Heyting algebra. (For open sets U, V in a topological space, $U \to V$ is the union of V and the interior of the complement of U.)

A LAWVERE–TIERNEY CLOSURE OPERATION (L–T for short) is an inflationary, idempotent, intersection-preserving operation:

$$x \leqslant \bar{x}\,, \qquad \bar{\bar{x}} = \bar{x}\,, \qquad \overline{x \wedge y} = \bar{x} \wedge \bar{y}\,.$$

(A *Kuratowski closure operation* is union-preserving.) The set of its values are called *closed elements*. The second lemma of the last section says that the closed elements of a L–T closure operation on a Heyting algebra forms an exponential ideal: $a \to b$ is closed whenever b is. Conversely, given any closure operation such that the closed elements form an exponential ideal, then the operation is L–T.

If the operation is double negation and if $b = \neg\neg b$, then $a \to b = a \to (\neg b \to 0) = (a \wedge \neg b) \to 0$, hence $a \to b = \neg\neg(a \to b)$. Consequently, $\neg\neg$ preserves intersection. The proof in [1.727] was just an adaptation of the proof in the last section.

1.859. Let **A** be a category with binary products. We say that an object B is BASEABLE if B^A exists for all A, (that is, if $(A \times -, B)$ is representable, all A). Let **B** be the full subcategory of baseable objects.

B is a full subcategory in a natural sense: given $f\colon B_1 \to B_2$ we can define, just as if all objects were baseable, the map $f^A\colon B_1^A \to B_2^A$.

The inclusion $\mathbf{B} \to \mathbf{A}$ is continuous but we will need only:

The inclusion $\mathbf{B} \to \mathbf{A}$ *preserves equalizers.*

BECAUSE: If $B_0 \to B_1$ is the equalizer of $f, g\colon B_1 \to B_2$, where $B_1, B_2 \in |\mathbf{B}|$ then $B_0{}^A$ is constructible as the equalizer of f^A, g^A.

(**B** is always an exponential category: if **B** is baseable, then B^A is baseable because $(B^A)^{A'} \simeq B^{A \times A'}$. We have no analysis concerning when **B** is pre-reflective.)

1.9. TOPOI

The composition of relations requires a notion of image (and the associativity of composition requires regularity). In any category with pullbacks, however, we can compose a map with a relation: given a map $f: A \to B$ and a table

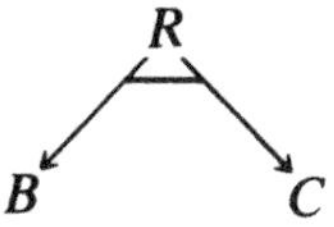

let

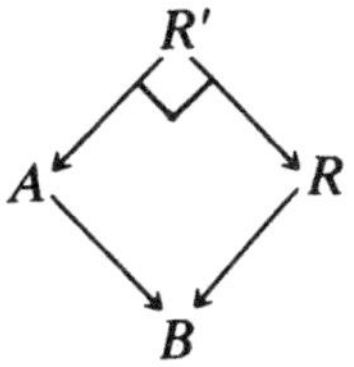

be a pullback;

then

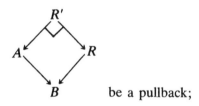

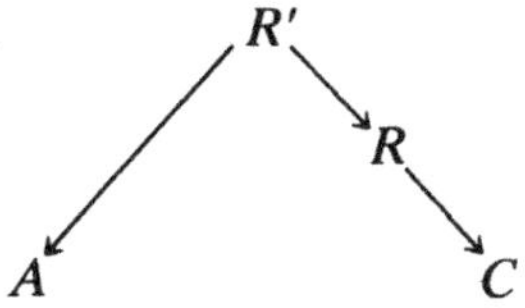

is still a monic pair;

we take it as the definition of fR.

We say that a relation U targeted at an object C is a UNIVERSAL RELATION if every relation R targeted at C is uniquely of the form fU. f is denoted ΛR. Given an object C we say that a POWER-OBJECT of C, denoted $[C]$, is an object which appears as the source of a universal relation targeted at C. We shall denote the universal relation as $\ni \subset [C] \times C$.

A TOPOS is a cartesian category in which each object has a power-object. We will show in this section that a topos is a positive effective transitive logos (hence, in particular, it is a pre-topos).

A topos is definable as a cartesian category in which

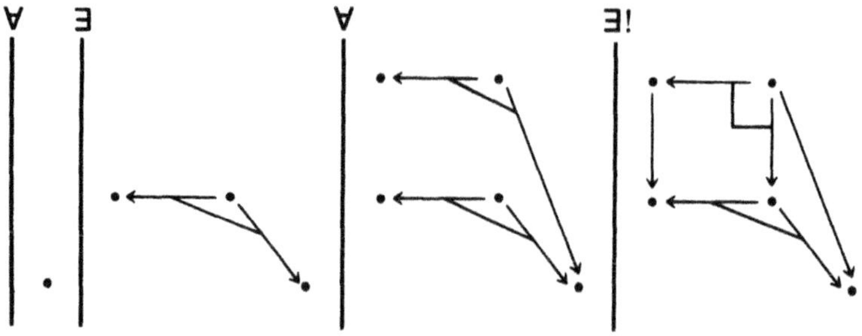

1.911. For any category with pullbacks and for any object B therein we can define a contravariant set-valued functor $\mathscr{R}el(-, B)$ that sends an object A to the set of relations from A to B. Given a map $f: A \to A'$, $\mathscr{R}el(f, B)$ sends R in $\mathscr{R}el(A', B)$ to fR in $\mathscr{R}el(A, B)$. (No regularity is needed to show that $g(fR) = (gf)R$, that is, $\mathscr{R}el(-, B)$ preserves composition.)

The existence of a power-object $[B]$ says that $\mathscr{R}el(-, B)$ is equivalent to the functor $(-, [B])$. And conversely: if $\eta: (-, [B]) \to \mathscr{R}el(-, B)$ is an isomorphism of functors, then $\eta_{[B]}$ carries $1_{[B]}$ to a universal relation targeted at B.

If **A** is regular, then it is a topos iff the graphing functor $\mathbf{A} \to \mathscr{R}el(\mathbf{A})$ [1.564] has a right-adjoint. In particular Grothendieck topoi are topoi [1.842]. In $\mathscr{S}$, $[B]$ is the set of subsets of B.

1.912. Relations from A to 1 correspond to subobjects of A, that is $\mathscr{R}el(-, 1) \simeq \mathscr{S}ub(-)$. The traditional notation for [1] is Ω. It is called a SUBOBJECT CLASSIFIER.

Translating entirely to subobjects (rather than relations) we may define a *universal subobject* as a monic map $\Omega' \rightarrowtail \Omega$ with the property that for any monic $A' \rightarrowtail A$ there exists a unique pullback

$$\begin{array}{ccc} A' & \rightarrowtail & A \\ \downarrow & \lrcorner & \downarrow \\ \Omega' & \rightarrowtail & \Omega. \end{array}$$

The map $A \longrightarrow \Omega$ is often denoted χ_A and called the CHARACTERISTIC MAP of $A' \subset A$. Note that its uniqueness forces the uniqueness of the map $A' \longrightarrow \Omega'$. Using the notion of inverse image, we may define a universal subobject as one such that for all subobjects $A' \subset A$ there is a unique $\chi: A \longrightarrow \Omega$ such that $\chi^{\#}(\Omega') = A'$.

In particular, for every object A there is a unique pullback

$$\begin{array}{ccc} A & \xrightarrow{1} & A \\ \big\downarrow & & \big\downarrow \\ \Omega' & \rightarrowtail & \Omega. \end{array}$$

Given any $A \longrightarrow \Omega'$ we may define $A \longrightarrow \Omega$ as $A \longrightarrow \Omega' \rightarrowtail \Omega$ to obtain a pullback. Hence for any object A there is a unique map from A to Ω'. That is, the universal subobject is a point. We henceforth drop the notation Ω'.

The universal subobject will be denoted $1 \xrightarrow{t} \Omega$.

1.913. *In any cartesian category with a subobject classifier all subobjects appear as equalizers.*

BECAUSE: $A' \rightarrowtail A$ is an equalizer of $\chi_{A'}$ and $A \longrightarrow 1 \xrightarrow{t} \Omega$.

Consequently,

In any cartesian category with a subobject classifier, covers coincide with epics.

1.914. The algebraic structure of Ω, that is, the maps of the form $\Omega^n \longrightarrow \Omega$ is of some interest. An n-ary operation, $g: \Omega^n \longrightarrow \Omega$, induces an n-ary operation on $\mathscr{Sub}(A)$ for each n: given $A_1, \ldots, A_n \subset A$ define $\hat{g}(A_1, \ldots, A_n) \subset A$ as the subobject whose characteristic map is

$$A \xrightarrow{\langle \chi_{A_1}, \ldots, \chi_{A_n} \rangle} \Omega^n \xrightarrow{g} \Omega .$$

Such yields an operation on the contravariant functor $\mathscr{Sub}(-)$: for $f: B \longrightarrow A$ it is the case that $\hat{g}(f^\#(A_1), \ldots, f^\#(A_n)) = f^\# \hat{g}(A_1, \ldots, A_n)$. Any such operation on $\mathscr{Sub}(-)$ so arises.

Let $g: \Omega^2 \longrightarrow \Omega$ be the characteristic map of $\langle t, t \rangle: 1 \longrightarrow \Omega \times \Omega$. Given $A_1, A_2 \subset A$ then $A' \subset \hat{g}(A_1, A_2)$ iff

$$A' \rightarrowtail A \xrightarrow{\chi_{A_i}} \Omega \quad = \quad A' \longrightarrow 1 \xrightarrow{t} \Omega, \quad i = 1, 2;$$

that is, iff $A' \subset A_i$, $i = 1, 2$. $\hat{g}(A_1, A_2) = A_1 \cap A_2$. Ω has, therefore, a semi-lattice structure. The unit is $1 \xrightarrow{t} \Omega$.

Now let $g: \Omega^2 \to \Omega$ be the characteristic map of $\langle 1, 1 \rangle: \Omega \to \Omega \times \Omega$. For any pullback

$$\begin{array}{ccc} E & \longrightarrow & A \\ \big\downarrow & & \big\downarrow \langle f,g \rangle \\ B & \xrightarrow[\langle 1,1 \rangle]{} & B \times B, \end{array}$$

E is the equalizer of f, g. $\hat{g}(A_1, A_2)$, therefore, is the equalizer of χ_{A_1}, χ_{A_2}. $A' \subset \hat{g}(A_1, A_2)$ iff $(A' \rightarrowtail A \xrightarrow{\chi_{A_1}} \Omega) = (A' \rightarrowtail A \xrightarrow{\chi_{A_2}} \Omega)$. Consider

$$\begin{array}{ccccc} A_i \cap A' & \longrightarrow & A_i & \longrightarrow & 1 \\ \big\downarrow & & \big\downarrow & & \big\downarrow t \\ A' & \longrightarrow & A & \xrightarrow[\chi_{A_i}]{} & \Omega. \end{array}$$

$A' \rightarrowtail A \xrightarrow{\chi_{A_i}} \Omega$ is the characteristic map of $A_i \cap A'$. Hence $A' \subset \hat{g}(A_1, A_2)$ iff $A_1 \cap A' = A_2 \cap A'$. Each $\mathscr{S}\!ub(A)$ has a Heyting double-arrow operation (and, of course, a Heyting single arrow operation: $x \to y = x \leftrightarrow (x \wedge y)$) and $\mathscr{S}\!ub(A)$ has a Heyting semi-lattice structure.

1.915. For any group G, $\mathscr{S}^G$ is easily seen to be a topos: the forgetful functor to $\mathscr{S}$ is a representation of topoi which tells us that the underlying set of $[B]$ is the usual set of subsets. $[B]$ becomes a G-set by defining for $X \subset B$ and $\sigma \in G$, $X^\sigma = \{x\sigma \mid x \in X\}$. In particular, Ω is the two-element G-set with trivial action.

1.916. Let $\mathbf{N}$ be the free monoid on one generator σ (i.e. $\mathbf{N}$ is the natural numbers under addition.) Construct Ω as the set whose elements are denoted: $(t-0), (t-1), \dots, (t-n), \dots, (t-\infty)$.

$(t-0)$ and $(t-\infty)$ will be fixed points, for all other elements we define $(t-n)^\sigma = t-n+1$. The universal point is $t-0$.

Given an $\mathbf{N}$-set A and a sub $\mathbf{N}$-set, $A' \subset A$, we define $\chi_{A'}(a) = t-n$ where $n = \inf\{j \mid a^{\sigma^j} \in A'\}$. If $a \in A'$ then $\chi_{A'}(a) = t-0$. If a is 'just outside' A', that is if $a \not\in A'$ but $a^\sigma \in A'$ then $\chi_{A'}(a) = t-1$. If a is 'permanently outside' A', that is, if for no finite j is a^{σ^j} in A', then $\chi_{A'} = t - \infty$.

Ω can usually be regarded as an object of 'truth-values'. In the case at hand, it may be regarded as the set of 'waiting times' for some event to occur.

For a general monoid M we may construct Ω as the set of right-ideals of M and define the action by $\mathscr{A}^\sigma = \{\alpha \mid \sigma\alpha \in M\}$. The entire ideal is the universal point. Given $A' \subset A$ define $\chi_{A'}(a) = \{\alpha \mid a^\alpha \in A'\}$. Note that for the free-monoid on two generators, Ω is of the order of the continuum.

1.917. In [1.84], $\mathscr{S}^{\mathbf{A}}$ was shown to be a topos. Given $T \in \mathscr{S}^{\mathbf{A}}$, the Yoneda lemma says that $[T](A) \simeq (H^A, [T]) \simeq \mathscr{S}\!ub(H^A \times T)$.

We may thus directly construct $[T]$: define $[T](A)$ as the set of subfunctors of $H^A \times T$. Given $f\colon A_1 \longrightarrow A_2$ in $\mathbf{A}$ and an element F in $[T](A_1)$, then $[T](f)$ sends F to F' where

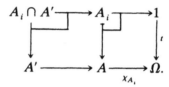

$$\begin{array}{ccc} F' & \rightarrowtail & H^{A_2} \times T \\ \downarrow & & \downarrow {\scriptstyle H^f \times 1} \\ F & \rightarrowtail & H^{A_1} \times T. \end{array}$$

We know of no use for this knowledge.

1.918. Similarly, and apparently as uselessly, we can give a direct description of power-objects in $\mathscr{S}h(X)$. Given a local homeomorphism $f\colon Y \longrightarrow X$ and an open subset $U \subset X$ then the sections $(U, (f))$ may be identified with the set of open subsets of $f^{\#}(U)$. Following [1.373, 1.374], therefore, the elements of $[f]$ in the stalk over $x \in X$ are named by open subsets of Y; two of them, Y_1, Y_2, name the same element over x iff there exists an open neighborhood U of x such that $Y_1 \cap f^{\#}(U) = Y_2 \cap f^{\#}(U)$. Given open Y' in Y and open U in X define $[Y', U]$ as the set of all elements over U named by Y'. The family of such subsets is a basis for the topology of $[f]$.

1.919. Let **E** be a cartesian category with a subobject classifier Ω. We can determine all the automorphisms of Ω, indeed, all of its monic endomorphisms.

For any endomorphism $g\colon \Omega \longrightarrow \Omega$ we will say that A' is a *g-large subobject of* A if $\hat{g}(A') = A$ (following the notational convention of [1.914]). A' is large iff its characteristic map factors through $g^{\#}(t)$. If g, g' yield the same large subobjects, then they are equal.

For monic $g\colon \Omega \rightarrowtail \Omega$ consider

$$\begin{array}{ccc} V & \overset{f}{\rightarrowtail} & \Omega \\ \downarrow & & \downarrow {\scriptstyle g} \\ 1 & \underset{t}{\rightarrowtail} & \Omega \end{array} \quad \text{and} \quad \begin{array}{ccc} U & \rightarrowtail & V \\ \downarrow & \lrcorner & \downarrow {\scriptstyle f} \\ 1 & \underset{t}{\rightarrowtail} & \Omega. \end{array}$$

A' is a g-large subobject of A iff there exists a pullback

$$\begin{array}{ccc} A' & \longrightarrow & A \\ \downarrow & \lrcorner & \downarrow \\ U & \longrightarrow & V. \end{array}$$

For A to have any large subobject it must be the case that $A = A \times V$. And in that case, the unique large subobject of A is $A \times U$.

The operation which sends A' to $(A \times V) \cap (A' \leftrightarrow A \times U)$ has exactly the same large subobjects, hence is the same operation.

Since g is monic $\hat{g}(A_1) = \hat{g}(A_2)$ implies $A_1 = A_2$. Viewing $\hat{g}$ as an operation on $\mathscr{S}ub(1)$ we have that $\hat{g}(V) = \hat{g}(1)$ hence $V = 1$.

Viewing $\hat{g}$ as an operation on $\mathscr{S}ub(A)$ we have that $g^2(A) = (A \leftrightarrow A \times U) \leftrightarrow A \times U$, that is, A is a g^2-large subobject of itself. The identity operation has exactly the same large subobjects, hence:

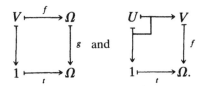

Every monic endomorphism of Ω is an involution.

The absence of non-automorphic monic endomorphisms on an object is a 'finiteness' condition. Ω need not satisfy the dual condition, the topos of **N**-sets yielding an example. If σ is the generator of **N** then its action on any **N**-set is a map of **N**-sets. According to [1.916] the action of σ on Ω is epic but not automorphic.

(The set $\mathcal{J} \subset \mathcal{S}ub(1)$ defined by $U \in \mathcal{J}$ iff $A' \leftrightarrow A \times U$ is an involution on each $\mathcal{S}ub(A)$ is a filter [1.634]. In $\mathcal{S}h(Y)$, $\mathcal{J}$ is the family of open sets whose complements are discrete.)

1.91(10). The axioms for topoi allow one slight relaxation. If a category has pullbacks then we can define power-objects whether it has a terminator or not. Alas, any discrete category has pullbacks, equalizers and power-objects.

Suppose however that **E** has binary products and equalizers (equivalently, binary products and pullbacks; equivalently, limits for all finite non-empty diagrams.) Suppose that **E** has power-objects. Finally suppose that **E** is non-empty. Then **E** does have a terminator and hence is a topos: for any pair of objects A, B let $M_{A,B}$ denote the relation tabulated by a product of A, B; for any $f: A' \longrightarrow A$, $fM_{A,B} = M_{A',B}$; hence for any A, B, $\Lambda M_{A,B}$ is a constant map, that is, $f(\Lambda M_{A,B}) = g(\Lambda M_{A,B})$ for all f, $g: A' \longrightarrow A$. For any B, $\Lambda M_{[B],B}$ is a constant endomorphism on $[B]$ (in particular, it is an idempotent); for every A there exists a map from A to $[B]$ namely $\Lambda M_{A,B}$; the equalizer of $1_{[B]}$ and $\Lambda M_{[B],B}$ is a terminator.

1.92. In a topos, $\Lambda 1: B \longrightarrow [B]$ is called a SINGLETON MAP (in $\mathcal{S}$ it sends b to the one-element set $\{b\}$). For any $f: A \longrightarrow B$, $f(\Lambda 1) \ni\; = f$ and $f(\Lambda 1)$ is a map, hence $f(\Lambda 1) = \Lambda f$. As a consequence, $\Lambda 1$ is monic.

A topos is exponential.

BECAUSE: First, any power object is baseable:

$$(A \times -, [B]) \simeq \mathcal{S}ub(A \times - \times B) \simeq \mathcal{R}el(-, A \times B)$$
$$\simeq (-, [A \times B]) .$$

That is, $[B]^A$ is constructible as $[A \times B]$.

For an object B, let $\chi: [B] \longrightarrow \Omega$ be the characteristic map of $\Lambda 1: B \rightarrowtail [B]$. Then B is the equalizer of χ and $[B] \longrightarrow 1 \xrightarrow{t} \Omega$. Since $[B]$ and Ω (which, of course, is [1]) are baseable, B is baseable [1.859].

1.921. F.W. Lawvere originally defined an *elementary topos* as a bicartesian exponential category with partial map classifiers (which will be defined in [1.934]). Whatever they are, Ω is a special case. Note that $\mathcal{R}el(\_, B) \simeq \mathcal{S}ub(\_ \times B) \simeq (\_ \times B, \Omega) \simeq (\_, \Omega^B)$, that is, $[B]$ is definable as Ω^B.

The first simplification was the discovery that partial map classifiers can be constructed just from cartesianness, exponentiation, and Ω. C.J. Mikkelsen discovered that the cocartesian structure is a consequence of the other axioms. A. Kock noticed that exponentiation is a consequence of power-objects (which greatly simplifies the proof, to come, of the slice lemma).

1.922. For a regular category **E** the right-adjoint of $\mathbf{E} \to \mathscr{Rel}(\mathbf{E})$, if it exists (that is, if **E** is a topos) is, of course, a covariant functor. Every topos, as will be shown, is regular. We will continue the brackets notation $[B]$ when the power-object functor is being viewed as covariant.

Ω^B denotes a contravariant functor on B. Given $g\colon B_1 \longrightarrow B_2$, Ω^g is the unique map such that

$$\begin{array}{ccc} (-, \Omega^{B_2}) & \xrightarrow{(-,\,\Omega^g)} & (-, \Omega^{B_1}) \\ \big\downarrow \wr & & \big\downarrow \wr \\ \mathscr{Sub}(- \times B_2) & \xrightarrow[(1\times g)^{\#}]{} & \mathscr{Sub}(- \times B_1). \end{array}$$

Another view of Ω^g: given a relation R in any category with pullbacks, not only can we define the composition fR for a map f, but we can (quite similarly) define Rg°. Ω^g is the unique map such that

$$\begin{array}{ccc} (-, \Omega^{B_2}) & \xrightarrow{(-,\,\Omega^g)} & (-, \Omega^{B_1}) \\ \big\downarrow \wr & & \big\downarrow \wr \\ \mathscr{Rel}(-, B_2) & \xrightarrow[-g^\circ]{} & \mathscr{Rel}(-, B_1). \end{array}$$

Hence,

$$\begin{array}{ccc} \Omega^{B_2} & \xrightarrow{\Omega^g} & \Omega^{B_1} \\ \big\downarrow \wr & & \big\downarrow \wr \\ [B_2] & \xrightarrow[{[g^\circ]}]{} & [B_1]. \end{array}$$

1.923. B^A was constructed above as a subobject of $[A \times B]$ via a pullback:

$$\begin{array}{ccc} B^A & \rightarrowtail & [A \times B] \\ \big\downarrow & & \big\downarrow f \\ 1 & \longrightarrow & [A]. \end{array}$$

In $\mathscr{S}$, f sends $R \subset A \times B$ to $\{a \mid \exists!_b \langle a, b\rangle \in R\}$. The map from 1 to $[A]$ sends 1 to the entire subobject.

1.924. Given $F, G \in |\mathscr{S}^{\mathbf{A}}|$, F^G is computable via the Yoneda lemma: $F^G(A) \simeq (H^A, F^G) \simeq (G \times H^A, F)$ [1.464]. Given a transformation $\eta\colon G \times H^{A_1} \longrightarrow F$ and a map $f\colon A_1 \longrightarrow A_2$ then $F^G(f)$ sends η to

$$G \times H^{A_1} \xrightarrow{1 \times H^f} G \times H^{A_2} \xrightarrow{\eta} F\,.$$

Suppose $\mathbf{A}$ has binary coproducts. Then

$$F^{H^A}(B) \simeq (H^B, F^{H^A}) \simeq (H^A \times H^B, F) \simeq (H^{A+B}, F) \simeq F(A+B)\,;$$

that is, $F^{H^A}(-) \simeq F(A + -)$. Thus

$$F^G(A) \simeq (H^A, F^G) \simeq (G, F^{H^A}) \simeq (G(-), F(A + -))\,.$$

$F^G(A)$ may be constructed as the set of transformations from G to 'F translated by A'.

We know of no use for these computations.

1.925. In the category of $\mathbf{N}$-sets, let A be the two-element object with a unique fixed point.

A^A is of the order of the continuum: the forgetful functor is represented by the regular representation, that is, $\mathbf{N}$ acting on itself; there are as many elements of A^A as there are maps from $\mathbf{N}$ to A^A; $(\mathbf{N}, A^A) \simeq (A \times \mathbf{N}, A)$; let $S \subset A \times \mathbf{N}$ be the complement (in $\mathscr{S}$) of the image of the action of σ, the generator of $\mathbf{N}$; for any function from S to A there is a unique extension to a map of $\mathbf{N}$-sets from $A \times \mathbf{N}$ to A: S is infinite.

Bearing in mind that, as far as we know, this is computation for its own sake, the structure of A^A can be fully explicated. Let B be an indecomposable $\mathbf{N}$-set (with respect to coproduct), with a fixed point and in which for every y there exist precisely two solutions to $x^\sigma = y$ (where σ generates $\mathbf{N}$). B is unique up to isomorphism. Define B_n as $B \times (\mathbb{Z}/n\mathbb{Z})$, where σ operates on $\mathbb{Z}/n\mathbb{Z}$ by adding 1. B_n is characterized by the fact that it is indecomposable, that it contains a copy of $\mathbb{Z}/n\mathbb{Z}$ and that for every y there are exactly two solutions of $x^\sigma = y$. $A \times B_n$ is characterized by the fact that it is indecomposable, that it contains a copy of $\mathbb{Z}/n\mathbb{Z}$ and that for every element in the image of the action by σ there are precisely four solutions of $x^\sigma = y$, precisely two of which are in the image of σ.

We leave it to the reader to find that A^A must be a coproduct of indecomposables of the form $A \times B_n$. For $n > 0$ the number of copies of B_n is $(1/n)\sum_{d \mid n} \mu(n/d)2^d$ (μ is the Möbius function). There are uncountably many copies of B_0.

1.926. In a topos, exponential structure restricts to the Heyting algebra structure on $\mathscr{S}\!ub(1)$.

1.93. The slice lemma for topoi:

If **E** *is a topos, B an object therein, then* **E**/B *is a topos and* $\Delta\colon \mathbf{E} \to \mathbf{E}/B$ *is a representation of topoi.*

BECAUSE: We show first that $\Delta[A]$ is a power-object for $\Delta(A)$:

$$\mathscr{S}ub_{\mathbf{E}/B}(- \times \Delta(A)) \simeq \mathscr{S}ub_{\mathbf{E}}(\Sigma(- \times \Delta(A)) \simeq \mathscr{S}ub_{\mathbf{E}}(\Sigma(-) \times A)$$
$$\simeq (\Sigma(-), [A]) \simeq (-, \Delta[A]) .$$

Of the four isomorphisms above, the first is immediate. The third is the definition of $[A]$. The fourth is the adjointness of Σ and Δ. Only the second isomorphism needs any analysis and it has nothing to do with the subobject functor. We need only that $\Sigma(- \times \Delta(A)) \simeq \Sigma(-) \times A$. Recall the construction of products in **E**/B: given an object $f\colon C \longrightarrow B$, its product with $\Delta(A)$ is constructed as the diagonal map in

$$\begin{array}{ccc} P & \longrightarrow & A \times B \\ \downarrow & & \downarrow \\ C & \xrightarrow[f]{} & B. \end{array}$$

P is easily verified to be $A \times C$. That is $\Sigma((f) \times \Delta(A)) \simeq \Sigma(f) \times A$.

We must now show that every object in **E**/B has a power-object. Note first that any object $C \in \mathbf{E}/B$ appears as a subobject of $\Delta(\Sigma(C))$. It sufficies to show, therefore, in an arbitrary cartesian category, that if $[A]$ exists, then $[A']$ exists for any $A' \subset A$.

The functor $\mathscr{S}ub(- \times A')$ is naturally embedded in $\mathscr{S}ub(- \times A)$ by inclusion. It is a retract of $\mathscr{S}ub(- \times A)$: the right-inverse, $\mathscr{S}ub(- \times A) \longrightarrow \mathscr{S}ub(- \times A')$, is defined via intersection. In any category in which idempotents split, retracts of representables are representable. In the case at hand, let $e\colon [A] \longrightarrow [A]$ be the unique map such that

$$\begin{array}{ccccc} (-,[A]) & & \xrightarrow{\quad (-,e) \quad} & & (-,[A]) \\ \downarrow \wr & & & & \downarrow \wr \\ \mathscr{S}ub(- \times A) & \longrightarrow & \mathscr{S}ub(- \times A') & \longrightarrow & \mathscr{S}ub(- \times A). \end{array}$$

e is an idempotent. Construct $[A']$ as the equalizer of $1_{[A]}$ and e. (An elementary construction: $e = \Lambda(\ni \cap ([A] \times A'))$.)

1.931. THE FUNDAMENTAL LEMMA OF TOPOI

For $f\colon A \longrightarrow B$ in any cartesian category $\mathbf{A}$, we define $f^{\#}\colon \mathbf{A}/B \to \mathbf{A}/A$ by pulling back; that is, an object $C \longrightarrow B$ is sent to $f^{\#}C \longrightarrow A$ where

$$\begin{array}{ccc} f^{\#}C & \longrightarrow & C \\ \downarrow & \lrcorner & \downarrow \\ A & \longrightarrow & B. \end{array}$$

$f^{\#}$ has a left-adjoint, $\Sigma_f\colon \mathbf{A}/A \to \mathbf{A}/B$ defined by composing; that is, by sending $D \longrightarrow A$ in $\mathbf{A}/A$ to $D \longrightarrow A \xrightarrow{f} B$ in $\mathbf{A}/B$.

For $f\colon A \longrightarrow B$ in a topos $\mathbf{E}$, $f^{\#}\colon \mathbf{E}/B \to \mathbf{E}/A$ has a right adjoint $\Pi_f\colon \mathbf{E}/A \to \mathbf{E}/B$.

BECAUSE: For any category, $\mathbf{A}/A$ is equivalent to $(\mathbf{A}/B)/(f)$. If $\mathbf{A}$ is cartesian, then

$$\begin{array}{ccc} \mathbf{A}/B & \xrightarrow{\Delta} & (\mathbf{A}/B)/(f) \\ & {\scriptstyle f^{\#}}\searrow \quad \swarrow{\scriptstyle \wr} & \\ & \mathbf{A}/A & \end{array}$$

$\mathbf{E}/B$ is a topos, therefore exponential, therefore $f^{\#}$ has a right-adjoint Π_f [1.854].

1.932. The fundamental lemma for topoi may be regarded as an improvement of the double-sharp property, most readily seen by showing that it implies double-sharps. Functors with left-adjoints preserve subterminators, hence we obtain, for any $f\colon A \longrightarrow B$ in a topos, a pair of adjoint maps between $\mathscr{V}al_{\mathbf{E}/B}$ and $\mathscr{V}al_{\mathbf{E}/A}$. $\mathscr{V}al_{\mathbf{A}/B} \simeq \mathscr{S}ub_{\mathbf{A}}(B)$ for any category, $\mathbf{A}$. The restriction of $f^{\#}$ to $\mathscr{S}ub(B)$ is easily seen to be the inverse image operation. Hence the restriction of Π_f to $\mathscr{S}ub(A)$ is the double-sharp operation [1.7]. Thus

The double-sharp axiom holds for topoi.

When topoi are shown to be pre-logoi, they will thus be shown to be logoi.

1.933. Double-sharps imply the preservation of unions of subobjects by inverse-images. The fundamental lemma implies more—the preservation of covering families by pullbacks:

Given $f\colon A \longrightarrow B$ and a cover $\{C_\iota \longrightarrow B\}_I$ in a topos **E**, *let*

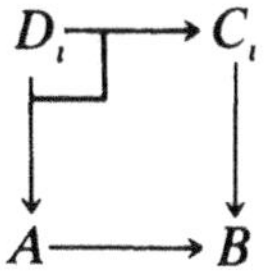

be a pullback for each i.

Then $\{D_i \longrightarrow A\}_I$ is a cover.

BECAUSE: Covers coincide with epics [1.913]. Any functor with a right-adjoint preserves epics. The given family $\{C_i \longrightarrow B\}$ may be regarded as a family of maps to the terminator in $\mathbf{E}/B$. Applying $f^{\#}$ we obtain a covering of the terminator in $\mathbf{E}/A$.

Consequently,

A topos is pre-regular.

1.934. For any cartesian category **C** we may define the category of partial maps, $\mathscr{P}\!ar(\mathbf{C})$: its objects are the same as those of **C**, the morphisms from A to B are pairs $\langle A', f\rangle$ where $A' \subset A$ and $f\colon A' \longrightarrow B$. The composition of $\langle A', f\rangle$ and $\langle B', g\rangle$ is $\langle f^{\#}(B'), fg\rangle$.

For a topos **E** *the natural inclusion of* **E** *in* $\mathscr{P}\!ar(\mathbf{E})$ *has a right-adjoint. That is, for any $A \in \mathbf{E}$ there exists $\tilde{A}$ such that $\mathscr{P}\!ar(-, A) \simeq (-, \tilde{A})$.*

An elementary translation:

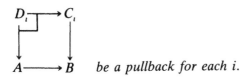

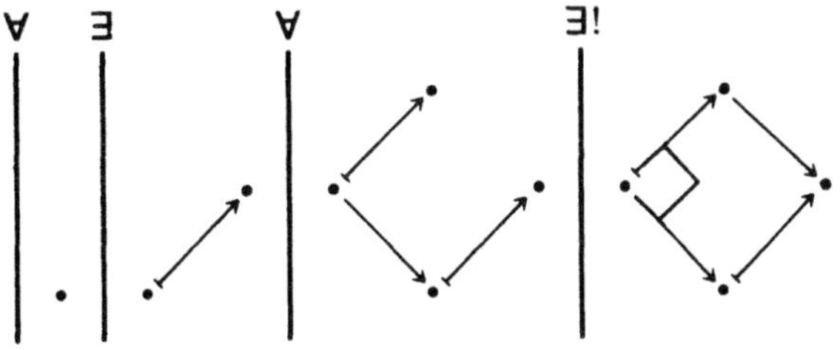

(In any boolean positive logos, $\tilde{A}$ is constructible as $A + 1$. Note that in a topos, $\tilde{1} = \Omega$.)

One may construct $A \mapsto \tilde{A}$ as $\Pi_t(A)$ where t is the universal point of Ω. ($\mathbf{A}/\Omega$ may be viewed as a category whose objects are pairs $(A', A\rangle$ where $A' \subset A$. $f\colon A \longrightarrow B$ is a map from $\langle A', A\rangle$ to $\langle B', B\rangle$ iff $f^{\#}(B') = A'$.)

1.935. Topoi are pre-regular and satisfy the slice condition [1.541]. The other two conditions needed for the capitalization lemma [1.54] are routine. Hence,

Every topos may be faithfully represented in a capital topos.

1.94. Given $F \subset [A]$, in a topos, we may view $\Gamma(F)$ as a subset of $\mathcal{S}ub(A)$. We will call this subset the family of SUBOBJECTS NAMED BY F. (F determines but is not determined by $\Gamma(F)$).

The INTERNALLY DEFINED INTERSECTION of F, $\bigcap F$, is constructed via the pullback diagram

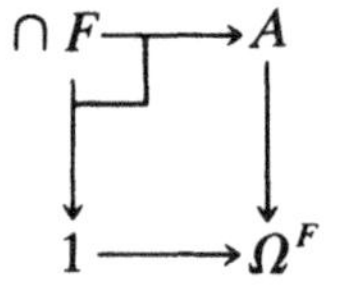

where $A \longrightarrow \Omega^F$ is the adjoint of $F \mapsto [A] \rightrightarrows \Omega^A$ and $1 \longrightarrow \Omega^F$ is the adjoint of $F \longrightarrow 1 \xrightarrow{t} \Omega$.

1.941. As we shall see, $\bigcap F$ is a lower bound of the subobjects named by F, but not in all cases, the greatest lower bound [1.94(10)]. It is preserved by representations of topoi. It will be characterized as the largest subobject of A that remains a lower bound under representations, that is, given any representation of topoi $T: \mathbf{A} \to \mathbf{B}$ then $T(\bigcap F)$ remains a lower bound of the subobjects named by $T(F)$.

The phrase 'internally defined' is from the public language (the public being the set of toponomers). Its use, however, is coextensive with the notion of being preserved by representations, which in turn, is coextensive with being constructible via a 'formula' in the partial operations of topos theory, which is coextensive in its turn, with its reasonably unique existence in any topos.

Note that in a complete topos the intersection of a family of subobjects $\{A_i\}_I$ in A is constructible as the equalizer of a pair of maps from A to Ω^I, one having characteristic maps as coordinates, the other having $A \longrightarrow 1 \xrightarrow{t} \Omega$ as each of its coordinates. If we view I as a copower of 1 then

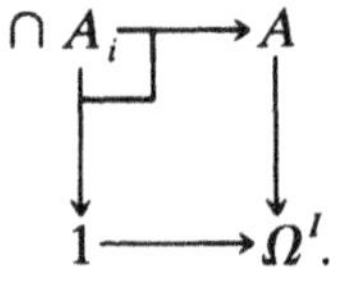

1.942. Given $A' \subset A$ the adjoint, $1 \mapsto \Omega^A$, of $\chi_{A'}$ is called the NAME OF A' and is denoted $\ulcorner A' \urcorner$. A' is named by F iff $\ulcorner A' \urcorner \subset F$.

The map $1 \longrightarrow \Omega^F$ that appears in the construction of $\bigcap F$ is the name of the entire subobject of F. For any $f: A \longrightarrow B$, Ω^f may be viewed as the internally defined inverse-image operation. In particular,

$\Gamma(\Omega^f)\colon \mathscr{Sub}(B) \longrightarrow \mathscr{Sub}(A)$ is the usual inverse-image operation. Hence

$$1 \xrightarrow{\ulcorner B \urcorner} \Omega^B \xrightarrow{\Omega^f} \Omega^A = \ulcorner A \urcorner .$$

In particular,

$$1 \longrightarrow \Omega^F \xrightarrow{\Omega^x} \Omega^1 = 1 \xrightarrow{t} \Omega \quad \text{for any } 1 \xrightarrow{x} F .$$

Our treatment of $\bigcap F$ rests on the following lemma:

For any subobject $A' \subset A$ named by F, $(A \longrightarrow \Omega^F \xrightarrow{\Omega^{\ulcorner A' \urcorner}} \Omega) = \chi_{A'}$.

BECAUSE: Consider the diagram

$$\begin{array}{ccc} (A, \Omega^F) & \xrightarrow{(A, \Omega^{\ulcorner A' \urcorner})} & (A, \Omega^1) \\ \downarrow \wr & & \uparrow \wr \\ (F, \Omega^A) & \xrightarrow[(\ulcorner A' \urcorner, \Omega^A)]{} & (1, \Omega^A). \end{array}$$

Starting with $A \longrightarrow \Omega^F$ in the NW corner and travelling counter-clockwise to the NE corner we come first, in the SW, to the inclusion map $F \subset \Omega^A$, then in the SE to the name of A' and then, in the NE, to the characteristic map of A'.

1.943. *$\bigcap F$ is contained in every subobject named by F.*

If F is well-pointed then $\bigcap F$ is the greatest lower bound of the subobjects named by F.

BECAUSE: Given $A' \subset A$, named by F, we wish to show, first, that $\bigcap F \subset A'$, equivalently, that

$$\begin{array}{ccc} \cap F & \longrightarrow & A \\ \downarrow & & \downarrow \chi_{A'} \\ 1 & \xrightarrow[t]{} & \Omega. \end{array}$$

But, using the last section,

$$(\bigcap F \longrightarrow A \xrightarrow{\chi_{A'}} \Omega) = (\bigcap F \longrightarrow A \longrightarrow \Omega^F \xrightarrow{\Omega^{\ulcorner A' \urcorner}} \Omega)$$
$$= (\bigcap F \longrightarrow 1 \longrightarrow \Omega^F \longrightarrow \Omega) .$$

We shall need a stronger statement than the one above:

If F is well-pointed then a map $B \longrightarrow A$ factors through $\bigcap F$ iff it factors through each of the subobjects named by F.

We wish to determine when

$$\begin{array}{ccc} B & \longrightarrow & A \\ \downarrow & -\!\!\!|\!\!\!- & \downarrow \\ 1 & \longrightarrow & \Omega^F \end{array} \quad \text{commutes.}$$

By hypothesis, the maps $\{1 \longrightarrow F\}$ are collectively epic. $\Omega^{(-)}$ is a contravariant functor with an adjoint on the right (itself). Any such functor carries epic families to monic families. The last mentioned square commutes therefore, whenever

$$(B \longrightarrow A \longrightarrow \Omega^F \xrightarrow{\Omega^{\ulcorner A' \urcorner}} \Omega^1) = (B \longrightarrow 1 \longrightarrow \Omega^F \longrightarrow \Omega^1)$$

for all $A' \subset F$.

Using the last section such is equivalent to

$$(B \longrightarrow A \xrightarrow{\chi_{A'}} \Omega) = (B \longrightarrow 1 \xrightarrow{t} \Omega) \quad \text{for all } A' \text{ named by } F,$$

which in turn is equivalent to f factoring through each A' named by F.

1.944. *A topos has a strict coterminator.*

BECAUSE: We may specialize to the case $A = 1$, $F = \Omega$. Then $\bigcap \Omega$ is a minimal subobject of 1. For any $p\colon B \longrightarrow 1$, $p^{\#}(\bigcap \Omega)$ is minimal in B (because $p^{\#}$ has a right-adjoint $p^{\#\#}$). In particular $B \longrightarrow \bigcap \Omega$ implies that B has no proper subobjects, which implies that $\bigcap \Omega$ is a strict coterminator [1.58].

1.945. *A topos is regular.*

BECAUSE: Since topoi are known to be pre-regular we need only construct images. Given $f\colon A \longrightarrow B$ consider

$$\begin{array}{ccc} F & \longrightarrow & \Omega^B \\ \downarrow & \lrcorner & \downarrow{\scriptstyle \Omega^f} \\ 1 & \xrightarrow[\ulcorner A \urcorner]{} & \Omega^A. \end{array}$$

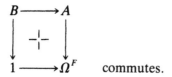

B' is named by F iff $f^{\#}(B') = A$, that is, iff f factors through B'. $\bigcap F$ is the image of f. We need only show that $f^{\#}(\bigcap F) = A$.

The capitalization lemma says that it suffices to specialize to a capital topos. F names the entire subobject, hence F has a point, therefore F is well-supported, and—in a capital topos—F is well-pointed. By [1.943] we are done.

1.946. *A topos is a logos.*

BECAUSE: We have everything except binary unions. Given $u_i\colon A_i \rightarrowtail A$, $i = 1, 2$, consider

$$\begin{array}{ccc} F_i & \longrightarrow & \Omega^A \\ \downarrow & & \downarrow{\scriptstyle \Omega^{u_i}} \\ 1 & \xrightarrow[\ulcorner A_i \urcorner]{} & \Omega^{A_i}. \end{array}$$

A' is named by F_i iff $A_i \subset A'$. Define $F = F_1 \cap F_2$. A' is named by F iff A contains both A_1 and A_2. $\bigcap F$ is the union of A_1, A_2. We need only show that $A_i \subset \bigcap F$.

Again we involve the capitalization lemma. F names the entire subobject, therefore has a point, therefore—in a capital topos—is well-pointed and by [1.943], $\bigcap F = A_1 \cup A_2$.

1.947. *A topos is a transitive logos.*

BECAUSE: It suffices to construct R^* [1.77] for reflexive R (since for any endo-relation R, $R^{\dagger}$ is constructible as $R(1 \cup R)^*$). Given a reflexive relation R on an object B consider the relation $R \times 1$ on $B \times B$. If Γ is applied to $[R \times 1]\colon [B \times B] \longrightarrow [B \times B]$ one obtains the endomorphism on $\mathscr{R}el(B, B)$ that sends S to RS. Let $F_1 \subset [B \times B]$ be the equalizer of $1_{[B \times B]}$ and $[R \times 1]$. S is named by F_1 iff $RS = S$. Since R is reflexive we can say, instead, that S is named by F_1 iff $RS \subset S$.

In the last section we constructed for any $A' \subset A$ a subobject of $[A]$ which names the subobjects that contain A'. Let $A = B \times B$ and A' be the diagonal $B \xrightarrow{\langle 1,1 \rangle} B \times B$. We obtain $F_2 \subset [B \times B]$ that names the reflexive relations on B.

S is named by $F = F_1 \cap F_2$ iff $RS \subset S$ and $1 \subset S$.

$\bigcap F$ is a lower bound of all reflexive relations such that $RS \subset S$. If $\bigcap F$, itself, is reflexive and if $R(\bigcap F) \subset \bigcap F$ then $\bigcap$F is the reflexive transitive closure of R [1.787].

It suffices to specialize to a capital topos. F names the entire subobject of $B \times B$, hence is well-supported, hence well-pointed. $\bigcap F$ is the greatest lower-bound of the family $\{S \subset B \times B \mid RS \subset S, 1 \subset S\}$. For ordinary reasons, therefore, $R(\bigcap F) \subset R(\bigcap F)$.

1.948. Let G be a group. G appears as an object in the topos of G-sets, $\mathscr{S}^G$. Let $F \subset [G]$ be the complement of $1 \xrightarrow{\ulcorner 0 \urcorner} [G]$. The only object named by F is the entire subobject; on the other hand, $\bigcap F$ is the empty subobject (assuming that G is non-trivial), most easily checked by applying the forgetful functor $U: \mathscr{S}^G \longrightarrow \mathscr{S}$. U is a faithful representation of topoi. UF is the family of non-empty subsets of UG. $U(\bigcap F) = \bigcap UF = \emptyset$.

1.949. Given $F \subset [A]$ the INTERNALLY DEFINED UNION $\bigcup F$ is constructed as the direct-image $\nearrow(\ni \cap (F \times A))$ where $\nearrow: [A] \times A \longrightarrow A$. ($\bigcap F$ can be constructed as $\nearrow^{\#\#}(\ni \cap (F \times A))$). The same sort of lemma holds for $\bigcup F$ as for $\bigcap F$: it is an upper bound for every subobject named by F and if F is well-pointed it is the least upper bound.

Say that $A' \subset A$ is a *permanent lower* (*upper*) *bound* of F if for all representations of topoi, T, it is the case that $T(A')$ is an lower (upper) bound of every subobject named by $T(F)$. Clearly $\bigcap F$ ($\bigcup F$) is a permanent lower (upper) bound of F. The capitalization lemma tells us that $\bigcap F$ is the greatest permanent lower bound and that $\bigcup F$ is the least of the permanent upper bounds.

1.94(10). The singleton map $\Lambda 1: A \longrightarrow [A]$ names the subobjects which are points. $\bigcup(\Lambda 1)$ is always entire. The only time that it is the least upper bound of the subobjects named by $\Lambda 1$ is when A is well-pointed.

We will say that a subobject of A is its WELL-POINTED PART if it is the least upper bound of the points of A, alternately, if it is the greatest well-pointed subobject of A. If the well-pointed part of A exists we denote it as $A^{\bullet}$.

There is no possible internal construction of $A^{\bullet}$. (The first paragraph of this section says that the obvious construction fails.) Recall that $\Delta(A) \in |\mathbf{A}/A|$ has a generic point, a point that lies in no $\Delta(A')$ for $A' \subsetneq A$. Hence if A is not well-pointed then $\Delta(A^{\bullet})$ is not $(\Delta A)^{\bullet}$.

A SOLVABLE TOPOS is a topos in which each object has a well-pointed part. Given F in a solvable topos we can construct $\bigcap \Gamma(F)$ and $\bigcup \Gamma(F)$ as $\bigcap(F^{\bullet})$ and $\bigcup(F^{\bullet})$. For every elementary definition on the subobjects of A there exists, in a solvable topos, a family $F \subset [A]$ that names precisely the subobjects satisfying the condition. Thus $\mathscr{Sub}(A)$ has least upper and greatest lower bounds for every elementarily definable family.

The definition of solvable topoi is elementary, but representations of solvable topoi are rare. As already observed, not even the slice lemma holds. (If $\mathbf{E}$ is solvable, then so is $\mathbf{E}/B$ but $\Delta: \mathbf{E} \to \mathbf{E}/B$ does not, in general, preserve well-pointed parts.)

Categories of micro-sheaves as described in [1.76] are topoi. We constructed in that section a micro-sheaf without a well-pointed part. Hence not all topoi are solvable. Every topos, however, may be faithfully represented in a solvable topos: capital topoi are solvable; $A^{\bullet}$ is either A or 0 depending on whether of not A is well-supported.

1.95. *A topos is a pre-topos.*

BECAUSE:

1.951. *A topos is effective.*

BECAUSE: Given an equivalence relation E on A factor ΛE as $A \xrightarrow{h} B \rightarrowtail [A]$. $h^\circ h = 1$. To see that $hh^\circ = E$ we note first the following lemma:

In a regular category, an equivalence relation E is distinguished by the pairs of maps $\langle f, g \rangle$ such that $fE = gE$.

BECAUSE: As may be readily verified in $\mathscr{S}$, $fE = gE$ iff $f^\circ g \subset E$. Any relation is of the form $l^\circ r$ for some pair of maps. Hence if $E \not\subset E'$ there exists $\langle l, r \rangle$ such that $lE = rE$ and $lE' \neq rE'$.

Returning to hh°. We wish to show that $fhh^\circ = ghh^\circ$ iff $fE = gE$. For any map h, $h = hh^\circ h$, hence $fhh^\circ = ghh^\circ$ iff $fh = gh$. Since h followed by a particular monic is ΛE we have $fhh^\circ = ghh^\circ$ iff $f(\Lambda E) = g(\Lambda E)$. But for any map f and relation R, $f(\Lambda R) = \Lambda(fR)$. Hence $fhh^\circ = ghh^\circ$ iff $\Lambda(fE) = \Lambda(gE)$ which of course is equivalent with $fE = gE$.

1.952. *A topos is positive.*

For any object A,

$$\begin{array}{ccc} 0 & \longrightarrow & 1 \\ \downarrow & & \downarrow{\scriptstyle \ulcorner 0 \urcorner} \\ A & \xrightarrow[\Lambda 1]{} & [A] \end{array} \qquad \text{is a pullback ,}$$

since ΛR factors through $\Lambda 1$ iff R is a map and it factors through $\ulcorner 0 \urcorner$ iff $R = 0$.

For any object A, therefore, there exists a disjoint union $A + 1$. Recalling the distributivity that holds in an exponential category we should expect to find a disjoint union of A and B inside $(A \times 1) \times (B \times 1)$. To be precise, $\langle \Lambda 1, \ulcorner 0 \urcorner \rangle \colon A \longrightarrow [A] \times [B]$ and $\langle \ulcorner 0 \urcorner, \Lambda 1 \rangle \colon B \longrightarrow [A] \times [B]$ are monic and their intersection is empty. (We have constructed $A + B$ as a subobject of $[A] \times [B]$. But, of course, $[A] \times [B] \simeq [A + B]$.)

1.953. In $\mathscr{S}$, the construction for the effectiveness of E [1.951] yields the set of equivalence classes. ΛE is the map that sends an element x to the set $\{y \mid xEy\}$.
The construction of $A + B$ [1.952] yields the set

$$\{\langle\{a\},\emptyset\rangle \mid a \in A\} \cup \{\langle\emptyset, \{b\}\rangle \mid b \in B\}\,.$$

1.954. *A topos has coequalizers.*

BECAUSE: Given

$$A \underset{g}{\overset{f}{\rightrightarrows}} B,$$

let $R = f°g$, an endorelation on B. Let $S = (R \cup R°)^*$, the equivalence closure of R [1.775, 1.947]. A topos is effective [1.951], so S is a level of a morphism $B \longrightarrow C$. It is readily verified that

$$A \underset{g}{\overset{f}{\rightrightarrows}} B \longrightarrow C \quad \text{is a coequalizer.}$$

1.955. *A topos is bicartesian.*

BECAUSE: We have just shown that it has coequalizers. A topos also has a coterminator and binary coproducts [1.952, 1.946].

1.96. A number of miscellaneous aspects of topoi

1.961. In any category, an object E is called an INJECTIVE if the functor $(-, E)$ carries monics into epics. An elementary version:

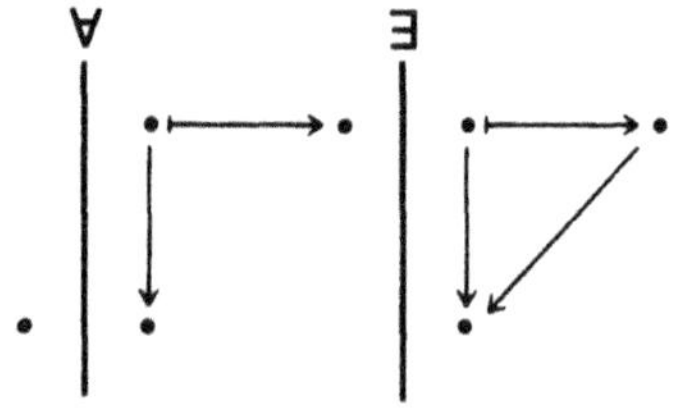

In a pre-topos, pushouts of monics exist and are monic [1.651], hence E is injective iff every monic $E \rightarrowtail A$ has a right-inverse [cf. projectives].
We say that an object E in an exponential category is INTERNALLY INJECTIVE if $E^{(-)}$ carries monics to epics. In a topos, Ω is internally injective; indeed, if $f\colon A \rightarrowtail B$ is monic then $\Omega^f\colon \Omega^B \longrightarrow \Omega^A$ is not only epic, it has a left-inverse $[f]\colon [A] \longrightarrow [B]$. (Recall that $\Omega^f = [f°]$, hence $\Omega^f[f] = [f°f]$. f is epic iff $f°f = 1$.)

In a topos, Ω is (externally) injective.

BECAUSE: Given monic $f: A \rightarrowtail B$ then Ω^f, as we have just seen, has a left-inverse, hence $\Gamma(f)$ has a left-inverse and so does (f, Ω):

$$\begin{array}{ccc} \Gamma(\Omega^B) & \xrightarrow{\Gamma(f)} & \Gamma(\Omega^A) \\ \downarrow \wr & & \downarrow \wr \\ (B, \Omega) & \xrightarrow[(f,\Omega)]{} & (A, \Omega). \end{array}$$

1.962. *If E is injective in an exponential category, then so is E^A for any A.*

BECAUSE: $(-, E^A) \simeq (- \times A, E)$. $- \times A$ preserves monics in any category.

(The theorem is also true for internal injectivity.)

Consequently, in a topos, Ω^A is injective, for all A. Since the singleton map embedds A into Ω^A, every object appears as a subobject of an injective.

1.963. Given A in a topos, let $\tilde{A}$ be as described in [1.934]. $\tilde{A}$ is injective: given monic $f: B_1 \rightarrowtail B_2$ then $(f, \tilde{A})$ is epic because $\mathscr{P}\!\mathit{ar}(f, A)$ is epic: to see the latter, it suffices to note that a partial map from B_1 to A may be construed as a partial map from B_2 to A.

1.964. We say that a category is VALUE-BASED if its values form a basis, as in $\mathscr{S}\!\mathit{h}(X)$ or in any capital topos.

In a value-based topos, Ω is a cogenerator.

BECAUSE: Given $f, g: A \longrightarrow B$, $f \neq g$, we wish to find $h: B \longrightarrow \Omega$ such that $fh \neq gh$. Since $(-, \Omega) \simeq \mathscr{S}\!\mathit{ub}(-)$ it suffices to find a subobject $B' \subset B$ such that $f^\#(B') \neq g^\#(B')$. Let $x: U \longrightarrow A$, $U \subset 1$ be such that $xf \neq xg$. Let $B' = \mathrm{Im}(xf)$. x factors through $f^\#(B')$. If it factored through $g^\#(B')$ then xg would factor through B'. Since maps from a subterminator are determined by their images, and $xf \neq xg$, a contradiction would result.

(For G a non-trivial group, Ω does not cogenerate in $\mathscr{S}^G$.)

1.965. An object C is an exponential category INTERNALLY COGENERATES if the functor $C^{(-)}$ is a contravariant embedding. If C is a cogenerator then it internally cogenerates: if $f \neq g$ then $\Gamma(C^f) \neq \Gamma(C^g)$, hence $C^f \neq C^g$.

In a topos, Ω internally cogenerates.

BECAUSE: Suppose $\Omega^f = \Omega^g$. Let **E** be a small subtopos containing f, g. We may faithfully represent **E** into a capital topos, in particular, a value-based topos, in which we have just seen, it must be the case that $f = g$.

1.966. A PROGENITOR is a object whose subobjects form a generating set. A topos is value-based iff the terminator is a progenitor. Any Grothendieck topos has a progenitor, constructed by taking a disjoint union of the objects in a generating set.

If G is a progenitor for a topos, then Ω^G is a cogenerator.

BECAUSE: $(-, \Omega^G)$ and $(G, \Omega^{(-)})$ are naturally equivalent functors, hence Ω^G cogenerates iff $(G, \Omega^{(-)})$ is an embedding. Given $f, g: A \longrightarrow B$, $f \neq g$ we have just seen that $\Omega^f \neq \Omega^g$. Let $h': G' \longrightarrow \Omega^B$, $G' \subset G$ be such that $h'\Omega^f \neq h'\Omega^g$. Ω^B is injective, therefore there exists $h: G \longrightarrow \Omega^B$ which extends h' and necessarily $h\Omega^f \neq h\Omega^g$. That is, there exists an element, h, of (G, Ω^B) on which the functions $(1, \Omega^f)$, $(1, \Omega^g)$ disagree.

1.967. *The following are equivalent conditions on a locally small topos*:

(a) *Arbitrary powers of objects exist.*

(b) *Arbitrary copowers of objects exist.*

(c) *Arbitrary copowers of* 1 *exist.*

Each implies that the topos is locally complete.

Note that (a) says that contravariant representable functors have adjoints, (b) that covariant representable functors have adjoints, and (c) that Γ has an adjoint.

(a) implies local completeness as follows: given a collection of subobjects $\{B_i\}_I$ of B let $f: B \longrightarrow \Pi_I \Omega$ be the map whose i-th coordinate is χ_{B_i}, let $g: B \longrightarrow \Pi_I \Omega$ be the map whose i-th coordinate is $B \longrightarrow 1 \xrightarrow{t} \Omega$. Then the equalizer of f, g is $\bigcap B_i$.

A locally small topos is well-powered (because $|\mathcal{S}ub(-)| = |(-, \Omega)|$) hence arbitrary intersections imply arbitrary unions.

(a) implies (b) because for any object A and set I we can obtain a family $\{u_i: A \longrightarrow \Pi_I (A + 1)\}$ such that $u_i u_i^\circ = 1$ and $u_i u_j^\circ = 0$, for all $i \neq j$. Let S be the union of the images of the u_i's. Then $\bigcup u_i^\circ u_i = 1_S$. S may be shown to be a copower in both the topos and in the category of its relations.

(b) implies (c) trivially.

(c) implies (a) because $\Pi_I A$ is easily constructible as $A^{\Sigma_I 1}$.

1.968. *A locally small topos is complete iff it is cocomplete.*

BECAUSE: If arbitrary coproducts exist then for any given collection $\{A_i\}$ there exists an object S and a collection of embeddings $\{u_i\colon A \rightarrowtail S\}$ (for instance, let $S = \Sigma_I A_i$). In the last section we saw that arbitrary powers exist. For each i consider

$$\begin{array}{ccc} P_i & \rightarrowtail & \Pi_I S \\ \downarrow & & \downarrow{\scriptstyle p_i} \\ A_i & \xrightarrow[u_i]{} & S \end{array}$$

Let $P = \bigcap P_i$. The collection $\{P \longrightarrow P_i \longrightarrow A_i\}$ is a product, as may be easily verified.

If arbitrary products exist, then a slight modification of the construction in the last section of copowers from powers yields a construction of coproducts from products.

1.969. The LAWVERE DEFINITION of a Grothendieck topos is a cocomplete topos with a generating set. The TIERNEY DEFINITION is a topos with a progenitor and arbitrary copowers of 1 (the last condition is more elegantly stated as 'with a geometric morphism to $\mathcal{S}$'). Using [1.967] we obtain arbitrary coproducts as follows: given $\{A_i\}_I$ let J be a set large enough so that (A_i, C) is embeddable in J for each i, where C is a cogenerator (the existence of which is insured by [1.966]). Let $B = \Pi_J C$. Since the obvious map $A_i \longrightarrow \Pi_{(A_i,C)} C$ is an embedding we obtain an embedding $A_i \rightarrowtail B$ each i. $\Sigma_I A_i$ may now easily be constructed as a subobject of $\Sigma_I B$.

1.96(10). Let **A** be the concrete category whose objects are of the form $\langle A, S, g\rangle$ where A and S are sets and g is a function from $A \times S$ to A. We will write $g\langle a, s\rangle$ as a^s. If $s \notin S$ we define a^s to be a. The forgetful operation forgets S and g. The source-target predicate is given by:

$$\langle A, S, g\rangle \xrightarrow{f} \langle A', S', g'\rangle \quad \text{iff} \quad f(a^s) = (fa)^s \quad \text{for all } a \in A \text{ and}$$
$$\text{for all } s \text{ in the universe.}$$

A satisfies all of the Giraud definition except for the existence of a generating set. **A** is not a topos: the only object with a power-object is 0. The forgetful functor is bicontinuous. If it were representable, then its representor would be a generator. It has, in fact, neither a left nor a right adjoint.

Let **B** be the full subcategory of **A** such that $\langle A, S, g\rangle \in \mathbf{B}$ iff S acts 'automorphically' on A, that is for each s the function $x \longrightarrow x^s$ is an automorphism. **B** is still bicomplete, still without a generating set, but it is a topos (and the forgetful functor is a representation of topoi).

B is one of the better counterexamples among topoi. The category of abelian groups in **B** has no injectives (other than the trivial group). Indeed, Ext(A, A) is a proper class for all $A \neq 0$.

An example due to Peter Johnstone: let **C** be the full subcategory of $\mathscr{S}^{\mathbb{Z}}$ such that $A \in \mathbf{C}$ iff for some $n > 0$ it is the case that $x^{\sigma^n} = x$, all $x \in A$ (where σ is the generator of $\mathbb{Z}$). **C** is a topos with a generating set and with arbitrary powers and copowers. It does not have a progenitor and it is not cocomplete.

1.96(11). The slice lemma for Grothendieck topoi:

If **E** *is a Grothendieck topos and B an object therein, then* $\mathbf{E}/B$ *is a Grothendieck topos, and the representation of topoi* $\Delta\colon \mathbf{E} \to \mathbf{E}/B$ *preserves progenitors.*

BECAUSE: Δ has a right adjoint [1.931], and thus it preserves coproducts. Every object in $\mathbf{E}/B$ is a subobject of some $\Delta(A)$, $A \in |\mathbf{E}|$.

1.97. Boolean topoi

1.971. Given a generating set **G** in a Grothendieck topos, we will say that an object is *small* if it appears as a subquotient of a **G**-object; that is, A is small if there exists $A \twoheadleftarrow G' \rightarrowtail G$, $G \in \mathbf{G}$.

In a boolean Grothendieck topos every object is a coproduct of small objects.

BECAUSE: Any object, A, is covered by its small subobjects. In a complete boolean algebra any cover may be refined to a partition. In the case at hand, suppose that $A = \bigcup_I A_i$ where A_i is small, each i. Well-order I and define $A'_i \subset A_i$ as the complement of $\bigcup_{j<i} (A_j \cap A_i)$. Then $\bigcup_I A'_i = A$ and $A'_i \cap A'_j = 0$ for $i \neq j$. Subobjects of small objects are small.

1.972. *In a boolean logos, if* 1 *is projective, then it is a progenitor.*

BECAUSE: If $x \neq y$ the complement E' of the equalizer of

$$A \underset{y}{\overset{x}{\rightrightarrows}} B$$

is nonzero, so its support U is nonzero [1.61]. Complemented subobjects of a projective are projective, so let $z\colon U \longrightarrow E'$. The equalizer of zx and zy is zero.

The following are equivalent conditions on a Grothendieck topos:

(1) *Boolean and* 1 *is projective.*

(2) *Boolean and value-based* (1 *is a progenitor*).

(3) *Boolean with a choice progenitor.*

(4) *Every object is a coproduct of subterminators.*

(5) *AC.*

BECAUSE: (1) implies (2) as a special case of the first assertion in this section.

(2) implies (3) because 1 is choice [1.57].

(2) implies (4) as a special case of [1.971].

(3) implies (5) because any small coproduct in the topos can be replaced by a well-ordered coproduct (by AC in $\mathscr{S}$), and any such coproduct of choice objects is choice, similarly to [1.662]. If G is a choice progenitor, then every object A is covered by a choice object $\Sigma_{G' \subset G, (G', A)} G'$, so it is itself choice.

(4) implies (5) because (4) says that every object is a subobject of $\Sigma_I 1$, for some I, and thus every object is decidable, so the topos is boolean [1.658]. As above, I can be well-ordered, thus $\Sigma_I 1$ is choice, and so are all of its subobjects.

(5) implies (1) because AC implies boolean [1.662].

1.973. A topos is IAC if $(\_)^A$ [1.853] preserves epics, for any A. (IAC stands for Internal Axiom of Choice).

If G is a group, the topos of G-sets $\mathscr{S}^G$ is IAC because the forgetful functor $\mathscr{S}^G \longrightarrow \mathscr{S}$ preserves exponentials, and preserves and reflects epics. Recall that $\mathscr{S}^G$ is AC iff G is trivial.

1.974. *A topos is* AC *iff it is* IAC *and* 1 *is projective.*

BECAUSE: Given an epic $f: A \twoheadrightarrow B$ in an IAC topos, $A^B \xrightarrow{f^B} B^B$ is epic. Pullbacks preserve epics, so

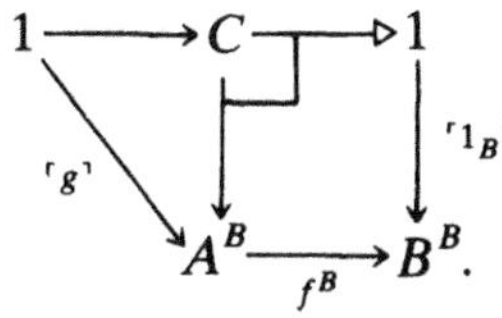

C is well-supported, and thus it has a point because 1 is projective. $gf = 1_B$.

Every functor from an AC topos preserves epics because it preserves left-invertibility. (This condition is in fact equivalent to AC by considering representable functors.)

1.975. By [1.93], $\Delta\colon \mathbf{A} \to \mathbf{A}/C$ is a representation of topoi. By [1.53], it is a faithful representation iff C is well-supported. We see from [1.974] that the slicing by C will create a left inverse for f in $\mathbf{A}/C$. Thus we say that 'a left inverse exists locally' (historically, in $\mathscr{Sh}(X)$).

1.976. *A small topos may be faithfully represented in an* AC *topos iff it is* IAC.

BECAUSE: Any small topos may be faithfully represented in a topos in which 1 is projective [1.935, 1.525]. Given such a faithful representation $T\colon \mathbf{A} \to \mathbf{B}$, replace $\mathbf{B}$ by the full subcategory $\mathbf{B}'$ of all subobjects of $T(\_)$. Given $A \in |\mathbf{A}|$, let $[A]^+ \subset [A]$ be the domain [1.662] of the universal relation. Let $C_A \subset A^{[A]^+}$ name [1.942] choice maps (that is, the maps contained in the universal relation). It is well-supported because of IAC as applied to the universal relation:

$$\begin{array}{ccc} & R & \\ \swarrow & & \searrow \\ [A]^+ & & A. \end{array}$$

Thus:

$$\begin{array}{ccc} T(C_A) & \simeq & C_{T(A)} \\ \downarrow & & \downarrow \\ T(1) & \simeq & 1, \end{array}$$

so the projectivity of 1 in $\mathbf{B}$ yields a point of $C_{T(A)}$. Thus $T(A)$ is choice, and $\mathbf{B}'$ is AC.

1.977. As usual, this has an immediate consequence:

Any universal sentence in the predicates of topoi which follows from AC *also follows from* IAC.

In particular:

An IAC *topos is boolean.*

1.978. A Grothendieck topos **A** is an ETENDUE if there is a well-supported B so that $\mathbf{A}/B$ is value-based [1.964].

A Grothendieck topos is a boolean etendue iff it is IAC.

BECAUSE: IAC implies that for G a progenitor, C_G is well-supported [1.976]. $\Delta: \mathbf{A} \to \mathbf{A}/C_G$ is a faithful representation of topoi that preserves progenitors [1.96(11)]. $\mathbf{A}/C_G$ is a boolean Grothendieck topos with a choice progenitor [1.977, 1.975], and thus it is value-based [1.972].

1.979. We will show in [2.542] that

For any topos **A** *there exists a boolean topos* **B** *and a faithful bicartesian representation* $\mathbf{A} \to \mathbf{B}$.

1.98. Arithmetic in topoi

A NATURAL NUMBERS OBJECT in a topos (NNO for short) is an object N together with morphisms $1 \xrightarrow{0} N$ and $N \xrightarrow{s} N$ such that for every $1 \xrightarrow{a} A \xrightarrow{t} A$ there exists a unique $N \longrightarrow A$ such that

$$\begin{array}{ccccc} 1 & \xrightarrow{0} & N & \xrightarrow{s} & N \\ & \searrow^{a} & \downarrow & & \downarrow \\ & & A & \xrightarrow[t]{} & A. \end{array}$$

If it exists, NNO is clearly unique up to isomorphism. In the category of sets, NNO is (a copy of) the set of natural numbers, $1 \xrightarrow{0} N$ is the number zero, and $N \xrightarrow{s} N$ is the successor function. The full subcategory of finite sets is a topos without a NNO.

1.981. *If* $1 \xrightarrow{0} N \xrightarrow{s} N$ *is a NNO, then for every* $A \xrightarrow{x} B \xrightarrow{t} B$ *there exists a unique* $A \times N \longrightarrow B$ *such that*

$$\begin{array}{ccccc} A & \xrightarrow{\langle 1_A, 0\rangle} & A \times N & \xrightarrow{1_A \times s} & A \times N \\ & \searrow^{x} & \downarrow & & \downarrow \\ & & B & \xrightarrow[t]{} & B. \end{array}$$

BECAUSE: A topos is exponential, so one may transpose a given $A \xrightarrow{x} B \xrightarrow{t} B$ to

$$1 \xrightarrow{\ulcorner x \urcorner} B^A \xrightarrow{t^A} B^A .$$

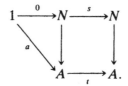

1.982. In $\mathscr{S}$, on may define functions on natural numbers by recursion, that is, given $g: A \longrightarrow B$ and $h: A \times N \times B \longrightarrow B$ there exists a unique $f: A \times N \longrightarrow B$ such that

$$f(a, 0) = g(a),$$

$$f(a, x + 1) = h(a, x, f(a, x)).$$

(Often A is a power of N and $B = N$.) Similarly in a topos:

1.983. Let $1 \xrightarrow{0} N \xrightarrow{s} N$ be a NNO. Then for any given $g: A \longrightarrow B$ and $h: A \times N \times B \longrightarrow B$ there exists a unique $f: A \times N \longrightarrow B$ such that

$$A \longrightarrow A \times 1 \xrightarrow{1_A \times 0} A \times N \xrightarrow{f} B \;=\; A \xrightarrow{g} B,$$

$$A \times N \xrightarrow{1_A \times s} A \times N \xrightarrow{f} B$$
$$= \; A \times N \xrightarrow{\langle p_1, p_2, f \rangle} A \times N \times B \xrightarrow{h} B.$$

BECAUSE: (Notation: For any object C, $C \xrightarrow{0} N$ means $C \longrightarrow 1 \xrightarrow{0} N$.) Let k be such that

$$
\begin{array}{ccc}
A & \xrightarrow{\langle 1_A, 0 \rangle} A \times N \xrightarrow{\quad 1_A \times s \quad} & A \times N \\
\langle 1_A, 0, g \rangle \searrow \quad \downarrow k & & \downarrow k \\
& A \times N \times B \xrightarrow{\langle p_1, p_2 s, h \rangle} A \times N \times B,
\end{array}
$$

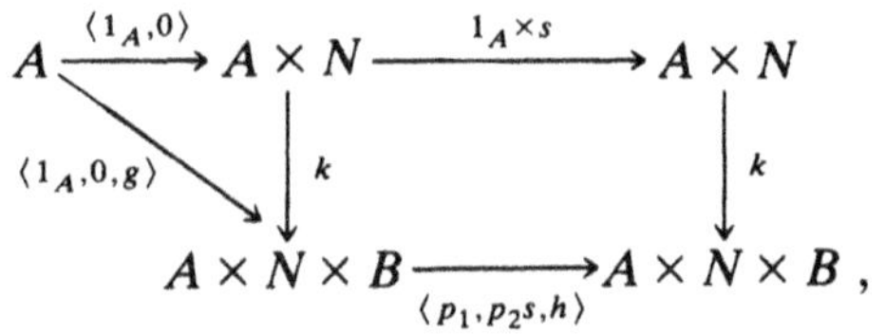

using [1.981]. We claim that one may let $f = kp_3$. Firstly, note that $kp_1 = p_1$ and $kp_2 = p_2$ because

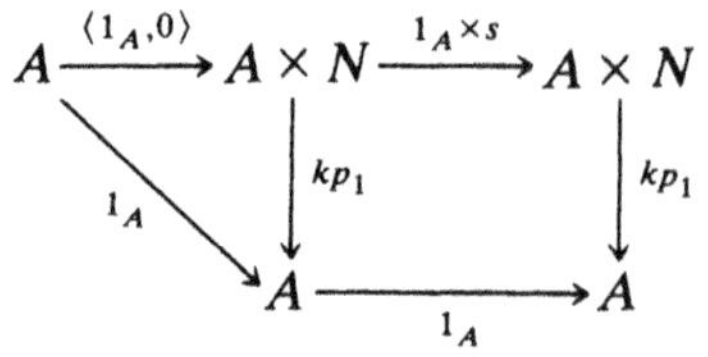

$$
\begin{array}{cc}
A \xrightarrow{\langle 1_A, 0 \rangle} A \times N \xrightarrow{1_A \times s} A \times N & \quad A \xrightarrow{\langle 1_A, 0 \rangle} A \times N \xrightarrow{1_A \times s} A \times N \\
1_A \searrow \downarrow kp_1 \quad \downarrow kp_1 \quad \text{and} & \quad 0 \searrow \downarrow kp_2 \quad \downarrow kp_2 \\
A \xrightarrow{\quad 1_A \quad} A & \quad N \xrightarrow{\quad s \quad} N,
\end{array}
$$

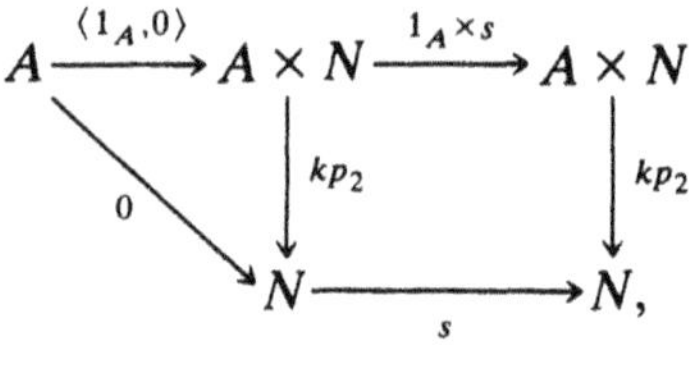

and using [1.981]. Now if $f = kp_3$, then $k = \langle p_1, p_2, f \rangle$, and therefore $\langle 1_A, 0 \rangle f = \langle 1_A, 0 \rangle kp_3 = \langle 1_A, 0, g \rangle p_3 = g$ and $(1_A \times s)f = (1_A \times s)kp_3 = k \langle p_1, p_2 s, h \rangle p_3 = kh = \langle p_1, p_2, f \rangle h$. For the uniqueness of f, suppose f has the required properties. Then $k = \langle p_1, p_2, f \rangle$ has the properties stated at the beginning of the argument. Apply the uniqueness part of [1.981].

1.984. Thus, for example, arithmetic operations addition, multiplication, and exponentiation can be defined on a NNO in any topos by the usual recursion equations. The reader may bear in mind that in [1.983], h could be given as $A \times N \times B \longrightarrow A \times B \longrightarrow B$. This is the case that applies here, with $A = B = N$.

Addition: $\langle 1_N, 0 \rangle\, a = 1_N$, $(1_N \times s)\, a = a\, s$.

Multiplication: $\langle 1_N, 0 \rangle\, m = 0$, $(1_N \times s)\, m = \langle p_1, m \rangle\, a$.

Exponentiation: $\langle 1_N, 0 \rangle\, e = 0\, s$, $(1_N \times s)\, e = \langle p_1, e \rangle\, m$.

1.985. We use matrix notation for morphisms from a coproduct [1.591].

Let $1 \xrightarrow{0} N \xrightarrow{s} N$ *be a NNO. Then*

$$\begin{array}{ccc} 1 & & N \\ & {}_0\searrow \quad \swarrow_s & \\ & N & \end{array}$$

is a coproduct and

$$N \underset{s}{\overset{1_N}{\rightrightarrows}} N \longrightarrow 1$$

is a coequalizer.

BECAUSE: Regarding the first conclusion, let

$$\begin{array}{ccc} 1 & & N \\ & {}_{u_1}\searrow \quad \swarrow_{u_2} & \\ & 1+N & \end{array}$$

be a coproduct. We will first show that

$$1 \xrightarrow{u_1} 1 + N \xrightarrow{\binom{0}{s} u_2} 1 + N \quad \text{is a NNO.}$$

Given $1 \xrightarrow{a} A \xrightarrow{t} A$, then

$$\begin{array}{ccccc} 1 & \xrightarrow{u_1} & 1+N & \xrightarrow{\binom{0}{s} u_2} & 1+N \\ & {}_a\searrow & \downarrow{\scriptstyle \binom{a}{f}} & & \downarrow{\scriptstyle \binom{a}{f}} \\ & & A & \xrightarrow[t]{} & A \end{array}$$

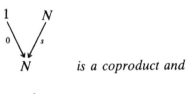

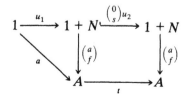

iff f is the unique morphism such that

$$\begin{array}{ccccc} 1 & \xrightarrow{0} & N & \xrightarrow{s} & N \\ & \searrow^{a} & \downarrow{\scriptstyle f} & & \downarrow{\scriptstyle f} \\ & & A & \xrightarrow[t]{} & A. \end{array}$$

Thus $1 \xrightarrow{u_1} 1 + N \xrightarrow{\binom{0}{s}} 1 + N$ is a NNO and $1 + N$ must be isomorphic to N. But this isomorphism must be $\binom{0}{s}$ by the calculation we just made (with $A = N$, $t = s$).

Regarding the second conclusion, let $f\colon N \longrightarrow A$ be any morphism such that $sf = f$. Then

$$\begin{array}{ccccc} 1 & \xrightarrow{0} & N & \xrightarrow{s} & N \\ & \searrow^{0f} & \downarrow{\scriptstyle g} & & \downarrow{\scriptstyle g} \\ & & A & \xrightarrow[1_A]{} & A \end{array}$$

both with $g = f$ and with $g = (N \longrightarrow 1 \xrightarrow{0f} A)$, so they must be equal because N is a NNO. This argument also shows that the factorization through 1 is unique (let $g = N \longrightarrow 1 \xrightarrow{a} A$ for a given a).

1.986. In the rest of section [1.98] we will show the converse of [1.985] and examine it in a broader setting. First we show that

Given $1 \xrightarrow{a} A$ and $A \xrightarrow{f} A$, there exists a subobject $A' \subset A$ and morphisms $1 \longrightarrow A'$ and $A' \longrightarrow A'$ such that

$$\begin{array}{ccccc} 1 & \xrightarrow{a'} & A' & \xrightarrow{f'} & A' \\ & \searrow^{a} & \downarrow & & \downarrow \\ & & A & \xrightarrow[f]{} & A \end{array}$$

and such that $1 + A' \xrightarrow{\binom{a'}{f'}} A$ is a cover.

BECAUSE: Given $1 \xrightarrow{a} A \xrightarrow{f} A$ in a topos **A**, let f^* be the reflexive-transitive closure of f [1.947, 1.77]. In the category of relations $\mathscr{Rel}(\mathbf{A})$ let $A'' = af^*$ and $f'' = A''f$. Observe that $f'' \subset af^*f \subset af^*f^* \subset af^* = A''$.

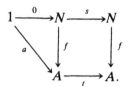

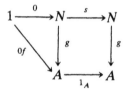

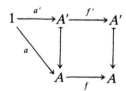

Clearly A'' determines a subobject of A that allows a, call it A', and f'' determines a map $f'\colon A' \longrightarrow A'$ so that the first conclusion holds. Regarding the second conclusion, note that we have already shown $a \cup f'' \subset A''$. We need to show $A'' \subset (a \cup f'')$. By [1.616] it suffices to show $f^* \subset (1 \cup f^*f)$. $1 \cup f^*f$ is reflexive, so by [1.787] it is enough to show $f(1 \cup f^*f) \subset (1 \cup f^*f)$. But $f(1 \cup f^*f) = f \cup ff^*f \subset f \cup f^*f^*f \subset f \cup f^*f \subset f^*f$; the latter because f^* is reflexive, so $f \subset f^*f$.

1.987. An object A together with morphisms $1 \xrightarrow{a} A$ and $A \xrightarrow{t} A$ has the PEANO PROPERTY iff any subobject $B \subset A$ that allows morphisms a and $t \restriction B = B \subset A \xrightarrow{t} A$ is entire.

The construction given in [1.986] in fact yields

Given $1 \xrightarrow{a} A$ *and* $A \xrightarrow{f} A$, *there exists a least subobject* $A' \subset A$ *that allows both* a *and* $A' \subset A \xrightarrow{f} A$. *The subobject* A' *has the Peano property.*

1.988. We will show in [2.542] that for any topos **A** there exists a boolean topos **B** and a faithful bicartesian functor $\mathbf{A} \longrightarrow \mathbf{B}$. Therefore the word 'boolean' will be removable from

In a boolean topos, let $1 \xrightarrow{a} A$ and $A \xrightarrow{t} A$ *be such that*

$$\begin{array}{ccc} 1 & & A \\ & {\scriptstyle a}\searrow \quad \swarrow{\scriptstyle t} & \\ & A & \end{array}$$

is a coproduct and

$$A \underset{t}{\overset{1_A}{\rightrightarrows}} A \longrightarrow 1$$

is a coequalizer.

Then $1 \xrightarrow{a} A \xrightarrow{t} A$ *has the Peano property.*

BECAUSE: If $A' \subset A$ and

$$\begin{array}{ccccc} 1 & \xrightarrow{a'} & A' & \xrightarrow{t'} & A' \\ & {\scriptstyle a}\searrow & \downarrow & & \downarrow \\ & & A & \xrightarrow[t]{} & A, \end{array}$$

we may assume that $1 + A' \longrightarrow A'$ is a cover (apply [1.986] to A' if necessary). Let A'' be the complement of A' in A. We must show that $A'' = 0$.

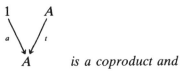

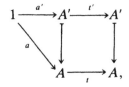

We claim that A'' allows $t \restriction A'' = A'' \subset A \xrightarrow{t} A$. If the topos were $\mathscr{S}$, this claim would easily follow just because $\binom{a}{t}$ is iso and $\binom{a'}{t'}$ is a cover. By [1.635] and [1.641], the claim holds in our topos. Let $A'' \xrightarrow{t''} A''$ be a (necessarily unique) morphism such that $A'' \xrightarrow{t''} A'' \subset A = A'' \subset A \xrightarrow{t} A$. $t = t' + t''$.

Let

$$A' \underset{t'}{\overset{1_{A'}}{\rightrightarrows}} A' \to C' \quad \text{and} \quad A'' \underset{t''}{\overset{1_{A''}}{\rightrightarrows}} A'' \to C''$$

be coequalizers. Because $A \underset{t}{\overset{1_A}{\rightrightarrows}} A \to 1$ is a coequalizer, $C' + C'' = 1$. As A' allows $1 \xrightarrow{a} A$, there exists $1 \longrightarrow C'$. But $C' \subset 1$, so $C' = 1$ and thus $C'' = 0$. By [1.944], $A'' = 0$. Therefore A' is entire and $1 \xrightarrow{a} A \xrightarrow{t} A$ has the Peano property.

1.989. It will be shown in [2.542] that the word 'boolean' is removable from

In a boolean topos, let $1 \xrightarrow{a} A$ and $A \xrightarrow{t} A$ *be such that*

$$\begin{array}{ccc} 1 & & A \\ & {\scriptstyle a}\searrow \; \swarrow{\scriptstyle t} & \\ & A & \end{array} \quad \textit{is a coproduct and } A \underset{t}{\overset{1_A}{\rightrightarrows}} A \to 1 \quad \textit{is a coequalizer.}$$

If $\binom{c}{u}: 1 + C \longrightarrow C$ *is a cover and*

$$\begin{array}{ccccc} 1 & \xrightarrow{c} & C & \xrightarrow{u} & C \\ & {\scriptstyle a}\searrow & \downarrow{\scriptstyle f} & & \downarrow{\scriptstyle f} \\ & & A & \xrightarrow[t]{} & A, \end{array}$$

then f is monic.

BECAUSE: We may assume that the topos is capital [1.935]. Let $K \subset C \times C$ be the level of f, K' the complement of the diagonal in K, and $A_1 \subset A$ the image of the morphism $K' \subset K \xrightarrow{p_1} C \xrightarrow{f} A$. Let A_2 be the complement of A_1. It is enough to show $A_2 = A$ because then $A_1 = 0$, so $K' = 0$ [1.944], and thus K is the diagonal and f is monic. We show that A_2 is entire by using the Peano property [1.988].

Because 1 is projective [1.525], A_2 allows $p: 1 \longrightarrow A$ iff there exists a unique $x: 1 \longrightarrow C$ such that $xf = p$. We claim that A_2 allows a. Suppose $xf = a$, $x \neq c$. As $\binom{c}{u}$ is a cover and 1 is projective, $x = yu$ for some $y: 1 \longrightarrow C$. But then $a = yuf = yft$, which is impossible because $\binom{a}{t}$ is iso. Thus A_2 allows a. A similar argument shows that if A_2 allows $p: 1 \longrightarrow A$ then A_2 allows pt. The image of $t \restriction A_2$ is well-pointed because it allows $1 \longrightarrow A$ and the topos is capital. Thus A_2 allows $t \restriction A_2$ and we are done.

1.98(10). *In any topos, let* $1 \xrightarrow{a} A$ *and* $A \xrightarrow{t} A$ *be such that*

$$1 \xrightarrow{a} A \xleftarrow{t} A \quad \textit{is a coproduct and} \quad A \overset{1_A}{\underset{t}{\rightrightarrows}} A \to 1 \quad \textit{is a coequalizer.}$$

Then $1 \xrightarrow{a} A \xrightarrow{t} A$ *is a NNO.*

BECAUSE: A has the Peano property [1.988, 2.542]. First note that the uniqueness condition for a NNO follows, that is, if $f, g: A \longrightarrow B$ were such that

$$\begin{array}{ccccc} 1 & \xrightarrow{a} & A & \xrightarrow{t} & A \\ & {\scriptstyle b}\searrow & \downarrow & & \downarrow \\ & & B & \xrightarrow[v]{} & B, \end{array}$$

then the equalizer E of f and g would allow both a and $E \subset A \xrightarrow{t} A$, and thus $E = A$.

For the existence condition, given $1 \xrightarrow{b} B$ and $B \xrightarrow{v} B$, apply [1.986] to $1 \xrightarrow{\langle a,b \rangle} A \times B$ and $A \times B \xrightarrow{t \times v} A \times B$ to obtain $1 \xrightarrow{c} C$ and $C \xrightarrow{u} C$, $\binom{c}{u}$ a cover, and morphisms $C \longrightarrow A$, $C \longrightarrow B$. It suffices to show that $C \longrightarrow A$ is iso.

Note that

$$\begin{array}{ccccc} 1 & \xrightarrow{c} & C & \xrightarrow{u} & C \\ & {\scriptstyle a}\searrow & \downarrow & & \downarrow \\ & & A & \xrightarrow[t]{} & A. \end{array}$$

$C \longrightarrow A$ is a cover because its image I allows both a and $t \restriction I$, so $I = A$

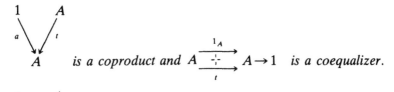

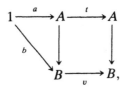

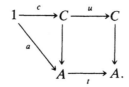

by the Peano property. But $C \longrightarrow A$ is also monic [1.989, 2.542] and thus it is iso [1.512].

1.98(11). *Let* $\mathbf{A}$ *and* $\mathbf{A}'$ *be topoi and* $T\colon \mathbf{A} \longrightarrow \mathbf{A}'$ *a bicartesian functor. If* $1 \xrightarrow{0} N \xrightarrow{s} N$ *is a NNO in* $\mathbf{A}$, *then* $1 \xrightarrow{T0} TN \xrightarrow{Ts} TN$ *is a NNO in* $\mathbf{A}'$.

BECAUSE: A NNO may be characterized in bicartesian terms [1.985, 1.98(10)].

1.98(12). Given an object A in a topos, an A-ACTION is an object B with morphisms $1 \xrightarrow{b} B$ and $A \times B \xrightarrow{v} B$. A FREE A-ACTION is an A-action $A \times 1 \xrightarrow{1_A \times e} A \times A^* \xrightarrow{s} A^*$ such that for any A-action $A \times 1 \xrightarrow{1_A \times b} A \times B \xrightarrow{v} B$ there is a unique morphism $A^* \xrightarrow{f} B$ so that

$$\begin{array}{ccccc}
A \times 1 & \xrightarrow{1_A \times e} & A \times A^* & \xrightarrow{s} & A^* \\
 & {\scriptstyle 1_A \times b}\searrow & \downarrow{\scriptstyle 1_A \times f} & & \downarrow{\scriptstyle f} \\
 & & A \times B & \xrightarrow[v]{} & B.
\end{array}$$

For any object A, a free A-action is clearly unique up to isomorphism. A NNO is a free 1-action.

In $\mathscr{S}$, A^* is the set of finite lists of elements of A. e is the empty list, and s puts an element of A at the beginning of a given list.

1.98(13). We note that the analogue of the bicartesian characterization of a NNO given in [1.985] and [1.98(10)] holds for a free A-action for any object A:

In any topos, for any object A, $A \times 1 \xrightarrow{1_A \times e} A \times A^* \xrightarrow{s} A^*$ *is a free* A*-action iff*

$$1 + A \times A^* \xrightarrow{\binom{e}{s}} A^* \quad \textit{is iso and}$$

$$A \times A^* \underset{s}{\overset{p_2}{\rightrightarrows}} A^* \longrightarrow 1 \quad \textit{is a coequalizer.}$$

BECAUSE: The reasoning is analogous to [1.985] and [1.98(10)].

1.98(14). *In a topos with a NNO, for any object* A *there exists a free* A*-action.*

BECAUSE: If $1 \xrightarrow{0} N \xrightarrow{s} N$ is a NNO, let

$$\begin{array}{ccc} 1 & & A \\ & {\scriptstyle u_1}\searrow \quad \swarrow{\scriptstyle u_2} & \\ & 1 + A & \end{array}$$

be a coproduct and consider the A-action given by the transpose $e: 1 \longrightarrow (1 + A)^N$ of $N \longrightarrow 1 \xrightarrow{u_1} (1 + A)$ and by the monic

$$A \times (1 + A)^N \xrightarrow{u_2 \times 1} (1 + A) \times (1 + A)^N,$$

whose target is isomorphic to $(1 + A)^{1+N}$ and thus to $(1 + A)^N$. Reason as in [1.986–7] to obtain A^* as the least subobject of $(1 + A)^N$ that allows e and $A \times A^* \subset A \times (1 + A)^N \longrightarrow (1 + A)^N$.

1(10). SCONING

A category with a terminator is EXACTING if the functor $\Gamma = (1, -)$ is exact, that is, if it preserves any finite colimits that may exist. (It always preserves any small limits that may exist.)

It is worth noticing that for an abelian category $\mathbf{A}$ and an object A therein, the representable functor $(A, -)\colon \mathbf{A} \to \mathscr{Ab}$ is exact iff A is projective. An exacting pre-logos is focal [1.733] because the pasting lemma [1.62] describes unions as pushouts. A Grothendieck topos is exacting iff 1 is a connected projective [1.(10)41]. This is not true for elementary topoi, counterexamples being topoi of microsheaves [1.94(10), 1.76].

$\mathscr{S}$ is the only exacting boolean Grothendieck topos. On the other hand, the arguments given by proof-theorists (relying on logical syntax) show that, in our terms, various free logoi and topoi are exacting. (We give a conceptual proof in [1.(10)3].) Thus every logos (topos) is covered by an exacting logos (topos).

The representations of a topos into exacting topoi cannot, in general, be collectively faithful. In chapter 2, we will give an example of a topos which has no exact functors into $\mathscr{S}$, and thus into no exacting topos [2.454]. On the other hand,

1.(10)1. *Every category* $\mathbf{A}$ *with a terminator is a slice of an exacting category* $\hat{\mathbf{A}}$. *If* $\mathbf{A}$ *is regular / a pre-logos / a pre-topos / a logos / cartesian-closed / a topos / a Grothendieck topos / a category with a natural number object, then so is* $\hat{\mathbf{A}}$.

BECAUSE: Given a category $\mathbf{A}$ with 1, its SCONE $\hat{\mathbf{A}}$ is defined by [1.2–1.22] as follows. Its objects are triples $\langle S, A, f\rangle$, where S is a set, $A \in |\mathbf{A}|$, and $f\colon S \to \Gamma(A)$ is a mapping. The proto-morphisms are pairs $\langle g, x\rangle$, where g is a mapping, and x is a morphism in $\mathbf{A}$. The source-target predicate $\langle S, A, f\rangle \to \langle S', A', f'\rangle$ is

$$\begin{array}{ccc} S & \xrightarrow{\;g\;} & S' \\ {\scriptstyle f}\downarrow & & \downarrow{\scriptstyle f'} \\ \Gamma(A) & \xrightarrow[\Gamma(x)]{} & \Gamma(A'). \end{array}$$

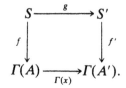

We write $\langle S, A\rangle$ for objects of the scone because the third coordinate is usually clear.

1.(10)11. $\langle \Gamma(1), 1\rangle$ is a terminator in the scone. It has a proper subobject $M = \langle \emptyset, 1\rangle$. All proper subterminators are included in M. Therefore, a terminator in the scone is not the union of any collection of proper subobjects. In particular, it is coprime [1.733].

The slice $\hat{\mathbf{A}}/M$ is equivalent to $\mathbf{A}$.

1.(10)12. The scone is exacting. In fact, $\Gamma\colon \hat{\mathbf{A}} \to \mathscr{S}$ preserves any small colimits that may exist, because

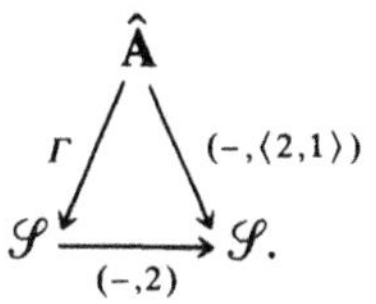

1.(10)13. $\hat{\mathbf{A}} \to \hat{\mathbf{A}}/M \to \mathbf{A}$ has both adjoints, so it preserves any limits and colimits that may exist. The left-adjoint is given by $\langle \emptyset, A\rangle$. The right-adjoint is given by $\langle \Gamma(A), A\rangle$.

1.(10)14. Additional structure on $\hat{\mathbf{A}}$ is easily determined by that on $\mathbf{A}$. For double-sharps, exponentiation, power-objects, and a natural number object use the formulae

$$\langle f, x\rangle^{\#\#}\langle S, A\rangle = \langle f^{\#\#}S, x^{\#\#}A\rangle\,,$$

$$\langle T, B\rangle^{\langle S, A\rangle} = \langle \hat{\mathbf{A}}(\langle S, A\rangle, \langle T, B\rangle), B^A\rangle\,,$$

$$[\langle S, A\rangle] = \langle \mathscr{S}\!ub_{\hat{\mathbf{A}}}\langle S, A\rangle, [A]\rangle\,,$$

$$N_{\hat{\mathbf{A}}} = \langle N_{\mathscr{S}}, N_{\mathbf{A}}\rangle\,,$$

and if $\mathscr{C} \subset |\mathbf{A}|$ is a generating set for $\mathbf{A}$, then $\{\langle 1, C\rangle : C \in \mathscr{C}\}$ is a generating set for $\hat{\mathbf{A}}$.

1.(10)2. The sconing construction for Heyting algebras was given in [1.72(10)].

The scone of the Heyting algebra of open sets of a topological space X is obtained by adding a focal point [1.733] to X: $\hat{X} = X + \{*\}$, $\mathscr{O}(\hat{X}) = \mathscr{O}(X) + \{\hat{X}\}$. X is a maximal proper open subset, and the inclusion $X \subset \hat{X}$ is open continuous.

How does this bear on the category $\mathscr{S}\!h(X)$? In $\mathscr{S}\!h(\hat{X})$, 1 is a coprime projective [1.733], and we will soon see that $\mathscr{S}\!h(\hat{X})$ is in fact exacting. $\mathscr{S}\!h(\hat{X})/X$ is equivalent to $\mathscr{S}\!h(X)$.

1.(10)21. We analyze the construction of $\mathscr{Sh}(\hat{X})$ from $\mathscr{Sh}(X)$ and describe it in purely categorical terms. What is a sheaf over $\hat{X}$? We have

$$\begin{array}{ccc} Z & \longrightarrow & Y \\ {\scriptstyle \psi}\downarrow & \lrcorner & \downarrow{\scriptstyle \varphi} \\ X & \subset & \hat{X}, \end{array}$$

where φ is a local homeomorphism. $\psi: Z \to X$ is a sheaf over X. The fibers of φ are $\varphi^{\#}(*)$ and the fibers of ψ. A germ at $*$ is a global section of φ (that is, a continuous map $\sigma: \hat{X} \to Y$ such that $\sigma\varphi = 1_{\hat{X}}$), thus an element of $\varphi^{\#}(*)$ together with the associated global section of ψ given by restricting σ to X. This gives the glueing conditions, i.e. a mapping from the stalk at $*$ to the global sections of ψ.

Given a sheaf ψ over X, a set S, and a mapping from S to the global sections of ψ, there is a sheaf φ over X such that ψ is obtained from it by pulling back to X, S is the stalk of φ at $*$, and the given mapping is a restriction of germs to X.

1.(10)3. In Appendix B we will give constructions of various free categories. If $\mathbf{A}$ is free, the representation $\mathbf{A} \to \hat{\mathbf{A}} \to \hat{\mathbf{A}}/M \to \mathbf{A}$ is the identity, so $\mathbf{A}$ is a RETRACT of $\hat{\mathbf{A}}$.

1.(10)31. *A retract of an exacting category is exacting.*

BECAUSE:

$$\mathbf{A} \xrightarrow{\Gamma} \mathscr{S} \text{ is a retract of } \mathbf{A} \to \hat{\mathbf{A}} \xrightarrow{\Gamma} \mathscr{S}$$

A retract of a weak-colimit preserving functor is again such.

It can also be verified directly that in the retract, 1 is again a coprime projective.

1.(10)32. Therefore,

Various free categories are exacting.

1.(10)4. An object A in a cocomplete abelian category $\mathbf{A}$ is a SMALL PROJECTIVE if the representable functor $(A, -): \mathbf{A} \to \mathscr{Ab}$ preserves all colimits, or equivalently, if it preserves sums.

A projective module is small iff it is finitely generated.

1.(10)41. *Let* $\mathbf{E}$ *be a Grothendieck topos,* $A \in |\mathbf{E}|$ *a connected projective. Then the representable functor* $(A, -): \mathbf{E} \to \mathscr{S}$ *preserves all small colimits.*

BECAUSE: It preserves finite coproducts, and therefore all coproducts: write $\Sigma_I A_i$ as $A_i + \Sigma_{j \neq i} A_j$, and recall that in a Grothendieck topos, coproducts are disjoint. Because A is projective, $(A, -)$ preserves images, and therefore all unions. Thus it preserves equivalence closures [1.775, 1.846], and hence equalizers.

CHAPTER TWO

ALLEGORIES

2.1. BASIC DEFINITIONS

Allegories, shortly to be defined, are to binary relations between sets as categories are to functions between sets. Consider, first, the one-object case, the basic model being the set of all binary relations on a fixed infinite set. Among the multitude of operations available on such a collection we consider only:

The constant, 1 ($x\,1\,y$ iff $x = y$),

The unary operation of RECIPROCATION, R° ($x\,R^\circ\,y$ iff $y\,R\,x$),

The binary operation of COMPOSITION, RS ($x\,RS\,y$ iff there exists z such that $x\,R\,z$ and $z\,S\,y$),

The binary operation of INTERSECTION, $R \cap S$ ($x\,R \cap S\,y$ iff $x\,R\,y$ and $x\,S\,y$).

The entire equational theory on these operations is decidable but not finitely axiomatizable [2.158]. Certain conditions, however, become our definition of a one-object allegory:

The monoid equations for composition:

$$1R = R = R1\,,$$

$$R(ST) = (RS)T\,.$$

The semi-lattice equations for intersection:

$$R \cap R = R\,, \qquad R \cap S = S \cap R\,,$$

$$R \cap (S \cap T) = (R \cap S) \cap T\,.$$

(Note that we have not required a unit for intersection.)

The anti-involution equations for reciprocation:

$$R^{\circ\circ} = R\,,$$

$$(RS)^\circ = S^\circ R^\circ\,, \qquad (R \cap S)^\circ = S^\circ \cap R^\circ\,.$$

Since intersection is commutative, the 'anti-' is misleading with respect to order. As usual, $R \subset S$ denotes the equation $R = R \cap S$. Note well that reciprocation preserves—not reverses—order:

$$R \subset S \quad \textit{implies} \quad R^\circ \subset S^\circ\,.$$

At this point no equation relates composition and intersection. We adjoin the equation of *semi-distributivity*:

$$R(S \cap T) \subset (RS \cap RT)\,.$$

This equation is equivalent with the Horn sentence:

$$S \subset T \quad \textit{implies} \quad RS \subset RT\,,$$

that is, composition preserves order.

We could describe the structure at this point as a semi-lattice ordered monoid with an anti-involution. We adjoin one more containment, the *law of modularity*:

$$(RS \cap T) \subset (R \cap TS^{\circ})S\,.$$

(Another ready example of a one-object allegory: the collection of normal subgroups of a group. $R \cap S, 1$ and RS are as standardly interpreted. $R^{\circ} = R$, or, if one insists, $R^{\circ} = R^{-1} = \{x^{-1} \mid x \in R\}$. Then the law of modularity above is precisely the classical modular identity.)

2.11. An ALLEGORY is a category with a unary operation denoted R° (using upper-case for morphisms) and a binary partial operation denoted $R \cap S$. $R \cap S$ is defined whenever $\Box R = \Box S$ and $R\Box = S\Box$. The equations:

$$\Box R^{\circ} = R\Box\,, \qquad R^{\circ}\Box = \Box R\,,$$

$$(\Box R)^{\circ} = \Box R\,, \qquad R^{\circ\circ} = R\,,$$

$$\Box(R \cap S) \succeq \Box R \qquad (\text{or } \Box(\mathrm{R} \cap \mathrm{S}) = \Box(\mathrm{RS}^{\circ})\Box\mathrm{S})\,,$$

$$R \cap R = R\,, \qquad R \cap S = S \cap R\,, \qquad R \cap (S \cap T) = (R \cap S) \cap T\,,$$

$$(RS)^{\circ} = S^{\circ}R^{\circ}\,, \qquad (R \cap S)^{\circ} = S^{\circ} \cap R^{\circ}\,,$$

$$R(S \cap T) \subset RS \cap RT \qquad (\text{or } R(S \cap T) = RS \cap R(S \cap T) \cap RT)\,,$$

$$RS \cap T \subset (R \cap TS^{\circ})S \quad (\text{or } RS \cap T = RS \cap T \cap (R \cap TS^{\circ})S)\,.$$

We have omitted $1^{\circ} = 1$ as an axiom because $1 = (1^{\circ})^{\circ} = (11^{\circ})^{\circ} = 1^{\circ\circ}1^{\circ} = 11^{\circ} = 1^{\circ}$

2.111. For any regular category **C** the category of relations is an allegory [1.563]. Among the most important examples are allegories constructed from formal theories as described in Appendix B. An example to keep in mind now is obtained from an arbitrary locale, $\mathscr{Z}$ [1.723]. A $\mathscr{Z}$-VALUED RELATION from a set I to a set J is defined as an $I \times J$ matrix of elements from $\mathscr{Z}$. If we write the (i, j)-th entry of such a matrix R as iRj and we think of $\mathscr{Z}$ as a set of 'truth-values', we may read iRj as 'the extent to which i is related to j'.

The rule for composition is obtained by the requirement that $(iRj) \wedge (jSk) \leqslant i(RS)k$, that is, we take the minimal such $I \times K$ matrix: $i(RS)k = \bigvee_j (iRj) \wedge (jSk)$. Note that the distributivity requirement in $\mathscr{Z}$ is necessary and sufficient for $R(ST) = (RS)T$. Intersection is defined by $i(R \cap S)j = (iRj) \wedge (iSj)$ and reciprocation by $jR^\circ i = iRj$. If $\mathscr{Z}$ is the two-element lattice, then we obtain the allegory of sets.

2.112. The modular identity $RS \cap T \subset (R \cap TS^\circ)S$ is equivalent, in the presence of the other axioms, to $RS \cap T \subset R(S \cap R^\circ T)$ and has an immediate consequence

$$R \subset RR^\circ R$$

(because $R \subset 1R \cap R \subset (1 \cap RR^\circ)R \subset RR^\circ R$).

2.113. Let M be a lattice-ordered commutative monoid and define $\Box x = x\Box = 1$, $x^\circ = x$. Except for the modular identity M is an allegory. Not even the above consequence $x \leqslant xx^\circ x$ need hold (consider the unit interval under multiplication).

Let $\mathscr{L}$ be a lattice and define

$$\Box R = R\Box = 0\,, \qquad RS = R \vee S\,,$$
$$R^\circ = R\,, \qquad R \cap S = R \wedge S\,.$$

Except for the modular identity, $\mathscr{L}$ is an allegory. Note that $R = RRR$. The modular identity is equivalent to the Horn sentence:

$$A \subset B \;\; \textit{implies} \;\; (A \cup X) \cap B = A \cup (X \cap B)\,.$$

That is, $\mathscr{L}$ is an allegory iff it is a MODULAR LATTICE.

Consider the non-modular lattice

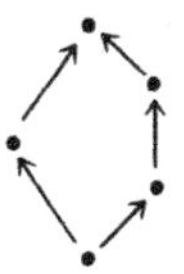

Every sublattice generated by two elements is modular, indeed distributive, and we may conclude that the modular identity is not a consequence of any consistent set of two-variable identities.

2.12. For an endomorphism R we recall the definitions:
R is REFLEXIVE if $1 \subset R$.
R is SYMMETRIC if $R^\circ \subset R$.
R is TRANSITIVE if $RR \subset R$.

Note that $R^\circ \subset R$ iff $R \subset R^\circ$ iff $R = R^\circ$.

Reflexive and transitive imply idempotent.

Symmetric and transitive imply idempotent:

$$R \subset RR^\circ R \subset R^3 \subset R^2 .$$

R is COREFLEXIVE if $R \subset 1$.

Coreflexive implies symmetric idempotent:

$$R \subset RR^\circ R \subset 1R^\circ 1 \subset R^\circ , \qquad R^2 \subset 1R \subset R .$$

Note that for a regular category **C**, the coreflexive relations on an object correspond to its subobjects.

If a morphism is reflexive, symmetric, and transitive, it is of course called an EQUIVALENCE RELATION. A symmetric idempotent may be viewed as a 'partial equivalence relation'.

2.121. *For coreflexive morphisms*, $AB = A \cap B$.

BECAUSE: $AB \subset A1 \subset A$ and $AB \subset 1B \subset B$ hence $AB \subset A \cap B$.

$$A \cap B \subset (A \cap B)^2 \subset AB .$$

2.122. The DOMAIN of a morphism R, $\mathscr{D}om\ R$, is defined as $1 \cap RR^\circ$. It is distinguished among coreflexive morphisms by:

For $A \subset \Box R$, $\mathscr{D}om\ R \subset A$ *iff* $R \subset AR$.

BECAUSE: If $(1 \cap RR^\circ) \subset A$ then $R \subset 1R \cap R \subset (1 \cap RR^\circ)R \subset AR$.
If $R \subset AR$ then $1 \cap RR^\circ \subset 1 \cap ARR^\circ \subset A(A^\circ \cap RR^\circ) \subset AA^\circ \subset A$.

2.123. *Consequently*, $\mathscr{D}om\ RS \subset \mathscr{D}om\ R$.

2.124. A lemma we will use repeatedly:

$$\mathscr{D}om(R \cap S) = 1 \cap SR^\circ .$$

BECAUSE: $\mathscr{D}om(R \cap S) \subset 1 \cap (R \cap S)(R \cap S)^\circ \subset 1 \cap RS^\circ$,

$$\begin{aligned} 1 \cap RS^\circ &\subset 1 \cap (1 \cap (1 \cap RS^\circ)) \subset 1 \cap (1 \cap (S \cap R)S^\circ) \\ &\subset 1 \cap (S \cap R)((S \cap R)^\circ \cap S^\circ) \subset \mathscr{D}om(R \cap S) . \end{aligned}$$

2.13. R is ENTIRE if $\mathscr{D}om\ R = 1$; equivalently if $1 \subset RR^\circ$.
R is SIMPLE if $R^\circ R \subset 1$.
R is a MAP if it is both entire and simple.

2.131.

$$R \textit{ and } S \textit{ are } \begin{Bmatrix} \textit{entire} \\ \textit{simple} \\ \textit{maps} \end{Bmatrix} \quad \textit{implies} \quad RS \textit{ is } \begin{Bmatrix} \textit{entire} \\ \textit{simple} \\ \textit{a map} \end{Bmatrix} .$$

From $\mathscr{D}om(RS) \subset \mathscr{D}om\ R$ we further obtain

RS entire implies R entire.

2.132. For an allegory **A** we shall denote the subcategory of maps by $\mathscr{M}ap(\mathbf{A})$ and reserve lower-case italics for morphisms therein. If **C** is a regular category then $\mathbf{C} \simeq \mathscr{M}ap(\mathscr{R}el(\mathbf{C}))$ [1.564].

2.133. The partial order when restricted to $\mathscr{M}ap(\mathbf{A})$ is discrete, that is:

$$f \subset g \quad \textit{implies} \quad f = g$$

BECAUSE: $f \subset g$ implies $f^\circ \subset g^\circ$ and $1g \subset ff^\circ g \subset fg^\circ g \subset f1$.

2.134. Reciprocation when restricted to $\mathscr{M}ap(\mathbf{A})$ is the partial operation that assigns inverses to isomorphisms. That is

If f and f° are maps then f is an isomorphism and $f^{-1} = f^\circ$.

BECAUSE: f° a map implies that ff° and $f^\circ f$ are maps. But $1 \subset ff^\circ$ and $f^\circ f \subset 1$, hence by the last section $1 = ff^\circ$ and $1 = f^\circ f$.

2.135. The isomorphisms in **A** and $\mathscr{M}ap(\mathbf{A})$ coincide. That is:

If R is an isomorphism then R is a map and $R^{-1} = R^\circ$.

BECAUSE: R is entire because RR^{-1} is.

R is simple because $R^\circ R \subset R^\circ RR^{-1}R \subset R^\circ RR^{-1}(R^{-1})^\circ R^{-1}R \subset R^\circ(R^{-1})^\circ \subset (R^{-1}R)^\circ \subset 1$.

2.136. $f(R \cap S) = fR \cap fS$. Indeed,

If F is simple then $F(R \cap S) = FR \cap FS$.

BECAUSE: $FR \cap FS \subset F(R \cap F^\circ FS) \subset F(R \cap S)$.

And, as always, $F(R \cap S) \subset FR \cap FS$.

2.14. A pair of maps f, g TABULATES a morphism R if

$$f^\circ g = R, \qquad ff^\circ \cap gg^\circ = 1.$$

We say that R is TABULAR if it has a tabulation and that an allegory **A** is a TABULAR ALLEGORY if every morphism in **A** is tabular.

If $\mathscr{V}$ is a *connected locale*, that is, if the only complemented elements are 0 and 1, then the allegory **A** composed of $\mathscr{V}$-valued relations between sets has very few maps. Indeed, $\mathscr{M}ap(\mathbf{A}) \simeq \mathscr{S}$. **A** is not tabular.

For any locale, however, there is a natural extension of the allegory composed of $\mathscr{V}$-valued relations that is tabular [2.168]. The same construction works for the allegories that arise from formal languages [Appendix B].

2.141. *If $ff^\circ \cap gg^\circ = 1$ in an allegory* **A** *then f, g is a monic pair in* $\mathscr{M}ap(\mathbf{A})$.

BECAUSE: If $hf = h'f$ and $hg = h'g$ then by the last section $h = h(ff^\circ \cap gg^\circ) = hff^\circ \cap hgg^\circ = h'ff^\circ \cap h'gg^\circ = h'(ff^\circ \cap gg^\circ) = h'$.

2.142. In a regular category, **C**, a pair f, g is monic iff $ff^\circ \cap gg^\circ = 1$ in $\mathscr{R}el(\mathbf{C})$ from which we see that $\mathscr{R}el(\mathbf{C})$ is tabular, indeed, the notions of tabulation in **C** and $\mathscr{R}el(\mathbf{C})$ coincide.

2.143. *If f, g tabulates R then $x^\circ y \subset R$ iff there exists (necessarily unique) h such that $x = hf$, $y = hg$.*

BECAUSE: If $x = hf$, $y = hg$ then $x^\circ y = (hf)^\circ(hg) \subset f^\circ h^\circ hg \subset f^\circ g$.

Conversely, if $x^\circ y \subset f^\circ g$ define $H = xf^\circ \cap yg^\circ$.

H is entire (using [2.124]) because

$$1 \subset 1 \cap (xx^\circ)(yy^\circ) \subset 1 \cap xf^\circ gy^\circ \subset \mathscr{D}om\ H.$$

H is simple because

$$H^\circ H \subset (fx^\circ \cap gy^\circ)(xf^\circ \cap yg^\circ)$$
$$\subset fx^\circ xf^\circ \cap gy^\circ yg^\circ \subset ff^\circ \cap gg^\circ \subset 1\,.$$

$Hf = x$ because $Hf \subset xf^\circ f \subset x$ and hence $Hf = x$ [2.133]. Similarly $Hg = y$.

2.144. Consequently, tabulations are unique up to unique isomorphism. That is,

If f, g and $f'g'$ both tabulate R then there exists a unique isomorphism u such that $f' = uf$, $g' = ug$.

2.145. *If a coreflexive morphism A is tabular in* **A** *then there exists monic $h \in \mathscr{Map}(\mathbf{A})$ such that $A = h^\circ h$.*

BECAUSE: If $A \subset 1$ and f, g tabulates A then $g \subset ff^\circ g \subset fA \subset f$ and $g = f$ by [2.133]. Clearly $gg^\circ = 1$.

2.146

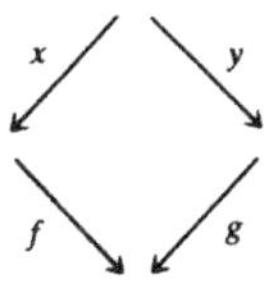
iff $x^\circ y \subset fg^\circ$.

BECAUSE: $xf = yg$ implies $x^\circ y \subset x^\circ ygg^\circ \subset x^\circ xfg^\circ \subset fg^\circ$.
$x^\circ y \subset fg^\circ$ implies $yg \subset xx^\circ yg \subset xfg^\circ g \subset xf$ and $yg = xf$ by [2.133].

2.147. *If* **A** *is a tabular allegory then $\mathscr{Map}(\mathbf{A})$ has pullbacks, equalizers, images and pullbacks transfer images.*

BECAUSE: The pullback of 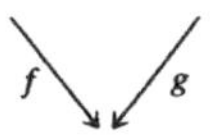 is constructible as a tabulation of fg° [2.143].

The equalizer of f, g is constructible as a tabulation of $\mathscr{Dom}(f \cap g)$.

The image of f is constructible as a tabulation of $\mathscr{Dom}(f^\circ)$. g is a cover iff $1 \subset g^\circ g$, that is, iff g° is entire.

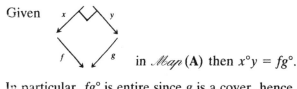

Given 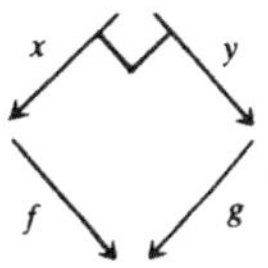 in $\mathscr{Map}(\mathbf{A})$ then $x^\circ y = fg^\circ$.

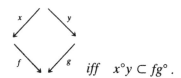

In particular, fg° is entire since g is a cover, hence $x^\circ y$ is entire and x is a cover.

2.148. *If* **A** *is a tabular allegory then* $\mathbf{A} \simeq \mathscr{Rel}(\mathscr{Map}(\mathbf{A}))$.

2.15. An object π in an allegory is a PARTIAL UNIT if 1_π is its maximum endomorphism.

λ is a UNIT if, further, every object is the source of an entire morphism targeted at λ.

An allegory is said to be UNITARY if it has a unit.

If **C** is a regular category then the (partial) units in $\mathscr{Rel}(\mathbf{C})$ coincide with the (sub)terminators in **C**.

The allegory composed of $\mathscr{V}$-valued relations between sets [2.111] is unitary.

2.151. For an object α let $\mathscr{Cor}(\alpha)$ be the semi-lattice of coreflexive morphisms on α. As in any category, (α, β) is the set of morphisms from α to β (it is a semi-lattice as well).

If π *is a partial unit then* $\mathscr{Dom}: (\alpha, \pi) \to \mathscr{Cor}(\alpha)$ *is an isomorphism of* (α, π) *onto an ideal of* $\mathscr{Cor}(\alpha)$.

BECAUSE: $\mathscr{Dom}$ is clearly an order-preserving function. Suppose $\mathscr{Dom}\ R \subset \mathscr{Dom}\ S$. Then $R \subset (\mathscr{Dom}\ S)R \subset SS^\circ R \subset S1$ (since 1 is maximum in (π, π)).

In particular, if $\mathscr{Dom}\ R = \mathscr{Dom}\ S$ then $R = S$.

Given $A \subset \mathscr{Dom}\ R$ then without using any property on π:

$$\mathscr{Dom}(AR) \subset \mathscr{Dom}\ A \subset A\,,$$

$$A \subset A(\mathscr{Dom}\ R)A \subset A \cap ARR^\circ A \subset \mathscr{Dom}(AR)\,.$$

2.152. Consequently:

If λ *is a unit in* **A** *then* $\mathscr{Dom}: (\alpha, \lambda) \to \mathscr{Cor}(\alpha)$ *is an isomorphism of semi-lattices.*

The unique entire morphism $p_\alpha: \alpha \to \lambda$*, being simple, is a map.*

λ *is a terminator in* $\mathscr{Map}(\mathbf{A})$.

If λ *is a unit then* $p_\alpha p_\beta{}^\circ$ *is maximum in* (α, β) *for any two objects* α, β.

BECAUSE: p_α is maximal in (α, λ), hence $Rp_\beta \subset p_\alpha$ any $R \in (\alpha, \beta)$ and $R \subset Rp_\beta p_\beta{}^\circ \subset p_\alpha p_\beta{}^\circ$.

2.153. Let **K** be a collection of partial endofunctions on the set of natural numbers $\mathbb{N}$. We make the following assumptions on **K**:

(i) **K** contains the identity,
(ii) **K** is closed under composition,
(iii) **K** contains two total functions ℓ and r such that for any φ, ψ in **K**, there exists θ in **K** defined on the common domain of φ and ψ, such that φ contains $\theta\ell$ and ψ contains θr:

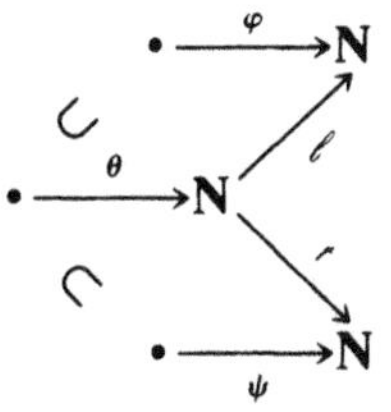

If $(-,-)\colon \mathbb{N} \times \mathbb{N} \to \mathbb{N}$ is a coding of pairs and $(n, k)\ell = n$, $(n, k)r = k$ for all numbers n, k, then in the condition (iii) one may let $\theta = (\varphi, \psi)$.

For example, **K** may be the collection of all partial recursive functions.

We consider the category **A** of ASSEMBLIES $A = (X, \{Y_n\}_{n \geq 0})$, where X is a set and $\{Y_n\}_{n \geq 0}$ a sequence of subsets $Y_n \subset X$ called CAUCUSES and written $A|_n$. Note that caucuses are not necessarily pairwise disjoint. The CARRIER of an assembly A is the set $|A| = \bigcup_n A|_n$. A morphism $f\colon A \to B$ is an ordinary function $f\colon |A| \to |B|$ for which there exists φ in **K** (a MODULUS of f) so that for every $n \in \mathbb{N}$, $x \in |A|$:

If $x \in A|_n$, then $\varphi(n)$ is defined and $f(x) \in B|_{\varphi(n)}$.

The category of assemblies is a positive pre-logos.

BECAUSE: The empty set is a coterminator. Given a pair of morphisms $f, g\colon A \to B$, their equalizer is obtained as $|E| = \{x \in |X| \colon f(x) = g(x)\}$ (that is, the ordinary equalizer of the maps f and g in the category of sets), and $E|_n$ is the ordinary intersection of $|E|$ and $A|_n$. A terminator is given by a one-element set 1 so that $1|_n = 1$. For binary products, let $(A \times B)|_n = A|_{n\ell} \times B|_{nr}$. Given a morphism $f\colon A \to B$, we wish to obtain the minimal subobject B' of B such that f factors through B'. Let $|B'|$ be the ordinary image of the map f and let $B'|_n = f(A|_n)$. B' is a subobject of B named by inclusion using the same modulus as f. (Note that we will have therefore shown that any subobject may be named by a monic given by inclusion with some modulus.) The required factor $A \to B'$ is obtained as f with the identity as its modulus. The minimality and the stability under pullbacks is left to the reader, as is the construction of disjoint unions.

The category of assemblies is not effective.

Note the functor $\nabla\colon \mathscr{S} \to \mathbf{A}$ given by $|\nabla X| = (\nabla X)|_n = X$, for all n; for $f\colon X \to Y$ let ∇f be f with the identity as modulus. ∇ preserves coterminator, equalizers, and finite products, but not unions.

In the allegory $\mathscr{Rel}(\mathbf{A})$, $R\colon A \to B$ may be named by a sequence $\{R_n\}$ of ordinary relations from $|A|$ to $|B|$ for which there exist φ, ψ in **K** such that for every $x \in |A|$, $y \in |B|$, and $n \in \mathbb{N}$: if $x\, R_n\, y$, then $\varphi(n)$ and $\psi(n)$ are defined, $x \in A|_{\varphi(n)}$, and $y \in B|_{\psi(n)}$. One of such sequences $\{R_n\}$ is included in another

$\{S_n\}$ iff there exists φ in **K** such that for each $n \in \mathbb{N}$, $x \in |A|$, $y \in |B|$, if $x\,R_n\,y$, then $\varphi(n)$ is defined and $x\,S_{\varphi(n)}\,y$.

The composition of relations R (from A to B) and S (from B to C) is the relation RS from A to C defined by: $x\,(RS)_n\,z$ iff there exists $y \in |B|$ such that $x\,R_{n\ell}\,y$ and $y\,S_{nr}\,z$.

For any object A the identity relation is named by: $x\,(1_A)_n\,x'$ iff $x = x'$ and $x \in A|_n$. Given a relation named by $\{R_n\}$, its reciprocal named by the sequence of ordinary reciprocals $\{(R_n)^\circ\}$.

$\mathscr{Rel}$(**A**) is a unitary tabular allegory, but not all equivalence relations split (as idempotents).

2.154. A (UNITARY) REPRESENTATION OF ALLEGORIES is a functor between allegories which preserves (units,) reciprocation and intersection. (As a functor, it preserves the category structure, and thus, in particular, composition.) The definitions of map and tabulation are equational, hence maps and tabulations are preserved by representations of allegories.

Given a representation of allegories T: **A** $\to$ **B** we obtain a representation of categories T': $\mathscr{Map}$(**A**) $\to$ $\mathscr{Map}$(**B**). If **A** and **B** are tabular, then T' preserves pullbacks, equalizers, and covers. If, further, **A** and **B** are unitary and T is unitary, then T' preserves terminators, and consequently T' is a representation of regular categories. T is an embedding iff T' is faithful.

Clearly, if T': $\mathbf{C}_1 \to \mathbf{C}_2$ is a representation of regular categories it induces a representation of unitary allegories $\mathscr{Rel}(\mathbf{C}_1) \to \mathscr{Rel}(\mathbf{C}_2)$, from which we obtain:

The category of small regular categories is isomorphic to the category of small unitary tabular allegories.

Consequently,

A small unitary tabular allegory may be faithfully represented in a power of the allegory of sets.

BECAUSE: This allegory is $\mathscr{Rel}$(**C**) for a small regular category **C**, so we may apply [1.55], and [1.52] to obtain a faithful representation $\mathbf{C} \to \Pi_I \mathscr{S}$, which yields a faithful representation of unitary allegories

$$\mathscr{Rel}(\mathbf{C}) \to \mathscr{Rel}(\Pi_I \mathscr{S}) \xrightarrow{\sim} \Pi_I\, \mathscr{Rel}(\mathscr{S})\,.$$

Tabular reflections of allegories will be considered in [2.16].

2.155. Let **B** be the subcategory of $\mathscr{Rel}(\mathscr{S})$ defined by $(A \xrightarrow{R} B) \in \boldsymbol{B}$ iff either $R = \emptyset$ or both R and R° are entire. **B** is not a suballegory of $\mathscr{Rel}(\mathscr{S})$: it is closed with respect to composition and reciprocation but not with respect to intersection. Nonetheless, each hom-set of **B** is easily seen to be a semi-lattice. All axioms for allegories hold except for the modular identity.

B is tabular (a tabulation that works in $\mathscr{Rel}(\mathscr{S})$ works in **B**). Hence the modular identity is not a consequence of tabularity. $\mathscr{Map}$(**B**) is not a regular category: it has equalizers and a terminator but it does not have products.

2.156. Let $\mathscr{L}$ be a modular lattice and suppose that $T: \mathscr{L} \to \mathscr{Rel}(\mathscr{S})$ is a representation of $\mathscr{L}$ viewed as an allegory, as described in [2.113, 2.154]. T picks a single object X in $\mathscr{S}$ and for each $R \in \mathscr{L}$ a binary relation on X. $T(R)$ is, in fact, an equivalence relation: it is reflexive because $0 \subset \mathrm{R}$ and $T(0) = 1_S$; it is symmetric because $R = R^\circ$; it is transitive because $R^2 = R$.

For each pair $R, S \in \mathscr{L}$, $T(R)$ and $T(S)$ commute (since R and S commute in $\mathscr{L}$). Hence T is a function from $\mathscr{L}$ to a family of commuting equivalence relations on X. T preserves intersection and when viewed as a function to the lattice of equivalence relations, it preserves $\cup$ (the smallest equivalence relation that contains a pair of commuting equivalence relations is their composition). Such a representation of $\mathscr{L}$ is called a *partition representation* or a *combinatorial representation*.

A *geometric representation* of $\mathscr{L}$ is a representation into the lattice of subspaces of a vector space over some skew field. Given such, we can replace each subspace with the congruence it induces on the ambient vector space and check that we thereby obtain a combinatorial representation.

2.157. A *projective plane* is a model of a certain two-sorted theory: the two sorts are *points* and *lines*; the unique predicate is an *incidence relation* which will be denoted as $x \in A$ where it is understood that lower-case italics refer to points and upper-case to lines; the axioms are:

for all x, y there exists A such that $x \in A$ and $y \in A$;

for all A, B there exists x such that $x \in A$ and $x \in B$;

for all x, y, A, B, $x \in A$, $x \in B$, $y \in A$, $y \in B$ imply $x = y$ or $A = B$.

An *interesting projective plane* is one such that each point is incident to at least three lines and each line is incident to at least three points. (It is easy to classify all uninteresting projective planes.)

For any skew field K we obtain a projective plane whose points are the 1-dimensional subspaces in K^3, whose lines are the 2-dimensional subspaces, and whose incidence relation is given by containment. Assuming, of course, that $0 \neq 1$ in K, the resulting plane is interesting. The classical question was: which interesting projective planes so arise?

For every projective plane there is an associated modular lattice (hence an associated allegory): let $\mathscr{L}$ be the disjoint union of the points and lines together with two new elements called 0 and 1. We partially order $\mathscr{L}$ by taking 0 as the minimum, 1 as the maximum and using the incidence relation in between. Conversely given any modular lattice in which all maximal chains have four elements we obtain a projective plane by taking its atoms as points and its coatoms as lines.

The *theorem of Desargues* says that if two triangles are 'in perspective' then their corresponding sides meet on a line. (That is, given P, a_1, a_2, b_1, b_2, c_1, c_2, u, v, w such that the triples $\langle p, a_1, a_2\rangle$, $\langle p, b_1, b_2\rangle$, $\langle p, c_1, c_2\rangle$, $\langle a_1, c_1, u\rangle$, $\langle a_2, c_2, u\rangle$, $\langle b_1, c_1, v\rangle$, $\langle b_2, c_2, v\rangle$, $\langle a_1, b_1, w\rangle$, $\langle a_2, b_2, w\rangle$ are colinear, then $\langle u, v, w\rangle$ is colinear. See Figure 1.)

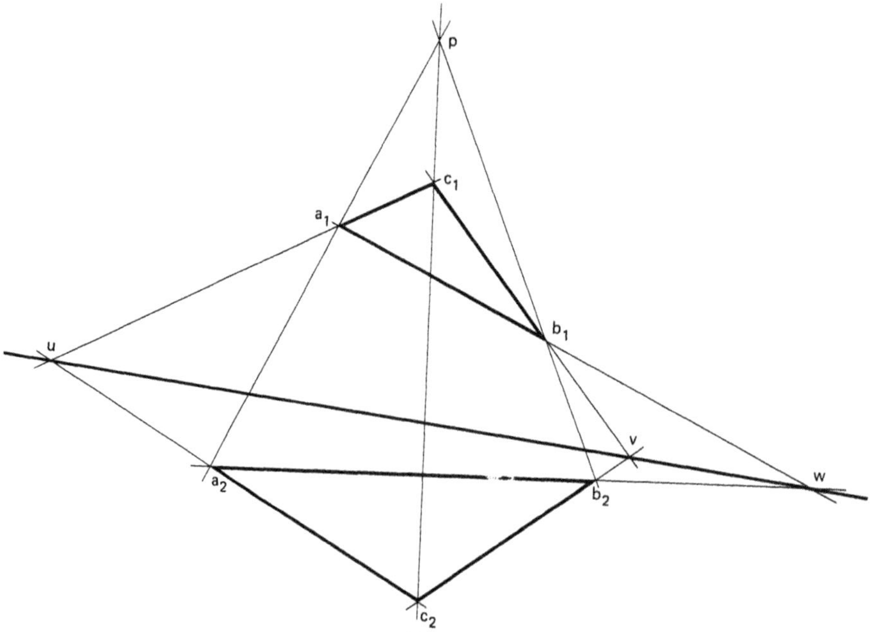

Fig. 1.

The theorem of Desargues in three-space is clear if the triangles are not coplanar (the line $\langle u, v, w\rangle$ is then the intersection of the planes determined by the triangles). If the triangles are coplanar, we may find a pair of non-coplanar triangles in perspective whose projection onto the relevant plane is the given pair of triangles. (Such is the traditional argument for the theorem of Desargues. We shall presently find quite a different argument.) One of the triumphs of 19th century mathematics provided the converse: every interesting projective plane that satisfies the theorem of Desargues is isomorphic to the projective plane arising from a skew field (necessarily unique up to isomorphism). Such planes may easily be embedded in a higher-dimensional projective space. (The skew field is a field iff the plane satisfies the classical theorem of Pappus. The original interest in Wedderburn's theorem that finite skew fields are fields seems to have been wholly geometric: it is equivalent to the theorem that all finite projective planes that satisfy Desargues also satisfy Pappus.)

Consider, now, the following Horn sentence for allegories:

$$(A_1A_2 \cap B_1B_2) \subset C_1C_2 \quad \textit{implies} \quad (A_1^\circ B_1 \cap A_2B_2^\circ) \subset (A_1^\circ C_1 \cap A_2C_2^\circ)(C_1^\circ B_1 \cap C_2B_2^\circ).$$

It is easily verified for $\mathscr{R}el(\mathscr{S})$. Starting with a projective plane, writing $a_1, a_2, b_1, \ldots$ as $A_1, A_2, B_1, \ldots,$ passing to the associated modular lattice, viewing such as an allegory, one will see that this Horn sentence is equivalent with the theorem of Desargues. (Note that Desargues implies modularity: given R, S, and T let $A_1 = R°$, $A_2 = T$, $B_1 = S$, $B_2 = 1$, $C_1 = 1$, $C_2 = S$.)

When Veblen and Wedderburn constructed (1907) a projective plane (with 91 points and 91 lines) that failed the theorem of Desargues, they provided us with an allegory (one object and 184 morphisms) which cannot be faithfully represented in $\mathscr{R}el(\mathscr{S})$.

2.158. In the next section we will discover existential conditions on allegories that insure faithful representations in $\mathscr{R}el(\mathscr{S})$. We pause here to show that no finite set of equations can suffice.

Fix a set of 'labels', $A, B, C, \ldots$. Let $\mathbf{G}$ be the category whose objects are directed edge-labelled graphs with a pair of distinguished vertices marked s and t. We allow multiple edges between vertices. We allow edges from a vertex to itself. We allow s, t to mark the same vertex. The maps in $\mathbf{G}$ are the obvious ones: they map vertices to vertices, edges to edges, they preserve incidence, direction, labels and s and t.

By the composition of two objects in $\mathbf{G}$ we mean the result of identifying the t-vertex of the first with the s-vertex of the second. Such is a bifunctor. The coproduct of two $\mathbf{G}$-objects is the result of identifying the s-vertex of the first with the s-vertex of the second, and then identifying the t-vertex of the first with the t-vertex of the second. Such, as always, is a bifunctor. Define the reciprocal of a $\mathbf{G}$-object as being the result of transposing the marks s and t. (Do not reverse the direction of the edges.) Such is clearly a functor.

Let $\hat{\mathbf{G}}$ be the poset associated with $\mathbf{G}°$; that is, we write $G_1 \leqslant G_2$ if there exists a map $G_2 \to G_1$, and we identify G_1 and G_2 if both $G_1 \leqslant G_2$ and $G_2 \leqslant G_1$. Note that the coproduct in $\mathbf{G}$, which, of course, is a product in $\mathbf{G}°$, becomes an intersection in $\hat{\mathbf{G}}$. The composition bifunctor becomes a binary operation on $\hat{\mathbf{G}}$ and the reciprocation functor becomes a unary operation on $\hat{\mathbf{G}}$. In this manner we obtain a one-object allegory. The identity morphism is the one-vertex graph with no edges.

For each label A define $[A]$ as the graph $s \cdot \xrightarrow{A} \cdot t$. Let $\bar{\mathbf{G}} \subset \hat{\mathbf{G}}$ be the suballegory generated by the $[A]$'s. The elements of $\bar{\mathbf{G}}$ are finite and connected. Note, however, that no matter how one directs or labels the edges, the graph

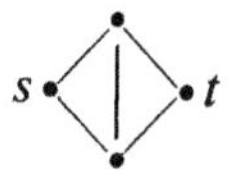

will fail to be in $\bar{\mathbf{G}}$. Indeed, the direction and labelling of the edges is always irrelevant: if a directed labelled graph is in $\bar{\mathbf{G}}$ then any redirecting and relabelling of the edges will yield a graph in $\bar{\mathbf{G}}$. (But as the example above shows, we cannot freely move s and t. A graph is in $\bar{\mathbf{G}}$ if it has a 'parallel-series decomposition' in the sense of circuit theory.)

By a *representable allegory* we mean an allegory which can be faithfully represented in a power of the allegory of sets. We will show that $\bar{\mathbf{G}}$ is the free one-object representable allegory on the given set of labels.

A set with a binary relation may be viewed, of course, as the set of vertices of a directed graph. A set S with a family of binary relations A, B, $C, \ldots$ may be viewed as the set of vertices of a labelled graph Γ_S. Given an allegorical term on A, B, $C, \ldots$ we may on one hand consider the resulting relation R on S and on the other hand, the resulting graph $[R]$ in $\bar{\mathbf{G}}$ (build from $[A]$, $[B], \ldots$). For $x, y \in S$ it is the case that xRy iff there exists a map $f\colon [R] \to \Gamma_S$ of directed labelled graphs such that $f(s) = x$ and $f(t) = y$.

Given S as above, we define $T\colon \hat{\mathbf{G}} \to \mathcal{R}el(\mathcal{S})$ by $x(TG)y$ iff there exists $f\colon G \to \Gamma_S$ such that $f(s) = x$ and $f(t) = y$. T may easily be seen to preserve the allegory operations. By restriction we obtain $T\colon \bar{\mathbf{G}} \to \mathcal{R}el(\mathcal{S})$. Since the $[A]$'s generate $\bar{\mathbf{G}}$, T is the unique representation of allegories such that $T[A] = A$, $T[B] = B, \ldots$.

Hence, given any representable one-object allegory $\mathbf{A}$ and specified morphisms A, $B, \ldots$ there exists a unique representation $T\colon \bar{\mathbf{G}} \to \mathbf{A}$ such that $T[A] = A$, $T[B] = B, \ldots$. To construct T let $\{F_i\colon \mathbf{A} \to \mathcal{R}el(\mathcal{S})\}$ be a faithful family of representations. For each i there exists $T\colon \bar{\mathbf{G}} \to \mathcal{R}el(\mathcal{S})$ such that $T_i[A] = F_i[A]$, $T_i[B] = F_i[B], \ldots$. We may lift the T_i's to obtain $T\colon \bar{\mathbf{G}} \to \mathbf{A}$.

$\bar{\mathbf{G}}$ itself is representable. Given $G_1, G_2 \in \bar{\mathbf{G}}$ such that it is not the case that $G_1 \leqslant G_2$ we wish to find a representation $T\colon \bar{\mathbf{G}} \to \mathcal{R}el(\mathcal{S})$ such that $TG_1 \not\subset TG_2$. Let S be the set of vertices of G_1 and interpret $A, B, \ldots$ as binary relations on S. The induced representation $\bar{\mathbf{G}} \to \mathcal{R}el(\mathcal{S})$ is as desired.

Note here that we have a decision procedure for equations. A containment holds in $\mathcal{R}el(\mathcal{S})$ iff it holds in $\bar{\mathbf{G}}$ iff there exists a map between two particular finite directed labelled graphs. The restriction to one object is not critical. Given a finite sequence of sets $S_1, S_2, \ldots, S_n$ and a finite family of binary relations we may take S as the disjoint union of the S_i's and reduce the problem to one about endo-relations.

There is a decision procedure, but there is no finite set of equations, true for $\mathcal{R}el(\mathcal{S})$, that account for all equations true for $\mathcal{R}el(\mathcal{S})$. We first recast the definition of allegories:

replace the general equation $R = R \cap R$ with the specific equations $A = A \cap A$, one for each variable A;

replace $(R_1 \cap R_2)S \subset (R_1S \cap R_2S)$ with $(R_1 \cap R_2)(S_1 \cap S_2) \subset (R_1S_1 \cap R_2S_2)$;

replace $(RS \cap T) \subset (R \cap TS^\circ)S$ with $R(S_1 \cap S_2) \cap T \subset (R \cap TS_1^\circ)S_2$.

In terms of graphs, the specific equations of the form $A = A \cap A$ simply say that two edges with the same label and same beginning and ending vertices may be identified. Note that the process of identifying such edges does not change the number of vertices.

Given a pair of terms, E_1, E_2, built from the allegory operations, we will say that a containment $E_1 \subset E_2$ is *separated* if each variable that appears at all appears exactly once in E_1 and exactly once in E_2. Except for the general containment $R \cap S \subset R$ and the specific containments $A \subset A \cap A$, all the containments, as now recast, are separated.

If a graph G is in $\bar{\mathbf{G}}$ and if G' is a connected subgraph of G containing s and t then one may show that $G' \in \bar{\mathbf{G}}$ and (not quite so easily) that the definition of allegory implies that $G \subset G'$. It is the onto functions between graphs that present the problem.

Consider the graph G_1:

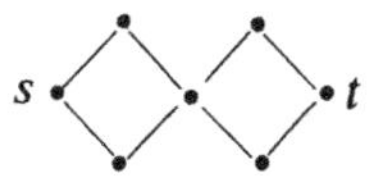

One may direct the edges at will. Label each edge differently (which as we will see means that we do not have to label them at all). Consider the result of identifying s and t, identifying the top two vertices, and identifying the bottom two vertices. The result is the graph G_2:

Both G_1 and G_2 are in $\bar{\mathbf{G}}$. The map from G_1 to G_2 yields a separated containment of the form

$$1 \cap (R_1 \cap R_4^\circ)(R_2 \cap R_3^\circ)(S_2^\circ \cap S_3)(S_1^\circ \cap S_4) \\ \subset (R_1R_2 \cap S_1S_2)(R_3R_4 \cap S_3S_4)\,.$$

G_2 was obtained from G_1 by identifying three pairs of vertices. If one identifies any one or any two of those pairs of vertices, the resulting graph is *not* in $\bar{\mathbf{G}}$.

Suppose that the above separated containment were a consequence of the definition of allegories. We would obtain a sequence of containments, each of which is a direct instance of one of the defining containments, which sequence would—by transitivity of containment—yield the above containment. Consequently, we would obtain a sequence of maps in $\mathbf{G}$ of the form $G_1 \to H_1$, $H_1 \to H_2, \ldots, H_n \to G_2$ *where each* H_i *is in* $\bar{\mathbf{G}}$. For each i the image of $G_1 \to H_i$ must be in $\bar{\mathbf{G}}$. Hence either $G_1 \to H_i$ is an inclusion or its image is isomorphic to G_2. There must exist i such that $G_1 \to H_i$ is an inclusion and the image of $G_1 \to H_{i+1}$ is isomorphic to G_2. The map $H_i \to H_{i+1}$, therefore, necessarily identifies at least three pairs of vertices. $H_i \to H_{i+1}$ is an instance of one of the defining containments. We reach a contradiction: each defining containment can yield graphs which identify at most one pair of vertices.

Suppose we adjoin a finite set of containments to the definition of allegory. For any such containment consider a graphical interpretation $H_1 \to H_2$. Factor $H_1 \to H_2$ as $H_1 \twoheadrightarrow H_3 \subset H_2$. The inclusion is a consequence of the original definition, and we may replace the given containment with the one which corresponds to $H_1 \twoheadrightarrow H_3$. Let H_1' be the result of relabelling H_1 so that each edge has a different label. Let H_3' be the result of identifying the vertices in H_1' in the same manner as they are identified by $H_1 \to H_3$. The containment that corresponds to $H_1' \twoheadrightarrow H_3'$ is separated. The given containment is easily seen to be an instance thereof. That is, we may assume that all new containments are separated.

Let n be an integer such that for each separated containment that has been added to the definition of allegory, it is the case that in its graphical interpretation

$H_1 \twoheadrightarrow H_3$ the number of vertices in H_1 is less than n plus the number of vertices in H_3. Let G_1 be of the form

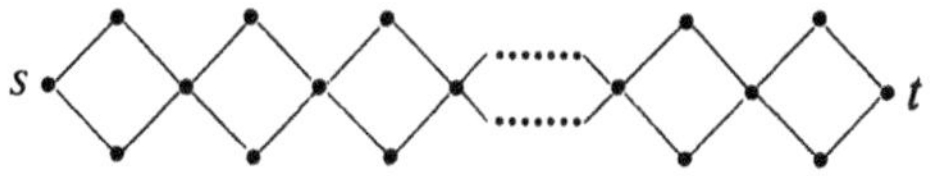

where the number of rhombi is n.

Let G_2 be the result of identifying s and t, identifying the top row of vertices, and identifying the bottom row of vertices. G_2 looks like

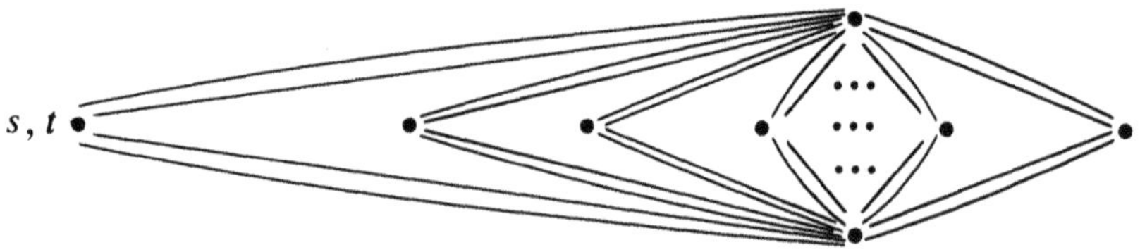

The corresponding containment is

$$1 \cap (R_0 \cap R_n^\circ)\left[\bigcap_{i=1}^{n-2} (R_i \cap R_{i+1}^\circ)(S_i^\circ \cap S_{i+1})\right](S_0^\circ \cap S_n)$$
$$\subset \prod_{i=0}^{n} (R_i R_{i+1} \cap S_i S_{i+1})\,.$$

In any proper factorization $G_1 \twoheadrightarrow G_3 \subset G_2$, G_3 must fail to be in **G**. A repetition of the argument for the original definition will work in this case. That is, the above containment is not a consequence of the given finite set of containments.

The equations chosen as the definition of allegory happen to be precisely those that account for all containments obtainable by identifying the vertices two at a time.

2.16. Tabular and effective reflections

2.161. Recall [2.14] that the allegory composed of $\mathscr{V}$-valued relations [2.111] is not tabular if $\mathscr{V}$ is a connected locale. For any locale, however, the allegory composed of $\mathscr{V}$-valued relations may be extended to a tabular allegory [2.168].

2.162. *If R, S splits a symmetric idempotent T then $S = R^\circ$*

BECAUSE: From $(RS)^\circ = RS$ and $SR = 1$ we wish to conclude that $S = R^\circ$.

$$R^\circ = (SR)R^\circ \subset SS^\circ(SR)R^\circ \subset SS^\circ R^\circ \subset S(RS)^\circ \subset (SR)S \subset S\,,$$

$$S^\circ \subset S^\circ(SR) \subset S^\circ(SR)R^\circ R \subset S^\circ R^\circ R \subset (RS)^\circ R \subset R(SR) \subset R\,.$$

Hence $S = R^\circ$.

2.163. Recall that both coreflexives and equivalence relations are symmetric idempotents [2.12]. As a consequence of the above, we have:

A coreflexive morphism A is a split idempotent iff there exists a map h such that $h^\circ h = A$, $hh^\circ = 1$; that is, iff A is tabular.

An equivalence relation E is a split idempotent iff there exists a map f such that $ff^\circ = E$, $f^\circ f = 1$; that is, iff E is effective.

2.164. For a class $\mathscr{E}$ of symmetric idempotents in an allegory **A**, the category $\mathit{Split}(\mathscr{E})$ [1.28] has a natural allegory structure:

$$(A \xrightarrow{R} B)^\circ = B \xrightarrow{R^\circ} A \quad \text{and} \quad (A \xrightarrow{R} B) \cap (A \xrightarrow{S} B) = A \xrightarrow{R\cap S} B.$$

($A(R \cap S) = R \cap S$ because $R \cap S \subset AR \cap S \subset A(R \cap A^\circ S) \subset A(R \cap S)$.) If $|\mathbf{A}| \subset \mathscr{E}$ then the full inclusion $\mathbf{A} \to \mathit{Split}(\mathscr{E})$ is a faithful representation of allegories.

If $\mathscr{E}$ includes all coreflexive morphisms then all coreflexive morphisms in $\mathit{Split}(\mathscr{E})$ split [1.281], hence are tabular. The universal property for $\mathit{Split}(\mathscr{E})$ says, therefore, that if Cor is the set of coreflexive morphisms and if **B** is any tabular allegory, $\mathbf{A} \to \mathbf{B}$ any representation of allegories, then there exists

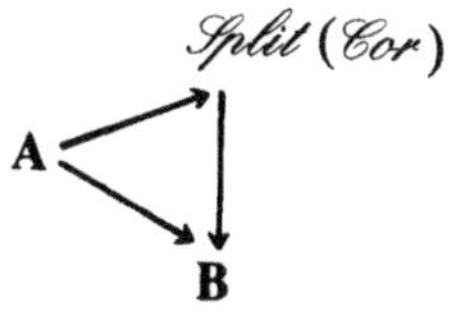

unique up to natural equivalence. We seek conditions for $\mathit{Split}(\mathit{Cor})$ to be tabular, in which case it is necessarily the tabular reflection of **A**.

2.165. We say that an allegory is PRE-TABULAR if every morphism is contained in a tabular morphism. If **A** is unitary then pre-tabularity is equivalent to the existence of tabulations for maximal morphisms. The most important examples [Appendix B] are the allegories arising from formal languages. $\mathscr{V}$-valued relations easily form a pre-tabular allegory.

*If **A** is pre-tabular, then $\mathit{Split}(\mathit{Cor})$ remains pretabular.*

BECAUSE: Given $A \xrightarrow{R} B$, suppose $R \subset f^\circ g$, $ff^\circ \cap gg^\circ = 1$. Let $C = \mathscr{Dom}(fA) \cap \mathscr{Dom}(gB)$. $C \xrightarrow{CfA} A$ and $C \xrightarrow{CgB} B$ are maps,

$$A \xrightarrow{R} B = (C \xrightarrow{CfA} A)^\circ (C \xrightarrow{CgB} B), \quad \text{and}$$

$$(CfA)(CfA)^\circ \cap (CgB)(CgB)^\circ = C\,.$$

2.166. *An allegory is tabular iff it is pre-tabular and all coreflexive morphisms split.*

BECAUSE: Given $R \subset f^\circ g$, $ff^\circ \cap gg^\circ = 1$ and the fact that coreflexives split, we wish to tabulate R.

$A = 1 \cap fRg^\circ$ is coreflexive. Let h be such that $h^\circ h = A$, $hh^\circ = 1$. Then $\langle hf, hg \rangle$ tabulates R:

$$(hf)^\circ(hg) \subset f^\circ h^\circ hg \subset f^\circ fRg^\circ g \subset R\,,$$

$$R \subset f^\circ g \cap R \subset f^\circ(1 \cap fRg^\circ)g \subset f^\circ h^\circ hg \subset (hf)^\circ(hg)$$

Using [2.136] and its dual,

$$(hf)(hf)^\circ \cap (hg)(hg)^\circ = hff^\circ h^\circ \cap hgg^\circ h^\circ = h(ff^\circ \cap gg^\circ)h^\circ$$
$$= hh^\circ = 1\,.$$

2.167. The last two sections yield:

If **A** *is a pre-tabular allegory,* $\mathscr{Cor}$ *its class of coreflexive morphisms, then* $\mathscr{Split}(\mathscr{Cor})$ *is its tabular reflection.*

A unit in **A** remains a unit in $\mathscr{Split}(\mathscr{Cor})$. Therefore a pre-tabular unitary allegory is faithfully and fully representable in an allegory of the form $\mathscr{Rel}(\mathbf{C})$, where $\mathbf{C} = \mathscr{Map}(\mathscr{Split}(\mathscr{Cor}))$ is a regular category. Apply [2.154].

2.168. A coreflexive $\mathscr{V}$-valued relation is a diagonal matrix. The objects in the tabular reflection may be redescribed as pairs $\langle I, \exists \rangle$ where $\exists$ is a function from I to $\mathscr{V}$. We read $\exists_i$ as 'the extent to which i exists'. The source-target predicate is

$$\langle I, \exists \rangle \xrightarrow{R} \langle J, \exists \rangle \quad \text{iff} \quad iRj \leqslant \exists_i \wedge \exists_j$$

that is, the extent to which i and j are related is bounded by the extent to which that both exist. R is entire iff $\exists_i = \bigvee_j iRj$ for all i and simple iff

$iRj \wedge iRj' = 0$ for all $i, j, j', j \neq j'$. R is a map, therefore, iff for each i the i-th row of R partitions $\exists_i$.

2.169. An EFFECTIVE ALLEGORY is one in which all equivalence relations split. As noted above [2.163] such is equivalent to the effectiveness of its category of maps.

If $\mathbf{A}$ is tabular (and unitary) and if $\mathscr{E}\mathit{q}$ is the class of all equivalence relations then $\mathscr{S}\mathit{plit}(\mathscr{E}\mathit{q})$ is effective and remains tabular (and unitary). It is the EFFECTIVE REFLECTION of $\mathbf{A}$, that is, given any effective $\mathbf{B}$ and any representation of allegories $\mathbf{A} \to \mathbf{B}$ there exists

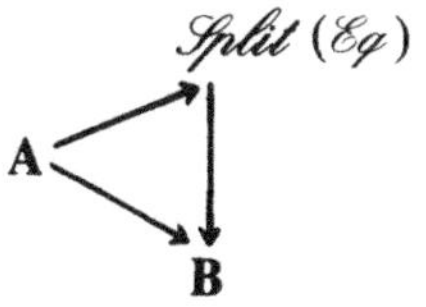

unique up to natural equivalence.

Starting with a regular category $\mathbf{C}$, it is now easy to construct its effective reflection in the category of regular categories: $\mathscr{M}\mathit{ap}(\mathscr{S}\mathit{plit}(\mathscr{E}\mathit{q}_{\mathscr{R}\mathit{el}(\mathbf{C})}))$.

Note that splitting all symmetric idempotents is equivalent to first splitting the coreflexive morphisms and then the equivalence relations.

2.16(10) A morphism R is SEMI-SIMPLE if there exist simple F, G [2.13] such that $R = F^\circ G$. (If only the containment $R \subset F^\circ G$ holds, then $R = F^\circ G'$, where $G' = (1 \cap FRG^\circ)G$; that is, if a morphism is contained in a semi-simple morphism, then it is semi-simple.) A SEMI-SIMPLE ALLEGORY is one in which every morphism is semi-simple. Thus a unitary allegory is semi-simple iff all maximal morphisms are semi-simple.

Let $\mathbf{A}$ *be an allegory,* $\mathscr{S}\mathit{id}$ *its class of symmetric idempotents.* $\mathscr{S}\mathit{plit}(\mathscr{S}\mathit{id})$ *is tabular iff* $\mathbf{A}$ *is semi-simple.*

BECAUSE: Suppose that $\mathscr{S}\mathit{plit}(\mathscr{S}\mathit{id})$ is tabular. Given $R \in \mathbf{A}$ let $(A \xrightarrow{F} \square R)$, $(A \xrightarrow{G} R\square)$ be a tabulation of $\square R \xrightarrow{R} R\square$. Then F and G are simple in $\mathbf{A}$ and $R = F^\circ G$.

Suppose that $\mathbf{A}$ is semi-simple. Given $A \xrightarrow{R} B$ in $\mathscr{S}\mathit{plit}(\mathscr{S}\mathit{id})$ let F, G be simple in $\mathbf{A}$ such that $R = F^\circ G$. Then $\square F \xrightarrow{FA} A$, $\square G \xrightarrow{GB} B$ are simple in $\mathscr{S}\mathit{plit}(\mathscr{S}\mathit{id})$ and $A \xrightarrow{R} B = (\square F \xrightarrow{FA} A)^\circ(\square G \xrightarrow{GB} B)$. That is, $\mathscr{S}\mathit{plit}(\mathscr{S}\mathit{id})$ remains semi-simple. We must show that a semi-simple allegory in which all symmetric idempotents split is tabular.

Given $R = F^\circ G$, where F, G are simple, note that $FF^\circ \cap GG^\circ$ is symmetric and transitive, hence idempotent [2.12]. Using [2.162], there exists H such that $HH^\circ = 1$ and $H^\circ H = FF^\circ \cap GG^\circ$. It is routine that HF and HG are maps and that they tabulate R.

(We thus have a representation theorem for small unitary semi-simple allegories. We will discover that the unit is unnecessary and that the condition of semi-simplicity can be further weakened [2.225].)

2.16(11). We say that a pair of idempotents in a category are *neighbors* if $ee'e = e$, $e'ee' = e'$.

If $\langle x, y\rangle$ splits e then $\langle e'x, ye'\rangle$ splits e' (because $(e'x)(ye') = e'ee' = e'$ and $(ye')(e'x) = (yx)ye'x(yx) = yee'ex = yex = yxyx = 1$). Hence if either of a neighboring pair of idempotents splits, then so does the other.

Suppose that A is an idempotent in an allegory. $A \cap A^\circ$ is symmetric and transitive, hence an idempotent. If $\langle R, S\rangle$ splits A then A and $A \cap A^\circ$ are neighbors:

$$\begin{aligned}(A \cap A^\circ)A(A \cap A^\circ) &\subset (RS \cap S^\circ R^\circ)RS(RS \cap S^\circ R^\circ)\\ &\subset S^\circ(SRS \cap R^\circ)RS(RSR \cap S^\circ)R^\circ\\ &\subset S^\circ(SRS)RS(RSR)R^\circ \subset S^\circ R^\circ \subset A^\circ\,,\end{aligned}$$

$$(A \cap A^\circ)A(A \cap A^\circ) \subset A^3 \subset A\,,$$

$$A \cap A^\circ \subset (A \cap A^\circ)^3 \subset (A \cap A^\circ)A(A \cap A^\circ)\,,$$

$$A \subset R(SRSR \cap 1)S \subset RS(RS \cap S^\circ R^\circ)RS \subset A(A \cap A^\circ)A\,.$$

We can, of course, split all idempotents if we forget the non-category part of the structure of an allegory. We have just seen, however, that if A is an idempotent in an allegory **A** and if there exists any faithful $T: \mathbf{A} \to \mathbf{B}$ where T is a representation of allegories such that TA splits in **B** then necessarily $(A \cap A^\circ)A(A \cap A^\circ) \subset A \cap A^\circ$ and $A \subset A(A \cap A^\circ)A$. These last two containments are the necessary and sufficient conditions for the existence of an extension of **A** in which A splits: the converse containments are automatic, hence A and $A \cap A'$ are neighbors; if $\mathcal{S}\!\mathit{id}$ is the class of all symmetric idempotents then $A \cap A^\circ$ splits in $\mathcal{S}\!\mathit{plit}(\mathcal{S}\!\mathit{id})$ hence A splits in $\mathcal{S}\!\mathit{plit}(\mathcal{S}\!\mathit{id})$. That is, *when we have split all the symmetric idempotents we have automatically split all idempotents that can ever be split in an allegory.*

If T is a partial ordering, that is, if $T^2 \subset T$ and $T \cap T^\circ = 1$ then the first of the above containments $(T \cap T^\circ)T(T \cap T^\circ) \subset T \cap T^\circ$ forces T to be the identity. Non-trivial partial orderings cannot split in an allegory.

If T is a strict dense partial ordering, that is, if $T^2 = T$ and $T \cap T^\circ = 0$ (for example, the strict order on the rational numbers) then the second of the above containments, $T \subset (T \cap T^\circ)T(T \cap T^\circ)$, forces T to be empty. Non-trivial strict dense partial orderings cannot split in an allegory.

2.16(12). We describe the allegory obtained by splitting the symmetric idempotent $\mathscr{V}$-valued relations. The objects are pairs $\langle I, R\rangle$, where R is a mapping from $I \times I$ to $\mathscr{V}$ such that $iRj = jRi$, and $iRj \wedge jRk \leqslant iRk$, for all $i, j, k \in I$. We read iRj as 'the extent to which i and j are equal'. The source-target predicate [1.2] is given as $T: \langle I, R\rangle \to \langle J, S\rangle$ iff $iTj \leqslant iRi \wedge jSj$, and $iRi' \wedge jSj' \wedge iTj \leqslant i'Tj'$, for all $i, i' \in I$, all $j, j' \in J$. This is the allegory of $\mathscr{V}$-VALUED SETS.

$T: \langle I, R\rangle \to \langle J, S\rangle$ is entire iff $iRi = \bigvee_{j\in J} iTj$, for all $i \in I$. It is simple iff $iTj \wedge iTj' \leqslant jSj'$, for all $i \in I$, all $j, j' \in J$.

2.16(13). Both the categories of recursive and primitive recursive functions [1.572, 1.573] are, in fact, pre-logoi, and their effective reflections are pre-topoi. We can make them effective, but in doing so we lose the axiom of choice: by [1.662] an AC pre-topos is boolean and neither effective reflection is boolean.

More generally,

If $\mathbf{C}$ *is an AC regular category, and if* $\hat{\mathbf{C}}$ *is its effective reflection, then* $\mathbf{C}$ *is equivalent to the full subcategory of projective objects in* $\hat{\mathbf{C}}$. *Hence if* $\mathbf{C}$ *is not effective then* $\hat{\mathbf{C}}$ *is not AC.*

BECAUSE: For each $B \in \hat{\mathbf{C}}$ there exists $A \in \mathbf{C}$ and a cover $A \twoheadrightarrow B$. If $C \in \mathbf{C}$ then it is projective in $\hat{\mathbf{C}}$: given $B \twoheadrightarrow C$ in $\hat{\mathbf{C}}$ choose $A \twoheadrightarrow B$ where $A \in \mathbf{C}$. Since $\mathbf{C}$ is full in $\hat{\mathbf{C}}$ and C is projective in $\mathbf{C}$ there exists $C \to A \twoheadrightarrow B \twoheadrightarrow C = 1$.

If B is projective in $\hat{\mathbf{C}}$ then choose $A \twoheadrightarrow B$ where $A \in \mathbf{C}$. There exists $(B \to A \to B) = 1$ in $\hat{\mathbf{C}}$. B is an equalizer of 1_A and $A \to B \to A$ forcing it to be isomorphic to a $\mathbf{C}$-object.

2.16(14). Let $\mathbf{A}$ be the category of assemblies [2.153]. The effective reflection $\mathbf{E}$ of $\mathscr{R}el(\mathbf{A})$ may be described as follows. The objects are pairs $\langle A, I\rangle$ (we will write A/I to suggest the formal quotient), where A is an assembly and $I: A \to A$ in $\mathscr{R}el(\mathbf{A})$ is an equivalence relation. A relation $R: A \to B$ in $\mathscr{R}el(\mathbf{A})$ is considered a morphism from A/I to B/J in $\mathbf{E}$ iff $IR = R = RJ$ in $\mathscr{R}el(\mathbf{A})$. Every assembly A may be considered as an object $A/1_A$ in $\mathbf{E}$.

The category $\mathbf{E}$ may also be presented as the result of splitting all symmetric idempotents of the full subcategory of $\mathscr{R}el(\mathbf{A})$ of those assemblies that have all caucuses equal (in other words, those assemblies that are simply sets). In fact, $\mathscr{R}el(\mathbf{A})$ is itself the result of splitting all coreflexive morphisms of this full subcategory.

2.2. DISTRIBUTIVE ALLEGORIES

Consider, again, the collection of binary relations on a fixed set. There is a relation 0 contained in every other relation. Given relations R, S there is a relation $R \cup S$ defined by $x(R \cup S)y$ iff xRy or xSy. Among the equations that hold are the usual distributive lattice equations for $\cup$ and $\cap$ together with $R0 = 0 = 0R$, $R(S \cup T) = RS \cup RT$ and $R \cap (S \cup T) = (R \cap S) \cup (R \cap T)$. These equations become our definition of a distributive allegory.

2.21. A DISTRIBUTIVE ALLEGORY is an allegory together with a unary operation denoted 0_R and a binary partial operation denoted $R \cup S$. $R \cup S$ is defined whenever $\Box R = \Box S$ and $R\Box = S\Box$.

The equations:

$$\Box 0_R = \Box R, \qquad 0_R\Box = R\Box,$$

$$\Box(R \cup S) \succeq (\Box R), \qquad (R \cup S)\Box \succeq S\Box,$$

$$R \cup R = R, \qquad R \cup S = S \cup R,$$

$$R \cup (S \cup T) = (R \cup S) \cup T,$$

$$0_R \cup S = (\Box R)S(R\Box),$$

$$R \cup (S \cap R) = R = (R \cup S) \cap R,$$

$$R0_S = 0_{RS}, \qquad R(S \cup T) = RS \cup RT,$$

$$R \cap (S \cup T) = (R \cap S) \cup (R \cap T).$$

2.211. If R and S have the same source and target, that is, if $\Box R = \Box S$ and $R\Box = S\Box$, then $0_R = 0_R \cup 0_S = 0_S$. For every pair of objects α, β therefore, the hom-set (α, β) is either empty or is a lattice with 0.

(Any allegory is a disjoint union of strongly connected components. A connected distributive allegory has a two-sided ideal of zero-maps.)

In any distributive allegory

$$(R \cup S)^\circ = S^\circ \cup R^\circ, \qquad (0_R)^\circ = 0_{R^\circ},$$

$$0_S R = 0_{SR} \quad \text{and} \quad (S \cup T)R = SR \cup TR.$$

2.212. If $\mathbf{C}$ is a pre-logos then $\mathscr{Rel}(\mathbf{C})$ is easily seen to be a distributive allegory [1.616].

If $\mathbf{A}$ *is a tabular unitary distributive allegory, then* $\mathscr{Map}(\mathbf{A})$ *is a pre-logos.*

BECAUSE: Under the hypothesis $\mathscr{Map}(\mathbf{A})$ is known to be a regular category [2.154]. The subobjects, in $\mathscr{Map}(\mathbf{A})$, of an object α correspond to the coreflexive morphisms on α, hence finite unions of subobjects exist.

Recalling that subobjects in $\mathscr{Map}(\mathbf{A})$ coincide with coreflexive morphisms in $\mathbf{A}$, we may construct $f^{\#}(A)$ as $1 \cap fAf^{\circ}$ for $A \subset 1$. Clearly, $f^{\#}(A \cup B) = f^{\#}A \cup f^{\#}B$ and $f^{\#}0 = 0$.

2.213. For any class of symmetric idempotents $\mathscr{E}$ in a distributive allegory, it is routine that $\mathscr{Split}(\mathscr{E})$ is a distributive allegory. Hence a pre-tabular distributive allegory may be completed, in the same manner as for allegories [2.167], to a tabular allegory by splitting the coreflexive morphisms. A semi-simple distributive allegory may be completed to an effective tabular distributive allegory by splitting all the symmetric idempotents. Both constructions yield faithful representations of distributive allegories. In particular, we can construct the effective reflection of a pre-logos as the category of maps of the effective reflection of its allegory of relations.

2.214. Suppose that $\mathbf{C}$ is a pre-logos and that A is the disjoint union of $A_1 \xrightarrow{u_1} A$, $A_2 \xrightarrow{u_2} A$. We obtain, in $\mathscr{Rel}(\mathbf{C})$, the equations

$$u_1u_1^{\circ} = 1, \qquad u_1u_2^{\circ} = 0,$$

$$u_2^{\circ}u_1 = 0, \qquad u_2u_2^{\circ} = 1,$$

$$u_1^{\circ}u_1 \cup u_2^{\circ}u_2 = 1.$$

This set of five equations is equivalent, in any distributive allegory, with the assertion that

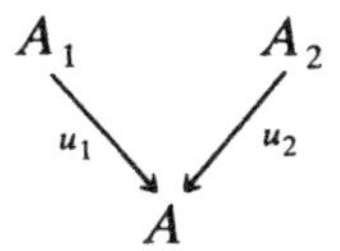

is a coproduct: given R_1, R_2 define $\begin{pmatrix} R_1 \\ R_2 \end{pmatrix}$ as $u_1^{\circ}R_1 \cup u_2^{\circ}R_2$.

Then clearly,

$$u_i\begin{pmatrix} R_1 \\ R_2 \end{pmatrix} = R_i, \quad i = 1, 2 .$$

For the uniqueness condition we need only verify that for any relevant R it is the case that

$$R = \begin{pmatrix} u_1 R \\ u_2 R \end{pmatrix} .$$

$$\left(R = 1R = (u_1^\circ u_1 \cup u_2^\circ u_2)R = u_1^\circ(u_1 R) \cup u_2^\circ(u_2 R) = \begin{pmatrix} u_1 R \\ u_2 R \end{pmatrix}.\right)$$

For the converse, suppose that $\langle U_1, U_2 \rangle$ is a coproduct in a distributive allegory. Define

$$p_1 = \begin{pmatrix} 1 \\ 0 \end{pmatrix}, \qquad p_2 = \begin{pmatrix} 0 \\ 1 \end{pmatrix} .$$

Clearly $U_1 p_1 = 1$, $U_1 p_2 = 0$, $U_2 p_1 = 0$, $U_2 p_2 = 1$. Let

$$R = p_1 U_1 \cup p_2 U_2 .$$

R is the unique morphism such that $U_i R = p_i$, $i = 1, 2$ hence $R = 1$. U_1 and p_1° are entire because $U_1 p_1$ and $p_1^\circ U_1^\circ$ are entire. Using $p_1 U_1 \subset p_1 U_1 \cup p_2 U_2 \subset 1$, we obtain $U_1 \subset 1U_1 \subset p_1^\circ p_1 U_1 \subset p_1^\circ 1$ and $p_1^\circ \subset 1\, p_1^\circ \subset U_1 U_1^\circ p_1^\circ \subset U_1(p_1 U_1^\circ) \subset U_1 1$. Thus $U_1 = p_1^\circ$ and similarly $U_2 = p_2^\circ$. Note that U_1, U_2 are maps since $U_i^\circ U_i \subset p_i U_i \subset 1$.

Consequently,

A pre-logos **C** *is positive iff* $\mathscr{R}el$(**C**) *has finite coproducts.*

2.215. Note that any allegory is isomorphic, via reciprocation, to its opposite allegory. Hence any allegory has coproducts precisely to the extent that it has products. Indeed, $\langle U_1, U_2 \rangle$ is a coproduct iff $\langle U_1^\circ, U_2^\circ \rangle$ is a product.

By a POSITIVE ALLEGORY we mean a distributive allegory with finite coproducts (equivalently with finite products.) If **A** is a tabular unitary positive allegory, then $\mathscr{M}ap$(**A**) is a positive pre-logos.

2.216. Let **A** be a distributive allegory. We construct its POSITIVE REFLECTION $\mathbf{A}^+$ by first formally creating a zero-morphism $\alpha \xrightarrow{0} \beta$ for every pair α, β such that (α, β) was empty. $\mathbf{A}^+$ is best described by [1.2–1.22]: objects are finite sequences, $\langle \alpha_1, \ldots, \alpha_n \rangle$, of **A**-objects, and proto-morphisms are matrices of **A**-morphisms. The source-target predicate is given by

$$\langle \alpha_1, \alpha_2, \ldots, \alpha_n \rangle \xrightarrow{R} \langle \beta_1, \beta_2, \ldots, \beta_m \rangle \quad \text{iff} \quad \Box R_{ij} = \alpha_i,\ R_{ij}\Box = \beta_j \quad \text{for all } i, j\,.$$

The operations are defined by

$$(RS)_{ik} = \bigcup_j R_{ij}S_{jk}\,,$$

$$(R^\circ)_{ij} = (R_{ji})^\circ\,,$$

$$(R \cap S)_{ij} = R_{ij} \cap S_{ij}\,,$$

$$(R \cup S)_{ij} = R_{ij} \cup S_{ij}\,,$$

$$(0_R)_{ij} = 0_{R_{ij}}\,.$$

(Note that the law of modularity in $\mathbf{A}^+$ requires the distributivity of $\cup$ and $\cap$. That is, if we dropped $(R_1 \cup R_2) \cap T = (R_1 \cap T) \cup (R_2 \cap T)$ from the definition of distributive allegory but added positivity then by defining $R = (R_1, R_2)$, $S = (1, 1)^\circ$ the law of modularity, $RS \cap T \subset (R \cap TS^\circ)S$, implies full distributivity.)

Define $\mathbf{A} \to \mathbf{A}^+$ by sending α to $\langle \alpha \rangle$. It is a full and faithful representation of distributive allegories, and $\mathbf{A}^+$ is positive.

If **A** is tabular, then so is $\mathbf{A}^+$: given R in $\mathbf{A}^+$ choose, for each i, j, a tabulation $\langle f_{ij}, g_{ij} \rangle$ of R_{ij}. Let γ_{ij} be the source of f_{ij}. The f_{ij}'s and g_{ij}'s appear as the entries (along with zero-morphisms) of a tabulation $\langle f, g \rangle$ of R with source $\langle \gamma_{11}, \gamma_{12}, \ldots, \gamma_{nm} \rangle$.

2.217. As repeatedly promised in chapter 1:

A pre-logos may be faithfully represented in a positive pre-logos.

BECAUSE: Starting with a pre-logos **C** we may construct its positive reflection as the category of maps in the positive reflection of the allegory of relations in **C**.

(A matrix R is entire iff for each i, $\bigcup_j R_{ij}R^\circ_{ji}$. It is simple iff R_{ij} is simple for each i, j and $R^\circ_{ij}R_{ij'} = 0$ for all i, j, j', $j \neq j'$. We may recast the last condition. $S^\circ T = 0$ iff $\mathscr{Dom}\ S$ and $\mathscr{Dom}\ T$ are disjoint:

$$(\mathscr{Dom}\ S) \cap (\mathscr{Dom}\ T) \subset (\mathscr{Dom}\ S)(\mathscr{Dom}\ T) \subset SS^\circ TT^\circ\,,$$

$$S^\circ T \subset S^\circ(\mathscr{Dom}\ S)(\mathscr{Dom}\ T)T \subset S^\circ[\mathscr{Dom}(S) \cap \mathscr{Dom}(T)]T\,.$$

Hence a matrix is simple iff each entry is simple and the domains of any two entries in the same row are disjoint. It is a map iff each entry is simple and domains of the entries on any row form a partition of 1.)

A pre-logos may be faithfully represented in a pre-topos.

BECAUSE: If **C** is a pre-logos its reflection in the category of pre-topoi is

$$\mathcal{M}ap(\mathcal{S}plit(\mathcal{E}q(\mathcal{R}el(\mathbf{C}))^{+})).$$

2.218. By combining the results [2.167, 2.16(10), 2.213, 2.217, 1.635] we obtain

A small pre-tabular or semi-simple unitary distributive allegory may be faithfully represented in a power of the allegory of sets.

2.219. Semi-simplicity for positive allegories is equivalent to an intriguing analogue arising from functional analysis. Recall that if R is an operator between Hilbert spaces that its *polarization* $R^\circ R$ is symmetric and positive semi-definite. In an allegory $R^\circ R = S$ is symmetric and, corresponding to positivity, $(\mathcal{D}om\ S) \subset S$. Every symmetric positive semi-definite operator on a Hilbert space is the polarization of something.

A positive allegory is semi-simple iff for every S such that $S^\circ = S$ and $(\mathcal{D}om\ S) \subset S$ there exists R such that $S = R^\circ R$.

BECAUSE: Given S as required let F and G be simple, $S = F^\circ G$. Take R as $(F \cup G)(\mathcal{D}om\ S)$.

Conversely, given arbitrary T consider

$$S = \begin{pmatrix} 1 & T \\ T^\circ & 1 \end{pmatrix}.$$

S is symmetric and reflexive, hence there exists R such that $S = R^\circ R$. We may write R as a matrix (F, G). F and G are easily verified to be simple and T is easily verified to be $F^\circ G$.

2.21(10). The equational theory of representable distributive allegories reduces easily to the equational theory of representable allegories. Because union distributes with the other operations every expression is equivalent to one of the form $E_1 \cup E_2 \cup \cdots \cup E_n$ where the E_i's are union-free.

If $E_1 \cup E_2 \cup \cdots \cup E_n \subset E'_1 \cup \cdots \cup E'_m$ in $\mathcal{R}el(\mathcal{S})$, where each E_i, E'_j is union-free then for each i there is a j such that $E_i \subset E'_j$ in $\mathcal{R}el(\mathcal{S})$; if not then there would be an i such for each j there exists a counterexample for $E_i \subset E'_j$; the cartesian product of the counterexamples would yield a counterexample for $E_i \subset E'_1 \cup E'_2 \cup \cdots \cup E'_m$. (Cartesian product yields a representation of allegories $\Pi_m \mathcal{R}el(\mathcal{S}) \to \mathcal{R}el(\mathcal{S})$. It is not a representation of distributive allegories.)

The equations true for $\mathscr{Rel}(\mathscr{S})$ are therefore known iff the union-free equations are known. Indeed, every equation is equivalent to a finite set of union-free equations. Using [2.158] we can therefore conclude that no finite set of equations in the operations of distributive allegories true for $\mathscr{Rel}(\mathscr{S})$ implies all equations true for $\mathscr{Rel}(\mathscr{S})$.

2.22. A distributive allegory is LOCALLY COMPLETE if (α, β) is a complete lattice for all α, β, and if composition and finite intersection distribute over arbitrary unions: that is, given R and $\{S_i\}_I$ one has $R(\bigcup S_i) = \bigcup RS_i$. For empty I we understand this to mean $R0 = 0$.

2.221. Let **A** be an allegory. A subset $\mathscr{A} \subset (\alpha, \beta)$ is a *downdeal* if for all $R \in \mathscr{A}$ and $S \subset R$ it is the case that $S \in \mathscr{A}$. Define $\bar{\mathbf{A}}$, the LOCAL COMPLETION of **A**, to be the allegory whose objects are the same as those of **A**, and whose morphisms are the downdeals. The composition of downdeals $\mathscr{A}$, $\mathscr{B}$ is defined as the downdeal generated by $\{RS \mid R \in \mathscr{A},\ S \in \mathscr{B}\}$. Reciprocation, intersection and union are defined in the obvious manners. $\bar{\mathbf{A}}$ is locally complete.

For $R \in \mathbf{A}$ let (R) denote its corresponding principal downdeal $\{S \mid S \subset R\}$. Then $\mathbf{A} \to \bar{\mathbf{A}}$, defined by sending R to (R), is a faithful representation of allegories. Consequently,

Any allegory may be faithfully represented in a locally complete distributive allegory.

2.222. Let **A** be a distributive allegory. We say that a subset $\mathscr{A} \subset (\alpha, \beta)$ is an *ideal* if it is a downdeal and if, further, it is closed under finite union. The last section can be repeated in terms of ideals. The allegory of ideals is a locally complete distributive allegory, and $\mathbf{A} \to \bar{\mathbf{A}}$ is a faithful representation of distributive allegories. Thus:

Any distributive allegory may be faithfully represented in a locally complete distributive allegory.

2.223. A locally complete distributive allegory is GLOBALLY COMPLETE if every indexed collection of objects has a disjoint union: that is, given $\{\alpha_i\}_I$ there exists $\{U_i\colon \alpha_i \longrightarrow \beta\}_I$ such that $U_iU_i^\circ = 1$ for all i, $U_iU_j^\circ = U_i^\circ U_j = 0$ for all $i \neq j$, and $\bigcup_i U_i^\circ U_i = 1$.

By the obvious extension of the argument in [2.214], we may easily see that disjoint unions in a locally complete distributive allegory coincide with coproducts (and with products).

2.224. Suppose that $\mathbf{A}$ is a locally complete distributive allegory. We construct its GLOBAL COMPLETION $\mathbf{A}^{\Sigma}$ in a manner entirely analogous to the construction of positive reflections of distributive allegories. The objects of $\mathbf{A}^{\Sigma}$ are indexed collections of $\mathbf{A}$-objects. The protomorphisms [1.2] are infinite matrices of $\mathbf{A}$-morphisms. Given a function $R: I \times J \to \mathbf{A}$, R satisfies the source-target predicate $\{\alpha_i\}_I \xrightarrow{R} \{\beta_j\}_J$ iff $R_{ij} \in (\alpha_i, \beta_j)$ for all i, j. The operations are defined by precisely the same formulas that appeared in [2.216].

$\mathbf{A}^{\Sigma}$ is a globally complete allegory, and $\mathbf{A} \to \mathbf{A}^{\Sigma}$, defined by sending R to the 1×1 matrix (R), is a full and faithful representation of locally complete distributive allegories. Hence:

Any locally complete distributive allegory may be faithfully represented in a globally complete allegory.

If $\mathscr{V}$ is a locale, then its global completion (viewing $\mathscr{V}$ as a one-object locally complete distributive allegory) is the allegory composed of $\mathscr{V}$-valued relations between sets [2.111].

2.225. The local completion of a tabular allegory is not, in general, tabular. It need not even be semi-simple. The local completion of a semi-simple allegory has the property that every morphism is the union of the semi-simple morphisms it contains. This property is maintained in the global completion. Better, a globally complete allegory with this property is necessarily semi-simple: given R let $\{F_i\}_I$, $\{G_i\}_I$ be collections of simple morphisms such that $R = \bigcup_I F_i^{\circ} G_i$. Let $\{U_i: \alpha_i \to \beta\}$ be a disjoint union, where α_i denotes the source of F_i (and of G_i). Define F as $\bigcup_I U_i^{\circ} F_i$ and G as $\bigcup_I U_i^{\circ} G_i$. Then F and G are simple and $R = F^{\circ} G$.

2.226. The SYSTEMIC COMPLETION of an allegory is the result of splitting the symmetric idempotents of its global completion. It is easily checked that splitting symmetric idempotents of a globally complete allegory maintains global completeness.

The systemic completion of a semi-simple allegory is tabular and effective (by the last section and [2.16(10)]. It is routine that a unit in a given allegory remains a unit in the systemic completion. In all but deliberately pathological cases, a unit, in fact, will be created in the systemic completion. Every object has a maximal endomorphism which is necessarily an equivalence relation. If M is a maximal endomorphism and if $ff^{\circ} = M$, $f^{\circ}f = 1$, then the target of f is a partial unit. Given any set of partial units we may apply this construction to their coproduct. Hence for every set of partial units there is a partial unit in which they may all be embedded. A systemic completion, therefore, has a unit iff there is only a set of isomorphism types of partial units.

Among the conditions on the original allegory that ensure the existence of a unit in the systemic completion is:

If an allegory **A**, *viewed as a category, has a generating set (for example, if* **A** *is small), then its systemic completion has a unit.*

BECAUSE: It is easy to check that a generating set [1.632] remains such in the systemic completion. We wish to show that there is a maximal partial unit. Suppose that π is such that for each α in the generating set there exists a map $f: \alpha \to \pi$. Then π is maximal: given $U: \pi \to \pi'$, suppose that U were not an isomorphism; since $UU^\circ = 1$ it must be the case that $U^\circ U \neq 1$; let α be an object in the generating set and let $F: \alpha \to \pi'$ be a morphism such that $FU^\circ U \neq F1$; we know that $F = AfU$ for $A = \mathscr{D}om\ F$; but then $FU^\circ U = AfUU^\circ U = AfU = F$.

2.227. *Let Y be a topological space, $\mathscr{O}(Y)$ the locale of open subsets thereof. The category of maps of $\mathscr{O}(Y)$-valued sets is equivalent to $\mathscr{S}h(Y)$.*

BECAUSE: The extent to which sections f, g of a local homeomorphism $X \to Y$ [1.373] are equal is given by $f\,R\,g = \bigcup \{U \subset Y \text{ open} \mid U \subset \Box f \cap \Box g,\ Uf = Ug\}$. The obtained $\mathscr{O}(Y)$-valued set [2.16(12)] is irredundant: $f\,R\,g = f\,R\,f = g\,R\,g$ implies $f = g$. It is in fact a maximal irredundant $\mathscr{O}(Y)$-valued set.

The locale $\mathscr{O}(Y)$ can also be regarded as a category whose objects are open subsets, and whose morphisms are the inclusions between open subsets. Given an $\mathscr{O}(Y)$-valued set $\langle I, R\rangle$ we obtain a left $\mathscr{O}(Y)$-set [1.273, 1.373] by considering the equivalence classes of pairs $\langle i, U\rangle$, where $U \subset (i\,R\,i)$, under the equivalence relation $\langle i, U\rangle \equiv \langle i', U'\rangle$ iff $U = U' \subset (i\,R\,i')$. Let $\Box\langle i, U\rangle = U$, and let $V\langle i, U\rangle = V$ if $V \subset U$. The associated sheaf of this left $\mathscr{O}(Y)$-set [1.373-4] can be made into a maximal irredundant $\mathscr{O}(Y)$-valued set $\langle \hat{I}, \hat{R}\rangle$ as above. In this case, $i\,R\,j = \langle i, (i\,R\,i)\rangle\, \hat{R}\, \langle j, (j\,R\,j)\rangle$ for all $i, j \in I$, and $f\,\hat{R}\,f = \bigcup_{i \in I} f\,\hat{R}\,\langle i, (i\,R\,i)\rangle$. Therefore, every map between $\mathscr{O}(Y)$-valued sets extends uniquely to a map between maximal irredundant $\mathscr{O}(Y)$-valued sets. But for any such $S: \langle \hat{I}, \hat{R}\rangle \to \langle \hat{J}, \hat{T}\rangle$, define a morphism of left $\mathscr{O}(Y)$-sets [1.271, 1.273, 1.374] as $F(f) =$ the unique g such that $g\,\hat{T}\,g' = f\,S\,g'$, for all g'.

2.228. We consider allegories with finite unions which distribute with composition, but not necessarily with intersection.

A tabular allegory with this property is distributive: given $R, S, T \in (\alpha, \beta)$ let $\langle f, g \rangle$ be a tabulation of $R \cup S \cup T$. The poset $\{Q \mid Q \subset f^\circ g\}$ is isomorphic to $\mathscr{Cor}(\Box f)$ (send Q to $1 \cap fQg^\circ$, send $A \in \mathscr{Cor}(\Box f)$ to $f^\circ Ag$). $\mathscr{Cor}(\Box f)$ is a distributive lattice since intersection and composition of coreflexives coincide.

A semi-simple allegory with the above mentioned property is distributive. Split the symmetric idempotents and obtain a tabular allegory with the same property.

An allegory with this property in which, further, every morphism is the union of semi-simple morphisms it contains [2.225] need not be distributive. For any group G consider the one-object allegory obtained by adjoining two new elements 0 and M, where M is understood to be the maximal element. (G retains the discrete ordering.) Composition is defined by using the group multiplication on G and defining $RM = MR = M$ for all R except 0. 0, of course, kills everything. Reciprocation is defined by using the inverse operation on G and taking 0 and M to be symmetric. This results in an allegory with finite unions which distribute with composition.

If G has more than one element, every morphism in the resulting allegory is the union of semi-simple morphisms it contains. If G has more than two elements, the resulting allegory is not distributive.

2.3. DIVISION ALLEGORIES

2.31. A DIVISION ALLEGORY is a distributive allegory with a binary partial operation denoted R/S defined iff $R\Box = S\Box$ such that

$$\Box(R/S) \succeq \Box R \qquad (\text{or } \Box(R/S) = \Box(R(S\Box)))\,,$$

$$(R/S)\Box \succeq \Box S \qquad (\text{or } (R/S)\Box = \Box(S(R\Box)))\,,$$

$$T \subset R/S \quad \text{iff} \quad TS \subset R\,.$$

The source-target information is summarized by

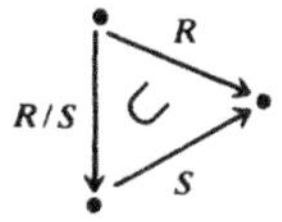

The diagram does not, in general, commute. It does 'semi-commute' (as indicated by the containment sign in the above triangle); that is, $(R/S)S \subset R$. R/S is maximal among morphisms that yield such semi-commutative triangles.

The double-Horn sentence in the above definition may be replaced with the following three containments:

$$(R_1 \cap R_2)/S \subset (R_1/S \cap R_2/S)\,,$$

$$T \subset (TS)/S\,,$$

$$(R/S)S \subset R\,.$$

The double-Horn sentence easily implies each of these three containments. Conversely: given $T \subset R/S$ then $TS \subset (R/S)S \subset R$; given $TS \subset R$ then $T \subset (TS)/S \subset (TS \cap R)/S \subset R/S$.

(The first containment may be replaced with an equality: $(R_1/S \cap R_2/S) \subset (R_1 \cap R_2)/S$ because $(R_1/S \cap R_2/S)S \subset (R_1/S)S \cap (R_2/S)S \subset (R_1 \cap R_2)$.)

2.311. Note that the defining containments for division do not make use of the zero or union operations. They do imply the distributivity of unions with composition. If $\{R_i\}_I$ is a given family of morphisms with a least upper-bound $\bigcup_I R_i$ then $\{R_i S\}$ has a least upper-bound, namely $(\bigcup R_i)S$. ($R_i S \subset T$ (all i) iff $R_i \subset T/S$ (all i) iff $\bigcup R_i \subset T/S$ iff $(\bigcup R_i)S \subset T$.)

2.312. When $\Box R = \Box S$ we may define $S\backslash R$ as $(R^\circ/S^\circ)^\circ$. Then $T \subset S\backslash R$ iff $ST \subset R$ (because $T \subset S\backslash R$ iff $T^\circ \subset R^\circ/S^\circ$ iff $T^\circ S^\circ \subset R$ iff $ST \subset R$):

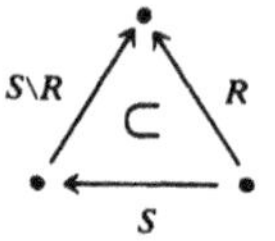

Either left or right division could have been chosen as primitive: each is definable in terms of the other. For relations on sets left-division looks more straightforward: $x(S\backslash R)y$ iff $\forall_z\, (z\, S\, x \Rightarrow z\, R\, y)$. We have chosen right-division as primitive because of its role in the definition of power-allegories [2.4].

2.313. For $\beta \xrightarrow{S} \gamma$ in a division allegory the function $(\alpha, \beta) \xrightarrow{(-)S} (\alpha, \gamma)$ has a right adjoint, namely $(\alpha, \gamma) \xrightarrow{(-)/S} (\alpha, \beta)$. We could have defined division allegories in this manner.

2.314. As observed in [1.784] $\mathscr{R}el(\mathbf{C})$ is a division allegory when $\mathbf{C}$ is a logos. The containments mentioned in (and near) that section are easily seen to hold for any division allegory:

$$(R/S)(S/T) \subset R/T\,,$$

$$R/(S_1S_2) = (R/S_2)/S_1\,,$$

$$1 \subset R/R\,,$$

$$(R/R)^2 \subset R/R\,.$$

Among other equations of note:

$$(R/R)R = R, \qquad R/1 = R\,,$$

$$R/(S_1 \cup S_2) = R/S_1 \cap R/S_2\,,$$

$$S\backslash(R/T) = (S\backslash R)/T\,.$$

The last equation allows us to write, unambiguously, $S\backslash R/T$. The result is the maximal morphism that can be inserted in

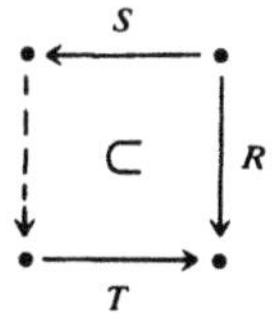

to maintain the indicated semi-commutativity.

Note that $(S\backslash R/T)^\circ = T^\circ\backslash R^\circ/S^\circ$.

2.315. Any locally complete distributive allegory is a division allegory: R/S is constructible as $\bigcup\{T \mid TS \subset R\}$. The process of local completion preserves division. Hence:

Any division allegory is faithfully representable in a locally complete distributive allegory, and thus in a globally complete allegory.

2.316. For an object α in a division allegory, $\mathscr{C}or(\alpha)$ is a Heyting algebra: given $A, B \in \mathscr{C}or(\alpha)$ we may construct $A \longrightarrow B$ as $1 \cap \mathrm{B}/\mathrm{A}$ (or $1 \cap \mathrm{A}\backslash\mathrm{B}$) using the fact that $AB = A \cap B$.

In a tabular unitary division allegory let $\ell: \gamma \longrightarrow \alpha$, $r: \gamma \longrightarrow \beta$ tabulate the maximal morphism from α to β. Then the poset (α, β) is canonically isomorphic to $\mathscr{C}or(\gamma)$ (send $R \in (\alpha, \beta)$ to $(1 \cap \ell R r^\circ) \in \mathscr{C}or(\gamma)$), hence (α, β) is a Heyting algebra.

Note that a Heyting algebra may be construed as a one-object division allegory: define

$$x \cap y = x \wedge y, \qquad xy = x \wedge y, \qquad x^\circ = x\,,$$

$$x \cup y = x \vee y, \qquad x/y = y \rightarrow x\,.$$

Indeed, any one-object division allegory in which the identity morphism is maximal is a Heyting algebra.

2.32. **A** *is a tabular unitary division allegory iff* $\mathscr{M}ap$(**A**) *is a logos.*

BECAUSE: One direction was shown in [1.784].

For the other direction we already know that $\mathscr{M}ap$(**A**) is a pre-logos. We need only verify the double-sharp axiom. Let $f: \alpha \longrightarrow \beta$ be a map. Note that

$$\begin{array}{ccc} \mathscr{Sub}(\beta) & \xrightarrow{f^{\#}} & \mathscr{Sub}(\alpha) \\ \wr\Big| & & \Big|\wr \\ \mathscr{Cor}(\beta) & \xrightarrow[1 \cap f(-)f^{\circ}]{} & \mathscr{Cor}(\alpha) \end{array}$$

where the lower function sends B to $1 \cap fBf^{\circ}$ (the domain of fB). To construct a right adjoint for $f^{\#}$ it suffices to construct one for $1 \cap f(-)f^{\circ}$.

$1 \cap fBf^{\circ} \subset A$ iff $B \subset f\backslash(1 \rightarrow A)/f^{\circ}$ where the arrow operation is the Heyting arrow on (α, α) (*not* $\mathscr{Cor}(\alpha)$) as just constructed.

2.33. The geometric representation theorem [1.74] now applies to countable tabular unitary division allegories. That is, any such allegory may be faithfully represented in a countable power of the allegory of $\mathscr{O}(R)$-valued sets [2.227], where $\mathscr{O}(R)$ is the locale of open sets of reals.

The Stone representation theorem [1.75] implies that any tabular unitary division allegory may be faithfully represented in the allegory of $\mathscr{V}$-valued sets, where $\mathscr{V}$ is the locale of open subsets of some Stone space [2.227].

2.331. It follows from the work of Ieke Moerdijk that:

Let X be a metrizable space without isolated points, $\mathscr{O}(X)$ the locale of open subsets thereof.

Any countable tabular unitary division allegory may be faithfully represented in a countable power of the allegory of $\mathscr{O}(X)$-valued sets.

Any countable logos may be faithfully represented in a countable power of $\mathscr{Sh}(X)$.

Any countable logos with a coprime terminator may be faithfully represented in $\mathscr{Sh}(X)$.

BECAUSE: It suffices to consider the allegory of $\mathscr{O}(2^*)$-valued sets, where $\mathscr{O}(2^*)$ is the locale of open subsets of the binary tree, and show that this allegory may be faithfully represented in the allegory of $\mathscr{O}(X)$-valued sets [1.742–7, 2.227]. We shall, in fact, give an embedding of complete Heyting algebras $\mathscr{O}(2^*) \longrightarrow \mathscr{O}(X)$ [1.723, 2.315]. The argument is due to Ieke Moerdijk.

Using topological facts whose proofs are given below, we can associate a nonempty, open subset $U_w \subset X$ to each finite binary word w by induction on length so that:

(i) X is assigned to the empty word,

(ii) For every n, if the length of w is at least n, then for each $x \in U_w$ there exists $y \in U_w - (U_{w*0} \cup U_{w*1})$ such that the distance between x and y is less than 2^{-n},

(iii) U_{w*0} and U_{w*1} are disjoint subsets of U_w,

(iv) $\mathscr{cl}(U_w) \subset (\mathscr{cl}(U_{w*0}) \cup U_{w*1}) \cap (U_{w*0} \cup \mathscr{cl}(U_{w*1}))$,

(v) $\mathscr{cl}(U_w) - U_w \subset \mathscr{cl}(U_w - (U_{w*0} \cup U_{w*1}))$.

Let $\mathbf{H} = \{\bigcup_{w \in A} U_w \mid A \subset 2^*\}$, and let us write U_A for an element $\bigcup_{w \in A} U_w$ of $\mathbf{H}$. We claim that $\mathbf{H}$ is closed under the complete Heyting algebra operations of $\mathcal{O}(X)$. The non-trivial cases are infinite meets and the arrow operation.

Let $\{A_\iota\}_I$ be a collection of subsets of 2^*. It is clear that $\mathrm{int}(\bigcap_I U_{A_\iota})$ contains $\bigcup\{U_w \mid U_w \subset U_{A_\iota}$; for all $i\}$. For the other direction, suppose that the ball of radius 2^{-n} around x is contained in U_{A_ι}, for all i. By (ii), for each i there exists $w \in A_\iota$, w of length at most n, so that $x \in U_w$. The containment in this direction is now clear.

In $\mathcal{O}(X)$, $U_A \Rightarrow U_B = \bigcup\{V \subset X \text{ open} \mid V \cap U_w \subset U_B \text{ all } w \in A\}$, so the result of the previous paragraph allows us to concentrate on the case when A consists of a single finite binary word w. We show that $U_w \Rightarrow U_B$ is included in $\bigcup\{U_v \mid U_v \cap U_w \subset U_B\}$; the other direction is obvious. It is appropriate to assume that each word in B is longer than w. Let $x \in (U_w \Rightarrow U_B) = \mathrm{int}((X - U_w) \cup U_B)$. If $x \in (X - \mathcal{cl}(U_w))$, the conclusion follows, because there is some U_v disjoint from U_w such that $x \in U_v$ (this is easily shown by induction on the length of w). Otherwise, $x \in \mathcal{cl}(U_B)$. If $x \in U_w$, there is a ball of small radius around x that is contained in both U_w and $(X - U_w) \cup U_B$, hence in U_B. If $x \in (X - U_w)$, then $x \in \mathcal{cl}(U_B) - U_w$, hence $x \in \mathcal{cl}(U_w) - U_w$ because the words in B are longer than w. This property, together with (v), implies that any neighborhood of x meets $U_w - (U_{w*0} \cup U_{w*1}) \subset U_w - U_B$, making $x \in \mathrm{int}(X - U_w) \cup U_B)$ impossible.

The description of the operations of $\mathbf{H}$ shows that it is isomorphic to $\mathcal{O}(2^*)$. It remains to show two topological facts needed in the argument above.

Let (X, d) be a metric space, δ a positive real. $A \subset X$ is δ-DENSE if for each $x \in X$ there exists $a \in A$ such that $d(x, a) < \delta$. $A \subset X$ is NOWHERE δ-DENSE if the distances between different elements of A are at least δ. For each $\delta > 0$, a dense set is δ-dense, and a nowhere δ-dense set is nowhere dense and closed.

For each $\delta > 0$ there exists a δ-dense, nowhere $\delta/2$-dense subset.

BECAUSE: The class of nowhere $\delta/2$-dense subsets is non-empty and closed under unions of chains, hence by Zorn's Lemma (equivalently, by the Axiom of Choice), there exists a maximal nowhere $\delta/2$-dense subset. It is δ-dense.

Let (X, d) be a metric space without isolated points and let U be a non-empty open subset thereof. Then for each $\delta > 0$ there exist disjoint, non-empty, open $V_0, V_1 \subset U$ such that:

(i) *$U - (V_0 \cup V_1)$ is δ-dense as a subset of U,*
(ii) $\mathcal{cl}(U) \subset (\mathcal{cl}(V_0) \cup V_1) \cap (V_0 \cup \mathcal{cl}(V_1))$,
(iii) $\mathcal{cl}(U) - U \subset \mathcal{cl}(U - (V_0 \cup V_1))$.

BECAUSE: (i) and (iii) will be satisfied if we let $A \subset U$ be δ-dense, nowhere $\delta/2$-dense, if we define a sequence $\{F_n\}_{\mathbb{N}}$ of subsets of U for which $U - \mathcal{cl}(\bigcup_{\mathbb{N}} F_n)$ is dense in U, and $\mathcal{cl}(U) - U \subset \mathcal{cl}(\bigcup_{\mathbb{N}} F_n)$, and if we define V_0 and V_1 so that $V_0 \cup V_1 \subset U - (A \cup \mathcal{cl}(\bigcup_{\mathbb{N}} F_n))$.

We write $B(x, \delta)$ for the ball of radius δ around x. Let $\{p_\beta\}_{\beta<\alpha}$ be a well-ordering of $\mathcal{cl}(U) - U$. We define the elements x_β^n of F_n by transfinite induction on $\beta < \alpha$. If $U \cap (B(p_\beta, 2^{-n}) - \mathit{cl}(\bigcup_{\sigma<\beta} B(x^n_\sigma, 2^{-n})))$ is non-empty, let x^n_β be any of its elements, and let $x^n_\beta = x^n_0$ otherwise. For any n:

(1) For each $x \in F_n$ there is $p \in \mathit{cl}(U) - U$ with $d(x, p) < 2^{-n}$,

(2) F_n is a closed, nowhere 2^{-n}-dense subset of U,

(3) For each $p \in \mathit{cl}(U) - U$ there is $y \in F_n$ with $d(p, y) < 2^{-n+1}$.

Let $F = \mathit{cl}(\bigcup_{\mathbb{N}} F_n)$. $\mathit{cl}(U) - U \subset F$ by (3). $U - (F \cup A)$ is a dense open subset of U, and $\bigcup_{\mathbb{N}} F_n \cup A$ a discrete subset of U, closed in U (if $p \in U$ and $\mu > 0$ are such that $B(p, \mu) \subset U$, and if $2^{-n} < \frac{1}{2}\mu$, then by (1), $B(p, \mu) \cap F_k = \emptyset$ for $k \geqslant n$; because A is a nowhere δ-dense, $B(p, \mu) \cap (F \cup A)$ has at most $n + 1$ elements for small enough μ, by (2)).

Let $W = U - (F \cup A)$. We now define disjoint open $W_0, W_1 \subset W$ such that $F \cup A \subset \mathit{cl}(W_0) \cap \mathit{cl}(W_1)$. Because $\bigcup_{\mathbb{N}} F_n \cup A$ is a discrete set closed in U and U is paracompact, for every $x \in \bigcup_{\mathbb{N}} F_n \cup A$ there is $\mu_x > 0$ such that $B(x, \mu_x) \cap B(y, \mu_y) = \emptyset$ whenever x, y are different element of $\bigcup_{\mathbb{N}} F_n \cup A$. Such a point x is not isolated, so there is a sequence $\{U(x, n)\}_{n \in \mathbb{N}}$ of disjoint non-empty open subsets of $B(x, \mu_x) - \{x\}$ such that for each neighborhood W_x of x there exists an n such that for each $k \geqslant n$, $U(x, k) \subset W_x$. Let

$$W^0_x = \bigcup_{\mathbb{N}} U(x, 2n), \qquad W^1_x = \bigcup_{\mathbb{N}} U(x, 2n + 1),$$

and define

$$W_i = \bigcup \{W^i_x \mid x \in \bigcup_{\mathbb{N}} F_n \cup A\}, \quad i = 0, 1.$$

We now let $V_0 = \mathit{int}(\mathit{cl}(W_0))$, $V_1 = W - \mathit{cl}(V_0)$. Because V_0 is regular open, $W - \mathit{cl}(V_1) = V_0$. V_0 and V_1 satisfy (ii): $(\mathit{cl}(V_0) \cup V_1) \cap (V_0 \cup \mathit{cl}(V_1))$ contains both W and $F \cup A$, and hence $\mathit{cl}(U)$.

2.34. *Let* **A** *be a division allegory,* $\mathscr{E}$ *a class of symmetric idempotents therein. Then* $\mathit{Split}(\mathscr{E})$ *is a division allegory. If* $|\mathbf{A}| \subset \mathscr{E}$ *then* $\mathbf{A} \to \mathit{Split}(\mathscr{E})$ *is a faithful representation of division allegories.*

BECAUSE: We already know that $\mathit{Split}(\mathscr{E})$ is a distributive allegory [2.213]. Given $A \xrightarrow{R} C$ and $B \xrightarrow{S} C$ in $\mathit{Split}(\mathscr{E})$ we may easily construct $(A \xrightarrow{R} C)/(B \xrightarrow{S} C)$ as $A \xrightarrow{A(R/S)B} B$. The double-Horn definition of division is readily verified.

Until now the forgetful operation $\mathit{Split}(\mathscr{E}) \to \mathbf{A}$ has preserved all partial operations except for the source and target operations. It does not preserve division. On the other hand, if A and B are identity maps in **A**, then $A(R/S)B = R/S$, hence $\mathbf{A} \to \mathit{Split}(\mathscr{E})$ preserves division.

2.341. If **A** is a pre-tabular division allegory we may therefore represent it in a tabular division allegory, namely $\mathit{Split}(\mathscr{Cor})$. If **A** is semi-simple we may faithfully represent it in $\mathit{Split}(\mathscr{Sd})$.

2.342. *If* **A** *is a division allegory then its positive reflection* $\mathbf{A}^+$ *is a division allegory.*

BECAUSE: Let I, J, K be finite sets, $\{\alpha_i\}_I$, $\{\beta_j\}_J$, $\{\gamma_k\}_K$ be objects in $\mathbf{A}^+$, let R be an $I \times K$ matrix from $\{\alpha_i\}$ to $\{\gamma_k\}$, S a $J \times K$ matrix from $\{\beta_j\}$ to $\{\gamma_k\}$.

Construct R/S as the $I \times J$ matrix such that $(R/S)_{ij} = \bigcap_k (R_{ik})/(S_{jk})$.

2.343. Finally, as repeatedly promised in chapter 1:

Every logos may be faithfully and fully represented in a positive effective logos.

The process is functorial. If **C** is a logos then

$$\mathbf{C} \to \mathscr{Map}(\mathscr{Split}(\mathscr{Eq}(\mathscr{Rel}(\mathbf{C}))^+))$$

is a reflection of **C** among positive effective logoi [2.32, 2.216, 2.169].

2.35. Analogously to the double-arrow operation on Heyting algebras we define in a division allegory a binary partial operation, SYMMETRIC DIVISION,

$$\frac{R}{S} = (R/S) \cap (S/R)^\circ$$

characterized by:

$$T \subset \frac{R}{S} \quad \text{iff} \quad TS \subset R \text{ and } T^\circ R \subset S\,.$$

(In $\mathscr{Rel}(\mathscr{S})$, $x\left(\frac{R}{S}\right)y$ iff x and y are related to the same things: $\forall_z\,(xRz \leftrightarrow ySz)$.)

We may easily verify the containments:

$$\frac{R}{S}\frac{S}{T} \subset \frac{R}{T}, \qquad \frac{R}{S}S \subset R\,,$$
$$\left(\frac{R}{S}\right)^\circ = \frac{S}{R}, \qquad \frac{R}{R}R \subset R\,.$$

2.351. For any R, $\frac{R}{R}$ is symmetric and characterized *among symmetric morphisms* by

$$T \subset \frac{R}{R} \quad \text{iff} \quad TR \subset R\,.$$

Clearly $1 \subset \frac{R}{R}$ and, by semi-cancellation,

$$\left(\frac{R}{R}\right)^2 \subset \frac{R}{R}\,.$$

That is, $\frac{R}{R}$ is an equivalence relation.

A morphism S is STRAIGHT if $\frac{S}{S} = 1$.

(In $\mathscr{R}e\ell(\mathscr{S})$, S is straight iff it relates different things to different things: $\forall_z\,(xSz \leftrightarrow ySz)$ implies $x = y$.)

Note that S is straight iff for every symmetric T such that $TS \subset S$ it is the case that T is coreflexive.

2.352. *If S is straight, and if $fS = gS$, then $f = g$.*

More generally,

If S is straight, F and G simple, and if $FS = GS$, then $(\mathscr{D}om\ F)G = (\mathscr{D}om\ G)F$.

BECAUSE: From $FS = GS$ we obtain $G°FS \subset G°GS \subset S$ and simillary $F°GS \subset S$. Hence if S is straight then

$$G°F \subset \frac{S}{S} \subset 1 \quad \text{and} \quad F°G \subset 1\,.$$

Thus

$$(\mathscr{D}om\ F)G \subset (\mathscr{D}om\ F)(\mathscr{D}om\ G)G \subset (\mathscr{D}om\ G)(\mathscr{D}om\ F)G$$
$$\subset (\mathscr{D}om\ G)FF°G \subset (\mathscr{D}om\ G)F\,.$$

Similarly, $(\mathscr{D}om\ G)F \subset (\mathscr{D}om\ F)G$.

2.353. For division allegories in which every morphism is the union of semi-simple morphisms it contains [2.225], we have a converse: given S such that $FS = GS$ implies $(\mathscr{D}om\ F)G = (\mathscr{D}om\ G)F$ for all simple F, G, then S is straight. It suffices to show that for all simple F, G such that $F°G \subset \frac{S}{S}$ it is the case that $F°G \subset 1$.

Given $F°G \subset \frac{S}{S}$ let $F' = (\mathscr{D}om\ G)F$ and $G' = (\mathscr{D}om\ F)G$.

Then $\mathcal{D}om\ F' = \mathcal{D}om\ G'$ and $F'^{\circ}G' = F^{\circ}G$. It suffices to show that $F' = G'$. But

$$G'S = F'F'^{\circ}G'S \subset F' \frac{S}{S} S \subset F'S .$$

Similarly $F'S \subset G'S$, hence by assumption on S, $F' = G'$.

In a tabular division allegory it suffices to verify the cancellation property on maps:

$$\text{if } \frac{S}{S} = \ell^{\circ}r \quad \text{then} \quad rS \subset \ell\ell^{\circ}rS \subset \ell \frac{S}{S} S \subset \ell S$$

and similarly, $\ell S \subset rS$ hence

$$\ell = r \quad \text{and} \quad \frac{S}{S} = r^{\circ}r = 1 .$$

2.354. *In an effective division allegory every morphism is of the form hS where h is a cover and S is straight.*

BECAUSE: Given R let $hh^{\circ} = \frac{R}{R}$, $h^{\circ}h = 1$. Define S as $h^{\circ}R$.

$$hS = hh^{\circ}R = \frac{R}{R} R = R .$$

S is straight because if T is symmetric and if $TS \subset S$ then

$$hTh^{\circ}R \subset hTS \subset hS \subset hh^{\circ}R \subset \frac{R}{R} R \subset R .$$

hTh° is symmetric, hence

$$hTh^{\circ} \subset \frac{R}{R} \subset hh^{\circ} .$$

Finally

$$T \subset h^{\circ}hTh^{\circ}h \subset h^{\circ}hh^{\circ}h \subset 1 .$$

2.355. *If SR is straight then S is straight.*

BECAUSE: Suppose T is symmetric and $TS \subset S$. Then clearly $T(SR) \subset SR$ and $T \subset 1$. In particular,

If S is right-invertible then it is straight.

2.356. *If S is straight then* $\frac{R}{S}$ *is simple.*

BECAUSE:

$$\left(\frac{R}{S}\right)^{\circ}\left(\frac{R}{S}\right) \subset \frac{S}{R}\frac{R}{S} \subset \frac{S}{S} \subset 1\,.$$

We could, of course, state this as a characterization of straightness: *S is straight iff* $\frac{R}{S}$ *is simple for all R.*

2.357. Note that $\frac{R}{1}$ is not equal to R, unless R is simple. $\frac{R}{1}$ is called the SIMPLE PART of R. Its domain is called the DOMAIN OF SIMPLICITY of R. Note that

$$\mathscr{D}om\left(\frac{R}{S}\right) = \mathscr{D}om(R/S \cap (S/R)^{\circ}) = 1 \cap (R/S)(S/R)\,.$$

Hence $\mathscr{D}om\left(\frac{R}{1}\right) = 1 \cap R(1/R)$.

$\frac{R}{1}$ *is the largest simple morphism of the form* AR, $A \subset 1$.

BECAUSE: If $A \subset 1$ and AR is simple then $AR1 \subset R$ and

$$(AR)^{\circ}R \subset R^{\circ}A^{\circ}AR \subset (AR)^{\circ}(AR) \subset 1 \quad \text{hence} \quad AR \subset \frac{R}{1}\,.$$

For the converse it suffices to show that $\frac{R}{1} = \left(\mathscr{D}om\,\frac{R}{1}\right)R$. $\left(\mathscr{D}om\,\frac{R}{1}\right)R \subset \frac{R}{1}$ because

$$\begin{aligned}\left(\left(\mathscr{D}om\,\frac{R}{1}\right)R\right)^{\circ}R &\subset R^{\circ}(1 \cap (1/R)^{\circ}R)R \\ &\subset R(1/R)^{\circ}((1/R) \cap R)R \\ &\subset (1/R\,R)^{\circ}1/R\,R\end{aligned}$$

and $\left(\mathscr{D}om\,\frac{R}{1}\right)R \subset R$. Clearly $\frac{R}{1} \subset R$, hence $\frac{R}{1} \subset \left(\mathscr{D}om\,\frac{R}{1}\right)R$.

2.4. POWER ALLEGORIES

A first attempt to define power allegories might be as an allegory together with two unary operations, denoted $\ni_R$, pronounced 'epsiloff', and $\Lambda(R)$, subject to equations (in which we write $(\Lambda(R))^\circ$ as $\Lambda^\circ R$):

$$\ni_R\Box = R\Box, \qquad \ni_R = \ni_{R\Box},$$

$$1 \subset \Lambda(R)\Lambda^\circ(R), \qquad \Lambda^\circ(R)\Lambda(R) \subset 1,$$

$$\Lambda(R) \ni_R = R.$$

That is: $\ni_R$ has the same target as R; $\ni_R$ depends only on the target of R (hence could be considered to be an operation from 'objects' to morphisms); $\Lambda(R)$ is a map; every morphism factors as $\Lambda(R)$ followed by $\ni_R$.

We are missing a uniqueness condition. We could add $\Lambda(f \ni_R) \succeq f$.

The venturi-tube is replaceable: $\Lambda(f \ni_R) = f(\Box \ni_R)$.

But such is not an equation, it is only a Horn sentence. If we avoid the notational convention of lower-case italics we are forced to write:

$$\textit{if } 1 \subset SS^\circ \textit{ and } S^\circ S \subset 1 \quad \textit{then} \quad \Lambda(S \ni_R) = S(\Box \ni_R).$$

Alternatively, we could impose the simpler but stronger condition:

$$\textit{if } F^\circ F \subset 1 \quad \textit{then} \quad F(\Box \ni_R) \subset \Lambda(F \ni_R).$$

We have just seen in [2.357] that in a division allegory this last Horn condition may be secured by replacing it with an equation:

$$\frac{R}{1}\left(\Box \ni_R\right) \subset \Lambda\left(\frac{R}{1} \ni_R\right).$$

Given the extra structure, however, of a division allegory, we find that Λ need not be posited as a primitive.

2.41. A POWER ALLEGORY is a division allegory with a unary operation, denoted $\ni_R$, such that:

$$\ni_R\Box = R\Box, \qquad \ni_R = \ni_{R\Box},$$
$$1 \subset (R/\ni_R)(\ni_R/R),$$
$$(\ni_R/\ni_R) \cap (\ni_R/\ni_R)^\circ \subset 1.$$

We have a name for the fourth containment—it says that $\ni_R$ is straight. We will say that the third containment says that $\ni_R$ is THICK.

2.411. In $\mathscr{R}el(\mathscr{S})$ the source of $\ni_R$ may, of course, be taken as the power-set of its target and $\ni_R$, itself, is then the reciprocal of the usual membership relation. The straightness condition is the axiom of extent: whenever a pair of subsets have the same elements they are equal. The thickness condition is the axiom of comprehension: for any binary relation R and element x there exists a subset A such that $y \in A$ iff xRy.

2.412. In almost all formulas the subscript on $\ni$ may be inferred and will be omitted.

Given any R, define $\Lambda(R)$ as $\frac{R}{\ni}$. Since $\ni$ is straight, $\Lambda(R)$ is simple [2.356]. The thickness condition is equivalent with the entireness of $\Lambda(R)$:

$$\mathscr{D}om\left(\frac{R}{\ni}\right) = \mathscr{D}om((R/\ni) \cap (\ni/R)^\circ)$$
$$= 1 \cap (R/\ni)(\ni/R) \qquad [2.357].$$

$\Lambda(R)$, therefore, is a map. Clearly $\Lambda(R)\ni \subset (R/\ni)\ni \subset R$. Finally,

$$R \subset (\mathscr{D}om(\Lambda(R)))R \subset (1 \cap (R/\ni)(\ni/R))R$$
$$\subset ((\ni/R)^\circ \cap (R/\ni))(\ni/R)R \subset \Lambda(R)\ni.$$

Hence, $\Lambda(R)\ni = R$.

The straightness of $\ni$ says, of course, that $\Lambda(R)$ is the unique map such that $\Lambda(R)\ni = R$. Indeed, if F is simple then $F \subset \Lambda(F\ni)$.

2.413. On several occasions we will wish to infer the thickness condition from other facts. Suppose we are given a division allegory with a unary operation $\ni_R$ such that $\ni_R\Box = R\Box$, $\ni_R = \ni_{R\Box}$ and such that for all R there exists map f such that $f\ni_R = R$. Then the thickness condition is a consequence: it suffices to show that $\frac{R}{\ni}$ is entire and for that it suffices to show that $f \subset \frac{R}{\ni}$; clearly $f \ni \subset R$; finally $f^\circ R \subset \ni$ because $f^\circ R \subset f^\circ f \ni \subset \ni$.

2.414. *If* **C** *is a topos then* $\mathscr{Rel}(\mathbf{C})$ *is a power allegory. Conversely, if* **A** *is unitary tabular power allegory then* $\mathscr{Map}(\mathbf{A})$ *is a topos.*

BECAUSE: If A is an object in a topos, **C**, let $\ni \subset [A] \times A$ be its universal relation. The uniqueness condition in the definition of universal relations forces $\ni$ to be straight [1.9]. Using the last section we may infer the thickness condition.

The converse should be clear.

2.415. For any object α, let $[\alpha]$ denote the source of $\ni_{1_\alpha}$. We will call $[\alpha]$ the POWER-OBJECT of α.

The SINGLETON MAP of α is defined as $\Lambda(1_\alpha)$. $\Lambda(1_\alpha)$ is monic:

$$\Lambda(1)\Lambda^\circ(1) \subset \frac{1}{\ni}\,\frac{\ni}{1} \subset \frac{1}{1} \subset 1\,.$$

For any map $\beta \xrightarrow{f} \alpha$, $\Lambda(f) = f\Lambda(1)$ (since $f\Lambda(1)$ is a map and $f\Lambda(1) \ni\ = f$).

2.416. As noted in [1.844], if **C** is a Grothendieck topos (Giraud definition) then $\mathscr{Rel}(\mathbf{C})$ is a complete effective distributive allegory. We used, there, the special adjoint functor theorem to show that $\mathscr{Rel}(\mathbf{C})$ is a power allegory [1.911, 1.842]. We here give a direct allegorical construction of $\ni$.

Let **A** be a complete effective distributive allegory with a progenitor γ [1.966]. (For this construction, **A** need not be semi-simple.) Fix an object α. Let I be the set of morphisms from γ to α and let $\Sigma_I\,\gamma$ be an I-fold copower of γ. There is an obvious morphism, T, from $\Sigma_I\,\gamma$ to α such that for every $R\colon \gamma \to \alpha$ there exists a map $u_\iota\colon \gamma \to \Sigma_I\,\gamma$ such that $R = u_\iota T$. Since **A** is effective there exists a factorization [2.354] $T = gS$ where S is straight. As we shall see, S is also thick.

S is a straight morphism such that for every $R\colon \gamma \to \alpha$ there exists a map f such that $R = fS$. We note, first, that S is maximal in the sense that if $S = hS'$, S' straight, then h is an isomorphism: since hS' is straight, h is straight; straight maps are monic (hh° is symmetric and $(hh^\circ)h \subset h$ hence $hh^\circ \subset 1$); it thus suffices to show that $h^\circ h = 1$; and for that it suffices to show that $Fh^\circ h \supseteq F$ for every simple morphism from γ; let f be such that $FS' = fS$; then $FS' = (fh)S'$ and by [2.352] we have $F = (\mathscr{Dom}\ F)fh$; hence $Fh^\circ h = (\mathscr{Dom}\ F)fhh^\circ h = (\mathscr{Dom}\ F)fh = F$.

Now let R; $\beta \to \alpha$ be arbitrary and let $[\alpha]$ denote the source of S. The morphism $\binom{S}{R}\colon [\alpha] + \beta \to \alpha$ may be straightened: that is, there exists a factorization $\binom{S}{R} = \binom{h}{h'}S'$ where h, h' are maps and S' is straight.

$S = hS'$ and by the above paragraph, h is an isomorphism. Hence $R = h'S' = h'h^{-1}S$ and $h'h^{-1}$ is a map. Thus every morphism targeted at α factors as a map followed by S, which insures that S is thick.

2.417. A generator (as opposed to a progenitor) for **A** is not good enough. In [1.96(10)] we described a category that satisfied all but the generating set condition of Giraud's definition but which failed to be topos. We make a modification in that construction to provide a complete effective distributive allegory with a generator but which is not a power allegory.

Let **C** be the category whose objects are quadruples $\langle S, s\colon S \to S, A, f\colon S \times A \to S\rangle$. We write $f(x, a)$ as x^a. If $a \not\in A$ we understand x^a to mean $s(x)$. In [1.96(10)] s was always taken as the identity map and hence never mentioned.) Given another such object $\langle S', s', A', f'\rangle$ we allow $g\colon S \to S'$ as a map iff $g(x^a) = (gx)^a$ for all $x \in S$ and all $a \in A \cup A'$. **C** satisfies all the Giraud conditions except for the existence of a generating set. And just as in [1.96(10)], the only object in **C** with a power-object is the coterminator.

$\mathscr{Rel}$(**C**) does have a generator: to wit $\langle G, s, \emptyset, \emptyset\rangle$ where $G = \{u, v\}$, $s(u) = s(v) = v$. Given $R, R'\colon S \to S'$ such that R is not contained in R' choose x, y such that xRy but not $xR'y$. Define $T\colon G \to S$ by wTz iff $w = v$ or $z = x$. Then $u(TR)y$ but not $u(TR')y$.

2.418. We recall the category of assemblies [2.153] and its effective reflection [2.16(14)].

Let **K** *be the collection of all recursive partial functions and let* **A** *be the corresponding category of assemblies. Then* $\mathscr{Map}(\mathscr{Split}(\mathscr{Eq}(\mathscr{Rel}\ \mathbf{A})))$ *is a topos.*

BECAUSE: We use the notation from [2.16(14)]. Given an object X/I, we construct its power object $[X]/E$. Let the assembly X' be such that $X'|_k = \{x \in |X|\colon x\, I_k\, x\}$. Let $\Phi(m, -)$ be the m-th partial recursive function. Let $[X]|_n$ consist of all assemblies $S \subset X'$ of modulus $\Phi(n_r, \_)$ such that $\Phi(n\ell, \_)$ is a modulus of $IS \subset S$. Let $S\, E_n\, S'$ iff $S \in [X]|_{n\ell\ell}$, $S' \in [X]|_{n\ell r}$, and $\Phi(n_r, \_)$ is a modulus of $S = S'$. The universal relation $\ni$ with target $[X]/E$:

$$S \ni_n x \quad \text{iff} \quad S \in [X]|_{n\ell} \text{ and } x \in S|_{n_r}\,.$$

This topos is called the REALIZABILITY TOPOS (more precisely, the recursive realizability topos, elsewhere: the Effective Topos). It was first studied from a somewhat different point of view by J.M.E. Hyland. In Appendix B we will discuss interpretations of higher-order logic related to topoi. The interpretation that corresponds to the Realizability Topos is the recursive realizability interpretation studied by S.C. Kleene, G. Kreisel, and A.S. Troelstra.

Consider the assembly **N** whose n-th caucus has n itself as the sole member. **N** is a proper subobject of ∇N. As constant 0 and the successor function s are in **K**, one has the morphisms $1 \to \mathbf{N}$ with modulus 0 and $\mathbf{N} \to \mathbf{N}$ with modulus s. It is easily shown by recursion in **K** that $1 \to \mathbf{N} \to \mathbf{N}$ is a natural numbers object in the Realizability Topos.

2.42. The splitting lemmas

If **A** *is a power-allegory then* $\mathscr{Split}(\mathscr{Cor})$ *is a power-allegory and* $\mathbf{A} \to \mathscr{Split}(\mathscr{Cor})$ *is a representation of power-allegories.*

BECAUSE: Given coreflexive A, let $[A]$ be the image of $\Lambda(\ni A)$ (that is, $[A] = \Lambda^\circ(\ni A)\Lambda(\ni A))$. Then $\ni_A$ is constructible as $[A]\xrightarrow{[A]\ni} A$. First, note that $[A]\ni\ = \Lambda^\circ(\ni A)\ni A$, hence that $[A]\ni\ = [A]\ni A$. Given symmetric $[A]\xrightarrow{S}[A]$ such that $S[A]\ni\ \subset [A]\ni$ we have $S\ni\ \subset S[A]\ni\ \subset [A]\ni\ \subset\ \ni$ hence $S \subset 1$ (because $\ni$ is straight). Thus $[A]\xrightarrow{S}[A] \subset [A]\xrightarrow{[A]}[A]$ (since $S = S[A]$) and $[A]\xrightarrow{[A]\ni} A$ is straight.

From the diagram

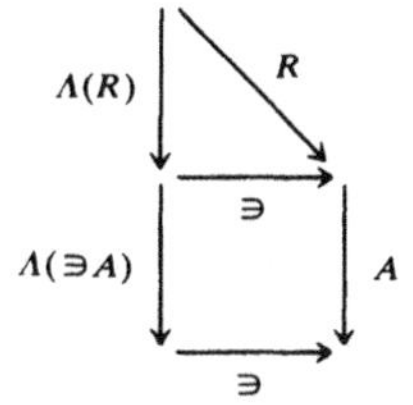

we may infer that

$$B\xrightarrow{R} A = B\xrightarrow{B\Lambda(R)\Lambda(\ni A)}[A]\xrightarrow{[A]\ni} A\,.$$

$B\xrightarrow{B\Lambda(R)\Lambda(\ni A)}[A]$ is a map. By [2.413] we are done.

2.421. *In a power allegory,* $\dfrac{R}{S} = \Lambda(R)\Lambda^\circ(S)$.

BECAUSE:

$$\Lambda(R)\Lambda^\circ(S) \subset \frac{R}{\ni}\,\frac{\ni}{S} \subset \frac{R}{S}\,,$$

$$\frac{R}{S} \subset \frac{R}{S}\,1 \subset \frac{R}{S}\,\Lambda(S)\Lambda^\circ(S) \subset \frac{R}{S}\,\frac{S}{\ni}\,\frac{\ni}{S} \subset \frac{R}{\ni}\,\frac{\ni}{S}$$
$$\subset \Lambda(R)\Lambda^\circ(S)\,.$$

2.422. In any division allegory,

$$E = \frac{E}{E} \quad \text{for any equivalence relation } E.$$

The last section, therefore, says that in a power allegory every equivalence relation is of the form ff°. Hence if coreflexives split then every equivalence relation is effective:

Let **A** *be a power allegory. Then* $\mathscr{Split}(\mathscr{Cor})$ *is an effective power allegory.*

2.423. *If* **A** *is a connected power allegory in which coreflexives split then it has a unit.*

BECAUSE: Given an object α define $M = 1_\alpha/0_\alpha$. M is clearly maximal in (α, α). Let f be a map such that $ff^\circ = M$, $f^\circ f = 1$. The target of f is a partial unit: given $R \subset f\Box$ then $R \subset f^\circ fRf^\circ f \subset f^\circ Mf \subset f^\circ ff^\circ f \subset f\Box$.

Suppose α is a power object. Connectivity (which in any allegory implies strong connectivity) says that every object has a map to α. Hence if π is a partial unit, as just constructed, such that there is a map from α to π, then π is a unit.

2.424. In particular, therefore:

If **A** *is a connected semi-simple power allegory then* $\mathscr{S}\!plit(\mathscr{C}\!or)$ *is a tabular unitary power allegory and* $\mathscr{M}\!ap(\mathscr{S}\!plit(\mathscr{C}\!or))$ *is a topos. Consequently,* $\mathscr{S}\!plit(\mathscr{C}\!or)$ *is also positive, effective and transitive.*

2.43. Pre-power allegories and the diagonal proofs

For a variety of technical reasons we define a PRE-POWER ALLEGORY as a division allegory in which each object appears as the target of a thick morphism.

2.431. It will be convenient to have a characterization of thickness that does not directly use division:

T is thick iff for all R such that $R\Box = T\Box$ there exists $\hat{R}$ such that

$$1 \subset \hat{R}\hat{R}^\circ, \qquad \hat{R}T \subset R, \qquad \hat{R}^\circ R \subset T\,.$$

BECAUSE: Given such $\hat{R}$, then $\hat{R} \subset \frac{R}{T}$ (that is, $\hat{R}T \subset R$ and $\hat{R}^\circ R \subset T$). Since $\hat{R}$ is entire, $\frac{R}{T}$ is entire, hence by definiton, T is thick.

Given the entireness of $\frac{R}{T}$ we may take $\hat{R}$ as $\frac{R}{T}$. Then

$$\hat{R}T \subset \frac{R}{T}T \subset R \quad \text{and} \quad \hat{R}^\circ R \subset \frac{T}{R}R \subset T\,.$$

(Note that the above containments for $\hat{R}$ imply that $\hat{R}T = R$ because $R \subset (\mathscr{D}\!om\ \hat{R})R \subset \hat{R}\hat{R}^\circ R \subset \hat{R}T$.)

2.432. *An effective pre-power allegory is a power allegory.*

BECAUSE: Given thick T may we factor it as hS where $S = h^\circ T$ is straight [2.354]. It suffices to show that S remains thick.

Given $R\Box = S\Box$ define $\hat{R}$ as $\frac{R}{T}h$. Then $\hat{R}$ is clearly entire (both $\frac{R}{T}$ and h are entire).

$$\hat{R}S = \frac{R}{T}hS \subset \frac{R}{T}T \subset R \quad \text{and} \quad \hat{R}^{\circ}R \subset h^{\circ}\frac{T}{R}R \subset h^{\circ}T \subset S\,.$$

2.433. *If* **A** *is a pre-power allegory,* $\mathscr{E}q$ *its class of equivalence relations then* $\mathscr{S}plit(\mathscr{E}q)$ *is a power allegory.*

BECAUSE: By the last section it suffices to show that $\mathscr{S}plit(\mathscr{E}q)$ is a pre-power allegory (since it is automatically effective [2.169]). Given an equivalence relation E let T be a thick morphism in **A** such that $T\Box = E\Box$. It suffices to show that $\Box T \xrightarrow{TE} E$ is thick in $\mathscr{S}plit(\mathscr{E}q)$. Given $E' \xrightarrow{R} E$ define $\hat{R} = E'\frac{R}{T}$. $\hat{R}$ is clearly entire.

$$\hat{R}TE \subset E'\frac{R}{T}TE \subset E'RE \subset R \quad \text{and}$$

$$\hat{R}^{\circ}R \subset \frac{T}{R}E'R \subset \frac{T}{R}RE \subset TE\,.$$

2.434. We are thus provided with the softest proof yet of the existence of power objects in a Grothendieck topos:

The systemic completion of a small locally complete distributive allegory is a power allegory.

BECAUSE: We may reduce to the case of a one-object locally complete distributive allegory **A**. The global completion is easily seen to be a pre-power allegory: given a set I let $[I]$ be the set of functions from I to **A** and let T be the $[I] \times I$ matrix defined via evaluation, that is, $T_{f,i} = f(i)$; given any $J \times I$ matrix R define $\hat{R}$ as the $J \times [I]$ matrix such that $\hat{R}_{j,f} = 1$ or 0 depending on whether or not $R_{j,i} = f(i)$ all $i \in I$. $\hat{R}$ is a map (each row has exactly one 1). $\hat{R}T = R$ and $\hat{R}^{\circ}R = \hat{R}^{\circ}\hat{R}T \subset T$ (because $\hat{R}$ is simple).

Hence the effective reflection of the global completion of **A** is a power allegory. Now split the coreflexive morphisms to obtain the systemic completion.

(2.435.) Cantor showed that for any set A, its power-object $[A]$ cannot appear as a sub-quotient of A. Except for the degenerate case, the same is true in any topos. We convert the problem to an algebraic one.

Suppose that in a power allegory there exists a morphism $F\colon A \to [A]$ such that $F^\circ F = 1$ (a partial map that covers $[A]$). Then $T = F \ni$ is thick: given R define $\hat{R}$ *as* $\frac{R}{\ni} F^\circ$; $\hat{R}$ is clearly entire;

$$\hat{R}T \subset \frac{R}{\ni} F^\circ F \ni \subset \frac{R}{\ni} \ni \subset R,$$

$$\hat{R}^\circ R \subset F \frac{\ni}{R} R \subset F \ni \subset T.$$

(We used only the thickness of $\ni$).

We would thus obtain an object with a thick endomorphism. But,

If a connected division allegory has a thick endomorphism, then it is equivalent to the one-object one-morphism allegory.

BECAUSE: It suffices to show that if T is a thick endomorphism on α, then $T = 1$, because then for any object β, the morphism $0\colon \beta \to \alpha$ factors as an entire morphism $\hat{0}$ followed by 1; that is 0 is entire; hence every object is a terminator.

The allegory of endomorphisms of α would be, by itself, a pre-power allegory. It clearly suffices to show:

(2.436). *The equational theory of one-object pre-power allegories is inconsistent.*

BECAUSE: We may recreate Cantor's proof by first defining R as $0/(1 \cap T)$, where T is thick. R is characterized by $S \subset R$ iff $S(1 \cap T) = 0$.

Let $\hat{R}$ be such that

$$1 \subset \hat{R}\hat{R}^\circ, \qquad \hat{R}T \subset R, \qquad \hat{R}^\circ R \subset T.$$

as insured by [2.431]. Then

$$\hat{R}(1 \cap T) \subset \hat{R}(1 \cap T)^2 \subset \hat{R}T(1 \cap T) \subset R(1 \cap T) \subset 0$$

hence $\hat{R} \subset R$. Thus

$$\hat{R} \subset \hat{R}(1 \cap \hat{R}^\circ \hat{R}) \subset R(1 \cap \hat{R}^\circ R) \subset R(1 \cap T) \subset 0.$$

The entireness of $\hat{R}$ now forces $1 \subset 0$, which, of course, forces any pair of morphisms to be equal.

2.437. The inconsistency of one-object pre-power allegories has two other classical theorems as corollaries.

Let **A** be the one-object allegory of recursively enumerable relations on the natural numbers. We may recursively enumerate all 'recipes' for recursive partial functions. That is, there exists a morphism T such that for any recipe for a recursively enumerable set A there exists n such that $A = \{m \mid nTm\}$. Given any description of a specific morphism R we may thus define a map $\hat{R}$ such that $\hat{R}T = R$ Clearly then

$$1 \subset \hat{R}\hat{R}^{\circ}, \ \hat{R}T \subset R \, , \ \hat{R}^{\circ}R \subset \hat{R}^{\circ}\hat{R}T \subset T \, .$$

Hence **A** cannot be a division allegory. If complements of recursively enumerable sets were also recusively enumerable then **A** would be boolean, hence a division allegory. Thus there are recursively enumerable sets which are not recursive (specifically, the set which corresponds to the coreflexive morphism $1 \cap T$).

2.438. Suppose that there were a recursive set of axioms for elementary number theory strong enough to yield Peano axioms [B.316–8] and strong enough so that we can prove that the set of Gödel-numbers of its provable consequences is a proper subset of the set of Gödel-numbers of well-formed sentences. Clearly such a proof does not really prove consistency. The remarkable theorem of Gödel is that, suitably interpreted, such a proof in fact shows inconsistency. A precise statement of the phase "suitably interpreted" is beyond the scope of this work. Suffice it to say that there is a formulation that allows an argument as follows.

Let **A** be the one-object allegory whose morphisms are named by recipes for recursively enumerable relations, two of them naming the same morphism if they are *provably* coextensive. We wish to show that **A** has only one morphism.

It is possible to show directly that **A** is a pre-power allegory but we shall, instead, show that it has just enough structure to allow the inconsistency demonstration of [2.436]. We may take T and the unary operation, $\hat{R}$, as described in the last section. It is still the case that

$$1 \subset \hat{R}\hat{R}^{\circ}, \qquad \hat{R}T \subset R \ \text{ and } \ \hat{R}^{\circ}R \subset T \, .$$

We do not need division in general, we need only show the existence of $0/(1 \cap T)$. By listing all proofs we can find a recipe for a morphism, R, such that mRn iff it is provably *not* the case that nTn. Clearly if $S(1 \cap T) = 0$ then $S \subset R$. The hypothesized existence of a simulated consistency proof directly allows us to prove that $R(1 \cap T) = 0$, hence that $R(1 \cap T) = 0$ *in* **A**.

Just as in [2.436] we can now show that $\hat{R} = 0$, hence that $1 = 0$.

2.44. The law of metonymy

The free power allegory on one object generator has a sequence of objects, say, $0, 1, 2, \ldots$ and a sequence of generating morphisms $(n+1) \xrightarrow{\ni_n} n$. The morphisms from n to m are named by expressions in the partial operations of division allegories, two such expressions naming the same morphism if, of course, the defining equations of power allegories force them to.

If we split the coreflexive morphisms a much less wieldy structure seems to emerge. It is an effective allegory but not tabular: free power allegories are not semi-simple [2.16(10), 2.225]. We shall discover that one further equation forces the semi-simplicity of the free power allegory on one object, that is, what used to be an existential condition becomes, in the presence of $\ni$, an equational condition.

2.441. As we shall see, the free power allegory on one object, though not positive, becomes positive when the coreflexive morphisms are split. First:

An allegory is PRE-POSITIVE if for every pair of objects $\langle \alpha, \beta \rangle$ there exists a pair of maps $\langle \ell, r \rangle$ such that

$$\ell\ell^\circ = 1_\alpha, \qquad rr^\circ = 1_\beta,$$

$$\ell r^\circ = 0 \qquad (\text{equivalently, } r^\circ \ell = 0).$$

Clearly, if an allegory is pre-positive then $\mathscr{S}plit(\mathscr{C}or)$ remains pre-positive. If the coreflexives split in a pre-positive allegory then it is positive ($\alpha + \beta$ is constructible as $\mathrm{Im}(\ell) \cup \mathrm{Im}(r)$). Hence a distributive allegory is pre-positive iff $\mathscr{S}plit(\mathscr{C}or)$ is positive.

A category is WELL-JOINED if for every pair of objects A, B there exists

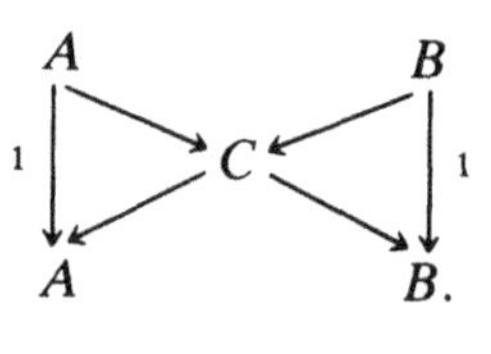

The free power allegory on one object is well-joined via its singleton maps and their reciprocals.

The following are equivalent properties for power allegories:

(1) *Pre-positive.*

(2) *Well-joined.*

(3) *For every pair of objects* α, β *there exists* $\alpha \xrightarrow{S_1} \gamma \xleftarrow{S_2} \beta$ *where* S_1 *and* S_2 *are straight.*

(4) *Connected and every morphism is of the form SF where S is straight and F is simple.*

BECAUSE: (1) obviously implies (2) (in any distributive allegory) and using the fact that right-invertibility implies straight [2.355], (2) implies (3) (in any division allegory).

To show that (3) implies (1) let $\alpha \xrightarrow{S_1} \gamma \xleftarrow{S_2} \beta$ be straight morphisms. We obtain maps $\alpha \xrightarrow{\Lambda(S_1)} [\gamma] \xleftarrow{\Lambda(S_2)} \beta$. But $\Lambda(S)$ is monic whenever S is straight:

$$\Lambda(S)\Lambda^\circ(S) = \frac{S}{\ni}\frac{\ni}{S} \subset \frac{S}{S}.$$

It suffices, therefore, to find $\ell, \mathscr{r}$ such that

$$\ell\ell^\circ = 1_{[\gamma]} = \mathscr{r}\mathscr{r}^\circ \quad \text{and} \quad \ell\mathscr{r}^\circ = 0.$$

For an arbitrary object α (e.g. $\alpha = [\gamma]$) let $\ni$ denote $\ni_\alpha$ and $\ni'$ denote $\ni_{[\alpha]}$.

$$\text{Define } \ell \text{ as } \frac{1}{\ni}\frac{1}{\ni'} \quad \text{and} \quad \mathscr{r} \text{ as } \frac{1/\ni}{\ni'}.$$

ℓ is clearly monic (being the composition of two monics). $1/\ni$ has a right-inverse $((1/\ni)\ni \subset 1$ and $1 \subset \frac{1}{\ni}\ni \subset (1/\ni)\ni)$ hence $1/\ni$ is straight and $\mathscr{r}$ is monic. Thus $\ell\ell^\circ = 1$ and $\mathscr{r}\mathscr{r}^\circ = 1$. To show that $\ell\mathscr{r}^\circ = 0$ it suffices to verify that

$$\frac{1}{\ni'}\frac{\ni'}{1/\ni} = 0.$$

But

$$\frac{1}{\ni'}\frac{\ni'}{1/\ni} \subset \frac{1}{1/\ni} \subset 1/(1/\ni)$$

and it suffices to show that $1/(1/\ni) = 0$. Clearly $\Lambda(0) \subset 1/\ni$ and $\Lambda(1) \subset 1/\ni$. Hence $1/(1/\ni) \subset 1/\Lambda(0) \cap 1/\Lambda(1)$. But for any map f, $R/f = Rf^\circ$, hence it suffices to show that $\Lambda^\circ(0) \cap \Lambda^\circ(1) = 0$, equivalently that $\Lambda(0) \cap \Lambda(1) = 0$. But

$$\mathscr{Dom}(\Lambda(0) \cap \Lambda(1)) \subset \Lambda(0)\Lambda^\circ(1) \subset \frac{0}{\ni}\frac{\ni}{1} \subset \frac{0}{1} \subset 0/1 \subset 0.$$

(In $\mathscr{Rel}(\mathscr{S})$, ℓ *sends* x to $\{\{x\}\}$ and $\mathscr{r}$ sends x to $\{\{x\}, \emptyset\}$. Following [2.357], $\frac{1/\ni}{1}$ is the simple part of $1/\ni$. But for every x both $x\ (1/\ni)\ \{x\}$ and $x\ (1/\ni)\emptyset$.)

(1) implies (4) in any division allegory: given R let $\ell, \mathscr{r}$ be such that $\ell\ell^\circ = \square R$, $\mathscr{r}\mathscr{r}^\circ = R\square$, $\ell\mathscr{r}^\circ = 0$. Define S as $\ell \cup R\mathscr{r}$ and F as $\mathscr{r}^\circ$. Clearly $SF = R$ and F is simple. S is straight because it has a right-inverse, ℓ°.

(4) implies (3) as follows: given α, β there exists a map $f: \alpha \to [\beta]$ because, by assumption, the allegory is connected and there exists a morphism $R: \alpha \to \beta$ and we may take f as $\Lambda(R)$. Factor f° as S_0F, where S_0 is straight and F is simple. Since f is entire, F° is entire, hence $F^\circ F = 1$. We take S_1 as F°; it is straight because it has a right-inverse. Define $S_2 = \Lambda(1)S_0$. S_2 is straight because for a monic g and straight S, gS is straight: if $T = T^\circ$ and $T(gS) \subset gS$, then $(g^\circ Tg)S \subset g^\circ gS \subset S$, and $g^\circ Tg \subset \frac{S}{S} \subset 1$, therefore $T \subset gg^\circ Tgg^\circ \subset gg^\circ gg^\circ \subset 1$.

2.442. Given an object α in a power allegory let $\ni$ denote $\ni_\alpha$ and $\supset$ denote $\ni/\ni$. $\supset$ is a transitive, reflexive endomorphism on $[\alpha]$. The straightness of $\ni$ says, directly, that $\supset$ is a partial ordering (not just a pre-ordering).

Let $\in$ denote $\ni^\circ$, $\ni'$ denote $\ni_{[\alpha]}$ and $\in'$ denote $(\ni')^\circ$. Define $\bigcup$ as $\Lambda(\ni'\ni)$ and define $\bigcap$ as $\Lambda(\in'\backslash\ni)$. Both $\bigcup$ and $\bigcap$ are maps from the second power object of α to its first.

(In $\mathscr{Rel}(\mathscr{S})$, given $\mathscr{F}' \in [[\alpha]]$ and $x \in \alpha$ then $\mathscr{F}(\ni'\ni)x$ iff there exists a set A such that $A \in \mathscr{F}$ and $x \in A$. $\bigcup\mathscr{F} = \{x \mid \exists_A\, x \in A \in \mathscr{F}\}$. In other words, the usual definition. $\mathscr{F}(\in'\backslash\ni)x$ iff for all $A \in \mathscr{F}$ it is the case that $x \in A$. $\bigcap\mathscr{F} = \{x \mid \forall A\, A \in \mathscr{F} \to x \in A\}$, again the usual definition.)

The LAW OF METONYMY is the containment

$$\supset \;\subset\; \bigcup{}^\circ \bigcap .$$

(In $\mathscr{Rel}(\mathscr{S})$ this says, simply, that for any pair of sets, one containing the other, there exists a family of sets whose union is the larger and whose intersection is the smaller.)

A pre-positive power allegory is semi-simple iff it obeys the law of metonymy.

BECAUSE: Given the law of metonymy we begin with the fact that $\supset$ is semi-simple. (Any morphism contained in a semi-simple one is itself semi-simple [2.16(10)]). If R is semi-simple and F is simple, then clearly RF is semi-simple. We may thus infer that $\ni$ is semi-simple by showing that $\ni = \supset \Lambda^\circ(1)$:

$$\sqsupset \Lambda^\circ(1) \subset (\ni/\ni)\frac{\ni}{1} \subset (\ni/\ni)\ni \subset \ni\,,$$

$$\ni \subset \ni\Lambda(1)\Lambda^\circ(1) \subset \sqsupset \Lambda^\circ(1) \qquad \text{since } \ni \Lambda(1) \subset \ni/\ni\,.$$

For any straight morphism S, $\Lambda^\circ(S)$ is simple since

$$\Lambda(S)\Lambda^\circ(S) = \frac{S}{\ni}\frac{\ni}{S} \subset \frac{S}{S}\,.$$

Hence $S = \Lambda(S)\ni$ is semi-simple.

By the last section, pre-positivity implies that every morphism is of the form SF where S is straight and F is simple. Since S is now known to be semi-simple we may conclude that every morphism is semi-simple.

The converse, which we will not need directly, is given in the next section.

2.443. If a connected power-allegory is semi-simple we know that $\mathcal{S}plit(\mathcal{C}or)$ is tabular and unitary, hence $\mathcal{M}ap(\mathcal{S}plit(\mathcal{C}or))$ is a topos. The capitalization lemma allows us, therefore, to specialize to the allegory of relations of a capital topos. The argument for the law of metonymy holding in $\mathcal{R}el(\mathcal{S})$ translates, without much difficulty, to this more general case.

Mostly for the exercise, but partly because we obtain a slightly sharper result, we have given an equational proof of

$$\sqsupset \subset \bigcup{}^\circ \bigcap \quad \textit{iff} \quad \sqsupset \textit{ is semi-simple iff } \sqsupset$$
is the union of semi-simple morphisms it contains.

BECAUSE: If $\sqsupset$ is the union of semi-simple morphisms it contains, then in $\mathcal{S}plit(\mathcal{C}or)$ it is the union of morphisms of the form $f^\circ g$. It suffices to show that $f^\circ g \subset \sqsupset$ implies $f^\circ g \subset \bigcup{}^\circ \bigcap$.

$f^\circ g \subset \sqsupset$ implies that $g\ni \subset ff^\circ g\ni \subset f(\ni/\ni)\ni \subset f\ni$ (and, conversely, if $g\ni \subset f\ni$ then $f^\circ g\ni \subset f^\circ f\ni \supset \ni$ hence $f^\circ g \subset \ni/\ni$). Given $g\ni \subset f\ni$ we wish to show that $f^\circ g \subset \bigcup{}^\circ \bigcap$.

Let $A = \Lambda^\circ(f \cup g)\Lambda(f \cup g) \subset 1$. We will show that $g\ni \subset f\ni$ implies $f^\circ g \subset \bigcup{}^\circ \bigcap$ by showing that $g\ni \subset f\ni$ implies $f = \Lambda(f \cup g)\bigcup$ and $g = \Lambda(f \cup g)\bigcap$. It suffices to show for arbitrary f, g that

$$\Lambda(f \cup g)\bigcup = \Lambda(f\ni \cup g\ni) \quad \text{and}$$

$$\Lambda(f \cup g)\bigcap = \Lambda(f\ni \cap g\ni)\,.$$

$\Lambda(f \cup g)\bigcup$ is a map, hence Λ of something, namely of

$$\Lambda(f \cup g)\bigcup\ni = \Lambda(f \cup g)\ni'\ni = (f \cup g)\ni = f\ni \cup g\ni\,.$$

Similarly, for the remaining equation it suffices to show that

$$\Lambda(f \cup g) \bigcap \ni = f \ni \cap g \ni .$$

Using that $R/f = Rf^{\circ}$ we obtain

$$\begin{aligned}\Lambda(f \cup g) \bigcap \ni &\subset \Lambda(f \cup g) \in' \backslash \ni \subset (\ni'/(f \cup g))^{\circ} \in' \backslash \ni \\ &\subset (\in / \ni' \ni'/(f \cup g))^{\circ} \subset (\in /(f \cup g))^{\circ} \\ &\subset (\in / f \cap \in / g)^{\circ} \subset (\in f^{\circ} \cap \in g^{\circ})^{\circ} \subset f \ni \cap g \ni .\end{aligned}$$

For the other containment, note first

$$\begin{aligned}\in' \Lambda^{\circ}(f \cup g)(f \ni \cap g \ni) &\subset (\Lambda(f \cup g) \ni')^{\circ}(f \ni \cap g \ni) \\ &\subset (f \cup g)^{\circ}(f \ni \cap g \ni) \subset f^{\circ} f \ni \cup g^{\circ} g \ni \\ &\subset \ni .\end{aligned}$$

Hence

$$\Lambda^{\circ}(f \cup g)(f \ni \cap g \ni) \subset \in' \backslash \ni .$$

Finally, therefore

$$\begin{aligned}f \ni \cap g \ni &\subset \Lambda(f \cup g) \Lambda^{\circ}(f \cup g)(f \ni \cap g \ni) \\ &\subset \Lambda(f \cup g) \in' \backslash \ni \subset \Lambda(f \cup g) \bigcap \ni .\end{aligned}$$

2.444. The law of metonymy is not a consequence of the other equations. Consider the one-object locally complete distributive allegory with three morphisms $0 \subset 1 \subset M$ $(M^2 = M)$. If its systemic completion were metonymic then it would be tabular. A tabulation for M would force M to be semi-simple in the global completion which would force M to be the union of semi-simple morphisms it contains in the original one-object allegory, which it clearly is not.

It is not necessary to reach all the way to the systemic completion. First consider the effective completion: it has two objects which we shall denote as α and λ; α has the original three endomorphisms $0, 1, M$; λ has two endomorphisms 0 and 1; there are just two morphisms from α to λ, 0 and p; there are just two morphisms from λ to α, 0 and p°;

$$p^{\circ}p = 1, \quad pp^{\circ} = M, \quad Mp = p, \quad p^{\circ}M = p^{\circ}, \quad M^2 = M ;$$

all other compositions involve either 0 or 1.

The objects of the positive reflection are of the form $m\alpha + n\lambda$. We may construct the power-object of $m\alpha + n\lambda$ as $(3^m - 2^n)\alpha + 2^{m+n}\lambda$. $\ni$ is characterized as the matrix with no repeated rows, in which every row with an M has at least one 1. Direct computation of $\supset$ on 2α yields a matrix in which M appears twice. Direct computation of $\cup \circ \cap$ yields a matrix with no M's. If we were now to impose the law of metonymy we would force $M = 1$ and obtain an allegory equivalent to the allegory of finite sets. But $\ni_\alpha = 3\alpha$ and if we maintain the straightness equation for $\ni$ we force $0 = 1$ and obtain the degenerate allegory.

2.445. We have just avoided introducing further algebraic operations that would insure tabularity and positivity for several reasons. Originally we were dissuaded by the non-canonicity of products and disjoint unions in $\mathscr{S}$. If we impose a particular binary operation, say, that delivers disjoint unions in $\mathscr{S}$, we complicate the representation theory of positive allegories—we must now preserve the specific choices, a task well beyond even the best representable functors.

We take the comparative simplicity of the free metonymic allegory on one object—and the fact that the simple process of splitting coreflexives yields a tabular positive allegory—as confirmation of the original decision.

(We could have, however, safely made connectivity in distributive allegories algebraic by defining 0 as a binary, rather than unary, operation, alternatively, by now adding a binary operation denoted, say, as ${}_R0_S$ with equations $\Box({}_R0_S) = \Box R$, $({}_R0_S)\Box = S\Box$, $0_{{}_R0_S} = {}_R0_S$.)

Tabularity is not a consequence of any reasonable set of equations on power-allegory operations. It suffices to exhibit a non-tabular algebraic substructure of a tabular power-allegory: e.g. the allegory of non-empty sets.

The same type of argument shows the same sort of thing for positivity: this time consider the allegory of all sets with other than three elements.

Pre-positivity requires more. Choose an uncountable cardinal κ such that $|I| < \kappa$ implies $|2^I| < \kappa$, e.g. the countable union of the iterated power-sets of the natural numbers. Arbitrarily divide the population into two classes, upper and lower: the lower class consisting of all those sets of cardinality less than κ. We now violate nature—by prohibiting almost all relations between different classes: if A and B are from different classes then $R: A \to B$ is legal iff it is *stilted*, that is, iff there exists a *finite* set C and a factorization $R = (A \to C \to B)$. Anything goes between sets from like class.

Stilted relations obviously form a two-sided ideal with respect to composition. Because power-objects of finite sets are finite every stilted relation is of the form fR where f has a finite target. Stilted relations do not form an ideal with respect to order (both the minimal and maximal morphisms between any two sets are stilted). They are closed, however, with respect to the boolean operations:

$$(R_1R_2 \cup S_1S_2) = (R_1, R_2)\binom{S_1}{S_2} \quad \text{and} \quad \neg(fR) = f(\neg R)\,.$$

We thus have a suballegory of $\mathscr{R}el(\mathscr{S})$.

Stilted relations form an ideal with respect to division ($(fR)/S = f(R/S)$, $R/(fS) = (R/S)f^\circ$) hence this allegory allows all sorts of division. Since each class is closed under power-set formation, we obtain an algebraic substructure of the allegory of sets. But sets of different class are prohibited, by law, from a common embedding.

A modification yields a completely distributive example: start with the allegory of relations between $\mathbb{Z}_2$-sets and define a stilted relation as one that factors through a *discrete* $\mathbb{Z}_2$-set. Then two objects from different classes are prevented from a common embedding (unless of course each is discrete).

2.446. We did not need the zero or union operations in defining power allegories. In fact they may be ommited entirely since the rest of the structure implies their existence and their equations:

Define 0_R as $(R/\ni)^\circ \backslash \ni$. Note that $R \subset S$ implies $0_S \subset 0_R$. $R/\ni$ is entire (since $(R/\ni)(\ni/R)$ is) hence $0_R \subset (R/\ni)(R/\ni)^\circ 0_R \subset (R/\ni)\ni \subset R$. Any contravariant deflationary operation on a semi-lattice is constant:

$$0_R \subset 0_{R\cap S} \subset R \cap S \subset S\,.$$

Define $R \cup S$ as $(\ni/R \cap \ni/S)\backslash \ni$. Note that both R and S appear in covariant positions: $R \subset R'$ and $S \subset S'$ imply $R \cup S = R' \cup S'$. $R \subset R \cup S$ because

$$R \subset (\ni/R)\backslash \ni \subset (\ni/R \cap \ni/S)\backslash \ni\,.$$

$R \cup R \subset R$ because

$$R \cup R \subset (\ni/R)\backslash \ni \subset ((R/\ni)(\ni R))((\ni/R)\backslash \ni) \subset (R/\ni)\ni \subset R\,.$$

Thus $R \subset T$, $S \subset T$ imply $R \cup S \subset T \cup T \subset T$.

The presence of the division operation forces most of the remaining equations for 0_R and $R \cup S$. (For example $0_R T \subset 0_{RT}$ because $0_R \subset 0_{RT}/T$.) Division, however, does not suffice for $(R_1 \cup R_2) \cap T \subset (R_1 \cap T) \cup (R_2 \cap T)$. [2.228].

In the proof of [2.441] we showed that for any object α there exist maps $\ell, \mathscr{r}$ such that $\ell\ell^\circ = \mathscr{r}\mathscr{r}^\circ = 1_\alpha$, $\ell\mathscr{r}^\circ = \mathscr{r}\ell^\circ = 0$. Given $R_1, R_2, T \in (\beta, \alpha)$ define $R = R_1\ell \cup R_2\mathscr{r}$, $S_1 = \ell^\circ \cup \mathscr{r}^\circ$ and apply the law of modularity to obtain $(R_1 \cup R_2) \cap T \subset RS \cap T \subset (R \cap TS^\circ)S \subset (R \cap TS^\circ)\ell^\circ \cup (R \cap TS^\circ)\mathscr{r}^\circ$. But,

$$(R \cap TS^\circ)\ell^\circ \subset R\ell^\circ \cap TS^\circ\ell^\circ \subset R_1 \cap T(\ell \cup \mathscr{r})\ell^\circ \subset R_1 \cap T\,,$$

and similarly $(R \cap TS^\circ)\mathscr{r}^\circ \subset R_2 \cap T$; hence $(R_1 \cup R_2) \cap T = (R_1 \cap T) \cup (R_2 \cap T)$.

The reverse containment, $(R_1 \cap T) \cup (R_2 \cap T) \subset (R_1 \cup R_2) \cap T$ holds, of course, in any lattice.

2.45. The Continuum Hypothesis

2.451. A boolean algebra is FREE on the set of generators S iff any map from S to any boolean algebra extends to a unique boolean homomorphism (that is, viewing boolean algebras as one-object division allegories, it extends to a representation of division allegories). For present purposes we will take the free boolean algebra on a set S to be the boolean algebra whose elements are named by boolean expressions built from the operators and constants of the theory of boolean algebras using S as a set of variable symbols, two expressions naming the same element iff the theory of boolean algebras says they must. We will need the following lemma:

Any collection of pairwise disjoint elements from a free boolean algebra is at most countably infinite.

BECAUSE: Define the support of an element as the subset of S needed for its name. It clearly suffices to show that for any natural number n there can only be finitely many pairwise disjoint elements with supports of size n. For that, of course, we may use induction, noting that the case $n = 0$ is immediate.

Elements with disjoint supports cannot be disjoint. Suppose there were an infinite collection of pairwise disjoint elements. Choose one element and note that for at least one s in its support there must be infinitely many elements in the collection whose supports also contain s. A given element in the free boolean algebra may be written in the form $(s \wedge P) \vee (\neg s \wedge Q)$ where the supports of P and Q are the result of removing s from the support of the given element. (Take the name of the given element and replace each occurrence of s with 1 to obtain P, and with 0 to obtain Q.) Thus we obtain either an infinite family of pairwise disjoint elements of the form $s \wedge P$ or of the form $\neg s \wedge Q$. In the first case we obtain a counter-example from the P's, in the second from the Q's. In either case it is a counter-example with smaller support size.

2.452. Let $\mathscr{B}$ be the boolean completion of a free boolean algebra, distinguished by the fact that the free algebra appears both as basis and cobasis. Note that $\mathscr{B}$ inherits the countability property established above.

If I and J are not isomorphic as sets, they remain non-isomorphic in the allegory of $\mathscr{B}$-valued sets. Indeed, recall that in any allegory, R is an isomorphism iff RR° and $R^\circ R$ are identity morphisms [2.135]. If R is a $\mathscr{B}$-valued relation between discrete sets, such is equivalent to the statement that each row and each column of R is a partition of unity. Suppose that I is infinite and of strictly smaller cardinality than J. Suppose that $R: I \to J$ is an isomorphism. For each $i \in I$ let J_ι consist of all $j \in J$ such that $R_{ij} \neq 0$. As just noted, each J_ι is countable. $\bigcup_I J_i$ is of cardinality strictly less than that of J, hence there must be a zero column.

2.453. In the late 19-th century Cantor was led to conjecture that every infinite set of reals is either in one-to-one correspondence with the integers or in one-to-one correspondence with the reals. (An equivalent version concerns the power-set of the integers instead of the reals.) This conjecture came to be known as the Continuum Hypothesis. In the late 1930's Gödel established its consistency with the axioms of set theory [B.5]. In the early 1960's P.J. Cohen proved its independence of those axioms.

Allegories in which the Continuum Hypothesis fails are inevitable. Any set I appears as a subobject of $[\Sigma_{\mathbb{N}}1]$ in the allegory of $\mathscr{B}$-valued sets, where $\mathscr{B}$ is the boolean completion of the free boolean algebra generated by the entries of an $I \times \mathbb{N}$ matrix which will be regarded as a "generic" morphism R from I to $\mathbb{N}$. We need to show that R is straight (because then $\Lambda(R)$ is the required embedding). The matrix R is straight iff

$$\bigcap_n \frac{R_{i,n}}{R_{i',n}} = 0, \quad \text{for all } i \neq i' .$$

A fortiori, it sufficies to show

$$\bigcap_n R_{i,n}/R_{i',n} = 0, \quad \text{for all } i \neq i' .$$

Suppose the intersection is >0. Because the elements of the free boolean algebra form a basis of $\mathscr{B}$, the free boolean algebra must have a non-zero element P such that $P \leqslant R_{i,n}/R_{i',n}$ for all n, equivalently $P \wedge R_{i'n} \leqslant R_{i,n}$ for all n. But if n is large enough so that the support of P [2.451] includes neither $R_{i'n}$ nor $R_{i,n}$, then the inequality cannot be a consequence of the theory of boolean algebras.

By the Stone representation theorem [1.75, 2.33] the free boolean algebra on a set of generators S is isomorphic to the boolean algebra of clopen subsets of the product space 2^S. Its boolean completion is isomorphic to the complete boolean algebra of regular open subsets of 2^S. If 2^S is regarded as a probability space the property in [2.451] that pairwise disjoint families are at most countable is immediate. The straightness of R seems to be more difficult from this point of view: each $\frac{R_{i,n}}{R_{i',n}}$ has probability $\frac{1}{2}$ and the family of all such is totally independent in the sense that each k-fold intersection has probability $(\frac{1}{2})^k$.

In [B.5] we will show how to interpret a formal language of axiomatic set theory in any boolean Grothendieck topos, and thus that the example just constructed yields the independence of the Continuum Hypothesis. (We note for the experts that the interpretation obtained in this example is the same as Cohen's.)

2.454. A topos is WELL-POINTED if the one-element set that consists of a terminator is a generating set.

We have just shown that for any set I there exists a value-based boolean AC Grothendieck topos $\mathbf{C}$ such that $\Sigma_I 1$ is embeddable in the power-object of $\Sigma_{\mathbb{N}} 1$ and in which $\Sigma_I 1 \simeq \Sigma_J 1$ iff $I \simeq J$ in $\mathscr{S}$.

We may take an ultra-filter $\mathscr{F} \subset \mathscr{V}al$ to obtain, $\mathbf{C}/\mathscr{F}$, a well-pointed AC topos that fails the Continuum Hypothesis.

$\mathbf{C}/\mathscr{F}$ does not, however, yield an entirely satisfactory model of set theory. If I is bigger than $[\mathbb{N}]$ in $\mathscr{S}$ then $\mathbf{C}$ fails to be a σ-topos: if it were a σ-topos then $\Gamma: \mathbf{C}/\mathscr{F} \to \mathscr{S}$ would be bicartesian, hence $\mathbf{C} \to \mathbf{C}/\mathscr{F} \xrightarrow{\Gamma} \mathscr{S}$ would be bicartesian. But:

There are no bicartesian functors from $\mathbf{C}$ *to* $\mathscr{S}$.

BECAUSE: Suppose $T: \mathbf{C} \to \mathscr{S}$ were bicartesian. Let $1 \xrightarrow{0} \mathbb{N}$ be the initial point of $\mathbb{N}$, $s: \mathbb{N} \to \mathbb{N}$ the successor map. Then $1 + \mathbb{N} \xrightarrow{\binom{0}{s}} \mathbb{N}$ is an isomorphism and $\mathbb{N} \to 1$ is the coequalizer of $1, s$. Such characterizes the natural numbers in $\mathscr{S}$. $\Sigma_{\mathbb{N}} 1$ in $\mathbf{C}$ possesses such structure, hence $T(\Sigma_{\mathbb{N}} 1) \simeq \mathbb{N}$.

T is a representation of boolean logoi, hence if R is straight then $T(R)$ is straight. Whereas $T(\Sigma_I 1)$ needn't be the I-fold copower of 1 it does have a family of distinct elements indexed by I (to wit: $\{T(1 \xrightarrow{u_i} \Sigma_I 1) \mid i \in I\}$). Since there is a straight morphism $T(R)$: $T(\Sigma_I 1) \to \mathbb{N}$ we would obtain an embedding $\Lambda(T(R))$ of $T(\Sigma_I 1)$ into $[\mathbb{N}]$, hence a family of distinct elements in $[\mathbb{N}]$ indexed by I, which, of coures, contradicts the choice of I.

We may, however, construct a well-pointed AC σ-topos that fails the Continuum Hypothesis. Let $\mathbf{C}$ be as described above and, by a Löwenheim–Skolem construction, let $\mathbf{C}' \subset \mathbf{C}$ be a countable elementary substructure. By [1.776] there does exists a bicartesian functor $T: \mathbf{C}' \to \mathscr{S}$. Let $\mathscr{F}' \subset \mathscr{V}al_{\mathbf{C}'}$ be the induced filter: $\mathscr{F}' = \{U \subset 1 \mid T(U) = 1\}$. Then $\mathbf{C}'/\mathscr{F}'$ is a well-pointed AC topos. It must be a σ-topos because $\Gamma: \mathbf{C}/\mathscr{F}' \to \mathscr{S}$ is faithful.

2.455. The bicartesian representation theory of $\mathbf{C}$ is more pathological than yet indicated. Any ideal of $\mathscr{B}$ closed under countable intersection is principal. The absence of atoms says, therefore, that no ultra-filter can be closed under countable intersections.

Let $\mathbf{C}$ *be a countably co-complete boolean logos in which there exists a well-supported object,* A, *such that* $\mathscr{S}ub(A)$ *has no ultra-filters closed under countable intersections. Then any cocartesian functor* $T: \mathbf{C} \to \mathscr{S}$ *is everywhere empty, that is,* $T = \emptyset$.

BECAUSE: Suppose there does exist cocartesian $T: \mathbf{C} \to \mathcal{S}$ not everywhere empty. For each well-supported object A we will construct an ultra-filter $\mathcal{F} \subset \mathcal{S}ub(A)$ which is closed under countable intersections.

Let B be such that $T(B) \neq \emptyset$. Since there exists $B \to 1$ it is necessarily the case that $T(1) \neq \emptyset$. We obtain a cocartesian functor $\mathbf{C} \to \mathcal{S}/T(1) \simeq \Pi_{T(1)} \mathcal{S}$ that preserves the terminator. Choose $x \in T(1)$ and replace T with $\mathbf{C} \to \Pi_{T(1)} \mathcal{S} \xrightarrow{p_x} \mathcal{S}$. T remains cocartesian. Moreover $T(1) = 1$.

Since T preserves finite coproducts and covers it preserves disjoint unions, images, and consequently, it preserves finite unions, complements, and hence finite intersection. Given $f: A \to B$ and a subobject $B' \subset B$, the inverse image $f^{\#}(B')$ is characterized, in any boolean logos, by $f(A') \subset B'$ and $f(\neg A') \subset \neg B'$. Hence T preserves inverse-images.

As in [2.454] we may infer that $T(\Sigma_{\mathbb{N}} 1) \simeq \mathbb{N}$, that is, T preserves the countable co-power of 1. Let $\{A_m\}$ be an arbitrary sequence of objects in $\mathbf{C}$ and let $g: \Sigma_{\mathbb{N}} A_n \to \Sigma_{\mathbb{N}} 1$ be the obvious map characterized by $g^{\#}(u_n) = A_n$. Such characterizes $\Sigma_{\mathbb{N}} A_n$ as a countable disjoint union in $\mathbf{C}$ and in $\mathcal{S}$. Thus T preserves arbitrary countable unions. And since it preserves complements, it preserves arbitrary countable intersections.

If A is well-supported then so is $T(A)$ and $T(A)$ cannot be empty. Choose $x \in T(A)$ and define $\mathcal{F} \subset \mathcal{S}ub(A)$ as $\{A' \mid x \in T(A')\}$. $\mathcal{F}$ is an ultra-filter closed under countable intersection.

2.5. QUOTIENT ALLEGORIES

A CONGRUENCE on an allegory is an equivalence relation on the morphisms that respects the allegory structure. (We shall be interested only in the case where different identity morphisms are never identified. Thus, R may be equivalent to S only if $\square R = \square S$ and $R\square = S\square$.) The QUOTIENT ALLEGORY is the allegory of equivalence classes with the obvious operations (which makes the assignment of equivalence classes into a representation of allegories).

2.51. *A quotient of a tabular (resp. unitary) allegory is such.*

BECAUSE: The equivalence class of an entire (resp. simple, tabular) morphism is such (in the quotient allegory). The equivalence class of a (partial) unit is a (partial) unit.

2.52. Any congruence on a distributive allegory respects zero. If a congruence on a distributive allegory respects binary unions, then the quotient allegory is distributive and the assignment of equivalence classes is a representation of distributive allegories. We consider two important examples.

2.521. The BOOLEAN QUOTIENT is obtained by identifying morphisms that are disjoint from the same things, i.e. identifying $R, S\colon \alpha \longrightarrow \beta$ wherever $R \cap T = 0$ iff $S \cap T = 0$, all $T\colon \alpha \longrightarrow \beta$. The congruence in question is the maximal congruence that does not identify nonzeros with zeros. The boolean quotient $\mathscr{Z}_{\neg\neg}$ of a locale $\mathscr{Z}$ is isomorphic to the boolean locale of regular elements of $\mathscr{Z}$, that is, of $R \in \mathscr{Z}$ for which $(R \to 0) \to 0 = R$. (Regular open subsets of a topological space are interiors of their closures.) The boolean quotient of the allegory of $\mathscr{Z}$-valued sets is equivalent to the allegory of $\mathscr{Z}_{\neg\neg}$-valued sets [2.551]. The category of maps of the boolean quotient of a tabular unitary division allegory is a boolean logos [2.536].

2.522. In a unitary distributive allegory with unit λ, let $U\colon \lambda \to \lambda$. The CLOSED QUOTIENT (with respect to U) identifies $R, S\colon \alpha \to \beta$ iff $R \cup p_\alpha U p_\beta^\circ = S \cup p_\alpha U p_\beta^\circ$ [2.152]. The congruence is the least one which identifies U with zero and which respects binary unions.

If the given allegory is a locale, this congruence identifies R and S iff $R \cup U = S \cup U$. Given an open subset U of a topological space X, the corresponding closed quotient of the locale $\mathcal{O}(X)$ of open subsets of X, is isomorphic to the locale $\mathcal{O}(X-U)$ given by the induced topology on the closed subset $X-U$. The corresponding closed quotient of the allegory of $\mathcal{O}(X)$-valued sets is equivalent to the allegory of $\mathcal{O}(X-U)$-valued sets [2.551].

2.53. For the rest of section 2.5, we will concentrate on AMENABLE CONGRUENCES, that is, congruences that respect binary unions and such that every congruence class $\bar{R}$ has the largest element, R^+. A quotient by an amenable congruence is an AMENABLE QUOTIENT. Closed quotients are amenable: $R^+ = R \cup p_\alpha U p_\beta^\circ$.

2.531. *If $R \subset S$, then $R^+ \subset S^+$.*

BECAUSE: $R \equiv R^+$ and $S \equiv S^+$, so $R \cup S \equiv R^+ \cup S^+$, and thus $R^+ \cup S^+ \subset (R \cup S)^+$. If $R \subset S$, then $R \cup S = S$, therefore $R^+ \subset R^+ \cup S^+ \subset S^+$.

2.532. $(R \cap S)^+ = R^+ \cap S^+$.

BECAUSE: $R \cap S \equiv R^+ \cap S^+$, so $R^+ \cap S^+ \subset (R \cap S)^+$. By [2.531], $(R \cap S)^+ \subset R^+ \cap S^+$.

2.533. It is now clear that

$\bar{R} \subset \bar{S}$ *in the quotient allegory iff* $R^+ \subset S^+$ *in the original allegory.*

2.534. *$T^+S^+ \subset (TS)^+$ and $(S^+)^\circ \subset (S^\circ)^+$.*

BECAUSE: $TS \equiv T^+S^+$ and $S^\circ \equiv (S^+)^\circ$.

2.535. It now follows from [2.533–4] that

If R is reflexive (resp. symmetric, transitive), so is R^+.

2.536. *An amenable quotient of a division allegory is a division allegory.*

BECAUSE: We construct $\bar{R}/\bar{S}$ as $\overline{R^+/S^+}$. $(R^+/S^+)S^+ \subset R^+$ implies $\overline{(R^+/S^+)}\bar{S} \subset \bar{R}$. On the other hand, if $\overline{TS} \subset \bar{R}$, then $(TS)^+ \subset R^+$ by [2.533], and thus $T^+S^+ \subset R^+$ by [2.534], so $T^+ \subset R^+/S^+$. Therefore $\bar{T} \subset \overline{R^+/S^+}$. (Note that $\overline{R^+/S^+} = \overline{R^+/S}$.)

In a tabular unitary division allegory the lattice (α, β) is a Heyting algebra and thus the boolean quotient is amenable: $R^+ = (R \to 0) \to 0$.

2.537. *An amenable quotient of an effective power allegory is an effective power allegory.*

BECAUSE: Effectivity is preserved by [2.535]. Thus it suffices to show that the quotient allegory is a pre-power allegory [2.432]. We show that $\overline{\ni_R}$ is thick. Note that $\ni_R = \ni_{R^+}$ [2.41]. From $1 \subset (R^+/\ni_R)(\ni_R/R^+) \subset (R^+/\ni_R)(\ni_R^+/R^+)$, we obtain $\bar{1} \subset (\bar{R}/\overline{\ni_R})(\overline{\ni_R}/\bar{R})$ as in the last section.

2.54. *Every coreflexive morphism of a quotient allegory is named by a coreflexive morphism of the given allegory.*

BECAUSE: If $\bar{S} \subset \bar{1}$ in the quotient allegory, let $R = \mathscr{D}om\ S = 1 \cap SS^\circ$ in the given allegory. Since $\bar{S}$ is a symmetric idempotent [2.12], we have $\bar{R} = \overline{1 \cap SS^\circ} = \bar{1} \cap \overline{SS^\circ} = \bar{1} \cap \bar{S} = \bar{S}$.

2.541. We now show that transitive closure in an allegory remains such in an amenable quotient.

$$(\bar{R})^* = \overline{R^*} \quad \textit{for any endomorphism } R\,.$$

BECAUSE: We may use the characterization of the transitive closure R^* as the least reflexive S such that $RS \subset S$ [1.787]. Note that $\overline{R^*}$ is reflexive. Also, $\bar{R}\overline{R^*} \subset \overline{R^*}$ because $RR^* \subset R^*$. On the other hand, if $\bar{S}$ is reflexive and $\bar{R}\bar{S} \subset \bar{S}$, then by [2.533–5], $1 \subset 1^+ \subset S^+$ and $RS^+ \subset R^+S^+ \subset S^+$, so $R^* \subset S^+$. By [2.531], $(R^*)^+ \subset S^{++} = S^+$. This yields $\overline{R^*} \subset \bar{S}$ by [2.533] and therefore $(\bar{R})^* = \overline{R^*}$.

2.542. *For any topos* **A** *there exists a boolean topos* **B** *and a faithful bicartesian representation* $\mathbf{A} \to \mathbf{B}$.

BECAUSE: Let Ω be a subobject classifier in **A** and $t\colon 1 \longrightarrow \Omega$ its universal subobject [1.912]. t is a subterminator in the slice topos $\mathbf{A}/\Omega$ and thus t is an endomorphism on a unit in the allegory $\mathscr{R}el(\mathbf{A}/\Omega)$ [2.151]. We will take two successive quotients. Firstly, consider the closed quotient $\mathbf{Q}_1$ of $\mathscr{R}el(\mathbf{A}/\Omega)$ with respect to t [2.522]. Secondly, let $\mathbf{Q}_2$ be the boolean quotient of this $\mathbf{Q}_1$ [2.521]. Finally, let $\mathbf{B} = \mathscr{M}ap(\mathbf{Q}_2)$. **B** is a boolean topos [2.537, 2.51, 2.414]. It is readily seen that the induced functor

$$F\colon \mathbf{A} \to \mathbf{A}/\Omega \to \mathscr{M}ap(\mathbf{Q}_1) \to \mathbf{B}$$

is a representation of positive pre-logoi [1.946, 1.952, 2.52, 2.212–7, 1.625] and thus F is a cartesian representation that preserves coterminator and binary coproducts. But F also preserves coequalizers because it preserves transitive closures [2.541, 1.954].

For the faithfulness of F, consider a proper subobject $A' \subset A$ in **A**. We have to show that $FA' \subset FA$ is proper in **B**. By [2.521], it suffices to find a subobject, P, of ΔA in $\mathbf{A}/\Omega$ that names a nonzero relation in $\mathbf{Q}_1$ disjoint from the relation named by $\Delta A'$. Because $\mathcal{S}ub_{\mathbf{A}/\Omega}(\Delta A) \simeq \mathcal{S}ub_{\mathbf{A}}(\Omega \times A)$ we may translate the condition on P as follows: to say that $P \subset \Omega \times A$ names a non-zero relation in $\mathbf{Q}_1$ is to say that P is not contained in $t \times A$; to say that P is disjoint from $\Delta A'$ is to say that $P \cap (t \times A') \subset t \times A$. The graph of the characteristic map of $A' \subset A$ will do.

2.55. *An amenable quotient of a locally (resp. globally) complete allegory is locally (resp. globally) complete.*

BECAUSE: $(\bigcup_i R_i)^+ \subset (\bigcup_i R_i^+)^+$ and $R_j^+ \subset (\bigcup_i R_i)^+$, for each j [2.531], so $\bigcup_i R_i^+ \subset (\bigcup_i R_i)^+$. Thus $(\bigcup_i R_i^+)^+ \subset (\bigcup_i R_i)^{++} = (\bigcup_i R_i)^+ \subset (\bigcup_i R_i^+)^+$. Therefore, if $S_i \equiv T_i$ for each i, then $\bigcup_i S_i^+ = \bigcup_i T_i^+$, so $(\bigcup_i S_i)^+ = (\bigcup_i S_i^+)^+ = (\bigcup_i T_i^+)^+ = (\bigcup_i T_i)^+$, that is, $(\bigcup_i S_i) \equiv (\bigcup_i T_i)$. Thus the quotient allegory is locally complete. Global completeness is readily preserved.

2.551. Disjoint unions in a globally complete allegory coincide with coproducts (and with products) [2.223, 2.214], so a congruence on a locale $\mathscr{V}$ naturally extends to its global completion, the allegory composed of $\mathscr{V}$-valued relations between sets [2.224, 2.111]. $R, S\colon I \longrightarrow J$ are identified iff $(iRj) \equiv (iSj)$ for all $i \in I$, $j \in J$. Thus one obtains such identification of morphisms between $\mathscr{V}$-valued sets [2.16(12)]. $R\colon \langle I, R\rangle \longrightarrow \langle I, R^+\rangle$ is an isomorphism in the quotient allegory. $(\_)^+$ is a representation of the quotient allegory in the allegory of $\mathscr{V}_{\equiv}$-valued sets, where $\mathscr{V}_{\equiv}$ is the quotient of $\mathscr{V}$. $(\_)^+$ yields an equivalence of categories.

2.56. Independence of the Axiom of Choice

We describe a 2-valued boolean Grothendieck topos with a sequence of non-zero subobjects $B_0, B_1, B_2, \ldots$ of 2^N (N a natural numbers object) whose product ΠB_n is zero. Such a topos is not IAC because $(-)^N$ fails to preserve epics. (Let $A = \Sigma\, B_n$ and let $f\colon A \longrightarrow N$ be the epic 'that sends B_n to n'. Then

$$\begin{array}{ccc} \Pi B_n & \longrightarrow & 1 \\ \downarrow & & \downarrow {}^{\ulcorner 1_N \urcorner} \\ A^N & \longrightarrow & N^N \\ & f^N & \end{array}$$

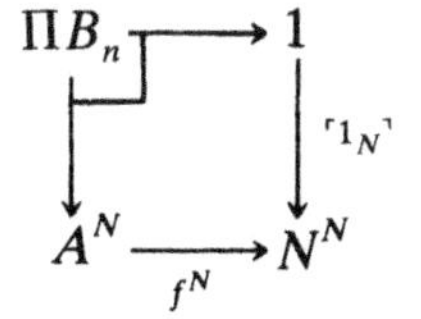

and because pullbacks preserve epics, f^N is not epic.) Furthermore, as we will show in Appendix B, this failure of IAC occurs in the part of the topos relevant for an interpretation of axiomatic set theory. Thus, as we will show in Appendix B, the Axiom of Choice [1.57] may not be inferred from the axioms of set theory.

The consistency of the Axiom of Choice with the axioms of set theory was established by Gödel in the late 1930's while the independence was first obtained by P.J. Cohen in the early 1960's soon after he showed independence of the Continuum Hypothesis from the axioms of set theory. While the interpretation of axiomatic set theory in the boolean Grothendieck topos discussed in [2.453] is essentially the same as Cohen's original interpretation in which the Continuum Hypothesis fails, the interpretation in the boolean Grothendieck topos discussed in this section is definitely different from Cohen's interpretation in which the Axiom of Choice fails. This fact is the point of departure of the monograph "Freyd's Models for the Independence of the Axiom of Choice" by A. Blass and A. Scedrov, Memoir of the Amer. Math. Soc. vol. 404, 1989.

2.561. Let **A** be the category whose objects are non-zero finite ordinals and whose proto-morphisms are arbitrary functions between them. (We use the convention that $n = \{0, 1, \ldots, n - 1\}$.) The source-target predicate $f: m \longrightarrow n$ means that $m \geqslant n$ and $jf = 1_n$ in $\mathscr{S}$, where $j: n \longrightarrow m$ is the inclusion map. Thus:

The objects of **A** *are non-zero finite ordinals and there exists a morphism* $m \longrightarrow n$ *iff* $m \geqslant n$.

2.562. The property just stated allows us to compute the subterminators in $\mathscr{S}^{\mathbf{A}°}$. The terminator is the functor all of whose values are one-element sets. Any subterminator U is determined by the set of all those n for which $U(n)$ is non-empty. If $U(n)$ is non-empty and $m \geqslant n$, then by [2.561] there exists a morphism $m \longrightarrow n$, so $U(m)$ must be non-empty. Thus any subterminator U is determined by the least n for which $U(n)$ is non-empty. Therefore the lattice of subterminators in $\mathscr{S}^{\mathbf{A}°}$ consists of a descending sequence that converges down to the empty functor.

Let $\mathscr{B}$ be the category of maps of the boolean quotient of $\mathscr{R}el(\mathscr{S}^{\mathbf{A}^\circ})$. $\mathscr{B}$ is two-valued.

BECAUSE: if U is not the empty subterminator, then for any subterminator V, $U \cap V$ is empty iff V is empty. Thus U is congruent to the entire subterminator [2.521].

2.563. We shall use some auxiliary notions and facts about the boolean congruence on an effective tabular unitary division allegory. Our arguments apply to any amenable congruence [2.53].

An object A is SEPARATED iff $1_A = 1_A^+$. A relation from A to B is DENSE iff it is congruent to the maximal relation from A to B.

We note in passing that for any object A there exists a separated object A' that becomes isomorphic to A in the quotient allegory ($1^+: A \longrightarrow A$ is an equivalence relation [2.535], so it splits as $1^+ = ff^\circ$, where $f: A \longrightarrow A'$ is a map such that $f^\circ f = 1_{A'}$ [2.163]; thus f is an isomorphism in the quotient allegory [2.134]).

If B is separated, then any map $A \longrightarrow B$ in the boolean quotient is named by a simple relation whose domain is a dense subobject of A.

BECAUSE: If $\bar{R}^\circ \bar{R} \subset \bar{1}$, then by [2.533–4], $(R^+)^\circ R^+ \subset 1^+$, but $1^+ = 1$. Also, if $\bar{R}$ is entire in the quotient, then $\overline{\mathscr{D}om(R^+)} = \overline{1 \cap R^+(R^+)^\circ} = \bar{1} \cap \overline{RR^\circ} = 1$. Let p be the terminal map from A in the given allegory. By [2.152], $\overline{\mathscr{D}om(R^+)p} = \bar{1}\bar{p} = \bar{p}$.

2.564. *If*

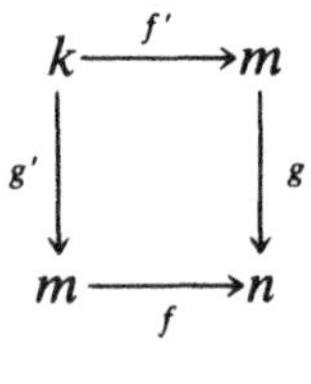

in $\mathbf{A}$, *then* $f = g$.

BECAUSE: Let $j: m \mapsto k$ be the inclusion in $\mathscr{S}$. In $\mathscr{S}$, $f = jg'f = jf'g = g$.

2.565. *In $\mathscr{S}^{\mathbf{A}^\circ}$, representable functors $H_n = (-, n)$ are separated.*

BECAUSE: We show the stronger claim that 1_{H_n} is complemented as a subobject of $H_n \times H_n$. If $f, g: m \longrightarrow n$ are distinct morphisms in $\mathbf{A}$, then for any $h: k \longrightarrow m$, [2.564] implies that hf are still distinct. Thus the assignment $X(m) = \{\langle f, g\rangle \in H_n(m) \times H_n(m) \mid f \neq g\}$ is a subfunctor of $H_n \times H_n$. Clearly, X is the complement of 1_{H_n}.

2.566. For any morphism $f: m \longrightarrow n$ in $\mathbf{A}$, let $H_f: H_m \longrightarrow H_n$ be the natural transformation that takes $k \longrightarrow m$ to $k \longrightarrow m \longrightarrow n$ [1.465].

If f and g are distinct morphisms from m to n in $\mathbf{A}$ then H_f and H_g have disjoint images in $\mathscr{S}^{\mathbf{A}^\circ}$.

BECAUSE: Let

$$\begin{array}{ccc} P & \longrightarrow & H_m \\ \big\downarrow & & \big\downarrow{\scriptstyle H_g} \\ H_m & \xrightarrow[H_f]{} & H_n . \end{array}$$

If P is not the empty functor, let k be such that $P(k)$ is a non-empty set. By the covariant Yoneda representation [1.465], consider a map $H_k \longrightarrow P$ in $\mathscr{S}^{\mathbf{A}^\circ}$. Then

$$\begin{array}{ccc} H_k & \longrightarrow & H_m \\ \big\downarrow & & \big\downarrow{\scriptstyle H_g} \\ H_m & \xrightarrow[H_f]{} & H_n \end{array}$$

so [1.465] and [2.564] imply $f = g$.

2.567. *Let $2 = 1 + 1$ in $\mathscr{B}$. For each n, there exists a jointly monic, countable collection of maps $H_n \longrightarrow 2$ in $\mathscr{B}$.*

BECAUSE: Fix n. For each $f: m \longrightarrow n$ in $\mathbf{A}$ consider the map $H_n \longrightarrow 2$ in $\mathscr{B}$ defined as the characteristic map of the image of the map $\overline{H_f}: H_m \longrightarrow H_n$ in $\mathscr{B}$. We show that this collection of maps $H_n \longrightarrow 2$ is jointly monic.

Let $\bar{R}, \bar{S}$ be distinct maps from Y to H_n in $\mathscr{B}$. By [2.563–5], they are named by simple morphisms R^+ and S^+ from Y to H_n in $\mathscr{Rel}(\mathscr{S}^{\mathbf{A}^\circ})$ whose domains are dense in Y. It is readily verified that the intersection of these domains, Q, is still dense in Y. In other words, Q is a coreflexive endorelation on Y such that $\bar{Q} = 1$. Let $u: Y' \longrightarrow Y$ be a monic in $\mathscr{S}^{\mathbf{A}^\circ}$ such that $Q = u^\circ u$ [2.145]. We obtain maps uR^+ and uS^+ from Y' to H_n in $\mathscr{S}^{\mathbf{A}^\circ}$. uR^+ and uS^+ must be distinct, for otherwise $u^\circ uR^+ = u^\circ uS^+$ and thus $\bar{R} = \overline{1R^+} = \overline{u^\circ uR^+} = \overline{u^\circ uS^+} = \overline{1S^+} = \bar{S}$, contrary to the assumption.

By [1.465], there exists m and a map $y: H_m \longrightarrow Y'$ such that $yuR^+ = yuS^+$. Their images are disjoint [2.566, 1.465]. Thus in $\mathscr{B}$, $\overline{yuR}$ and $\overline{yuS}$ have disjoint images (calculate the domains of the reciprocals). Let

$\bar{\rho}: H_n \longrightarrow 2$ be the characteristic map of the image of $\overline{yuR}$. Then $\overline{R\rho} \neq \overline{S\rho}$ because $\overline{R\rho} = \overline{S\rho}$ would imply $\overline{yuR\rho} = \overline{yuS\rho}$, which is impossible.

2.568. H_n's *from a generating set in* $\mathscr{B}$.

BECAUSE: Given two distinct maps $\bar{R}$, $\bar{S}$ from F to G in $\mathscr{B}$, let

$$E \rightarrowtail F \overset{\bar{R}}{\underset{\bar{S}}{\rightrightarrows}} G$$

be the equalizer. Let E' be the complement of E in F, as computed in $\mathscr{B}$. If E' were the empty functor, then E' would be the minimal subobject of F in $\mathscr{B}$ as well, so E would be entire, impossible. Thus there exists n and a map $f: H_n \longrightarrow E'$. Clearly $\overline{fR} \neq \overline{fS}$.

2.569. Let $N: \mathbf{A}^\circ \to \mathscr{S}$ be the constant functor whose value is the set of natural numbers and let $c_0: 1 \longrightarrow N$ and $c_s: N \longrightarrow N$ be the 'constant' natural transformations whose values are zero and successor, respectively. It is easily seen that N, c_0, and c_s constitute a NNO in $\mathscr{S}^{\mathbf{A}^\circ}$ [1.98]. Because the quotient functor $\mathscr{S}^{\mathbf{A}^\circ} \longrightarrow \mathscr{B}$ is bicartesian [2.52, 2.541], it preserves NNO [1.98(11)]. Furthermore, the quotient functor preserves disjoint sums [2.55, 2.223, 1.967] so the countable coproduct of 1 is a NNO in $\mathscr{B}$. Thus [2.567] implies

In $\mathscr{B}$, *each* H_n *is a subobject of* 2^N.

2.56(10). *Given* $f: m \longrightarrow n$ *and* $g: m \longrightarrow (n+1)$ *in* $\mathbf{A}$ *there exist* $h, h': (m+1) \longrightarrow m$ *such that* $hf = h'f$ *and* $hg = h'g$.

BECAUSE: We may let $h(m) = n$, $h'(m) = f(n)$, and $h(i) = h'(i) = i$, for each $i < m$.

2.56(11). *Any simple morphism* $H_n \longrightarrow H_{n+1}$ *in* $\mathscr{Rel}(\mathscr{S}^{\mathbf{A}^\circ})$ *is empty.*

BECAUSE: Let $R: H_n \longrightarrow H_{n+1}$ be simple and let $v: F \longrightarrow H_n$ be a monic map such that $v^\circ v = \mathscr{Dom}(R)$ [2.145]. $vR: F \longrightarrow H_{n+1}$ is a map. Suppose that F is not the empty functor. Choose m such that $F(m)$ has an element, t, and let $t: H_m \longrightarrow F$ be the corresponding map [1.465]. Let H_f be the composite $H_m \longrightarrow F \longrightarrow H_n$ and H_g the composite $H_m \longrightarrow F \longrightarrow H_{n+1}$. Let h, h' be as described in [2.56(10)]. Because v is monic, we obtain $H_h t = H_{h'} t$. Thus $H_h tvR = H_{h'} tvR$, that is, $H_h H_g = H_{h'} H_g$. This contradicts [2.56(10)].

2.56(12). *In $\mathscr{B}$, the product $P = \Pi H_n$ is the empty functor.*

BECAUSE: Suppose P is not the empty functor. By [1.465], let n be such that there exists a map $H_n \longrightarrow P$ in $\mathscr{S}^{\mathbf{A}^\circ}$ and thus there exists a map $H_n \longrightarrow H_{n+1}$ in $\mathscr{B}$. By [2.563–5], this map is named by a simple morphism $H_n \longrightarrow H_{n+1}$ in $\mathscr{R}el(\mathscr{S}^{\mathbf{A}^\circ})$ with a dense domain, which contradicts [2.56(11)].

APPENDICES

APPENDIX A

In [1.75–1.77] we use the following facts from topology:

A1. *Any two countable dense linearly ordered sets without endpoints are isomorphic.*

BECAUSE: Each of them can be enumerated by natural numbers, so an isomorphism can be defined by induction, using Cantor's back-and-forth argument. At even stages, pick the first element in the given enumeration of the second set beyond those already specified, so that the assignment preserves strict order. At odd stages, pick the first such element of the first set.

A.2. *A G_δ set in a complete metric space has a complete metric with the same topology.*

BECAUSE: A new metric can be defined on $\bigcap_{i=1}^{\infty} U_i$, U_i open, by letting $f_i(x)$ be the inverse of the distance from x to the complement of U_i, and adding

$$\sum_{i=1}^{\infty} 2^{-i} \frac{|f_i(x) - f_i(y)|}{1 + |f_i(x) - f_i(y)|}$$

to the old distance between x and y.

A.3. *Any compact Hausdorff space without isolated points and with a countable base of clopen sets is homeomorphic to the countable power of the two-point discrete space (with the product topology on the power).*

Any complete metric space without isolated points and with a countable base of clopen sets, and with the further property that no nonempty clopen set is compact, is homeomorphic to the countable power of the discrete space of natural numbers (with the product topology on the power).

BECAUSE: Both homeomorphisms will be given by defining nonempty clopen subsets Y_w of the space, X, for each finite word w, so that the clopens of the form Y_{wa} form a partitioning of Y_w. (For the empty word, λ, $Y_\lambda = X$). If X is compact we understand that the finite words are words on $A = \{0, 1\}$. In the second case we understand that they are words on $A = \{0, 1, 2, \ldots\}$.

We will define $f\colon A^{\mathbf{N}} \to X$ by defining $f(a_0, a_1, a_2, \ldots)$ to be the unique element in $Y_{a_0} \cap Y_{a_0a_1} \cap Y_{a_0a_1a_2} \cap \cdots$. We must insure, of course, that there is a unique element in the set in question.

If X is compact with a basis $U_1, U_2, U_3, \ldots$ of clopens we will require for each n and w of length n that $Y_{w0} = Y_w \cap U_n$, $Y_{w1} = Y_w - U_n$ whenever $\emptyset \subsetneq Y_w \cap U_n \subsetneq Y_w$. This not only insures that f is well defined but forces it to be continuous since the Y_w's can easily be seen to form a basis. The absence of isolated points is used in insuring that we may choose Y_{w0}, Y_{w1} to be a partitioning of Y_w into non-empty subsets when either $\emptyset = Y_w \cap U_n$ or $Y_w \subset U_n$.

In the second case we will inductively choose the Y_w's so that the diameter of each Y_w is bounded by the inverse of the length of w. Since Y_w is complete but not compact we know that it is not "totally bounded", that is, we know that for all small $\varepsilon > 0$ there is no finite cover of Y_w by sets of diameter less than ε. On the other hand Y_w does have a countable basis of clopens and we know that it is covered by a sequence of clopens of diameter less than ε, from which we may obtain a partitioning Y_{w0}, $Y_{w1}, Y_{w2}, \ldots$ into non-empty clopens of diameter less than ε. The completeness of X now insures that f is well-defined. The continuity of f follows from the fact that the Y_w's form a basis (each point of X lies in sets of the form Y_w of arbitrarily small diameter).

In each case, f is a homeomorphism because the canonical basis of $A^{\mathbf{N}}$ is sent to the manufactured basis of the Y_w's.

Note that the second condition is satisfied by the space of irrationals with the topology induced by the standard topology on the real line.

A.4. *Any two countable atomless boolean algebras are isomorphic.*

BECAUSE: Their Stone spaces [1.389] satisfy the first condition of [A.3].

A.5. *Any countable dense subset of Cantor space is homeomorphic to the rationals.*

BECAUSE: We begin by taking Cantor space to be the space of the topological group $\mathbf{Z}_2^{\mathbf{N}}$ where $\mathbf{Z}_2 = \{0, 1\}$ is the two-element additive group. Let $D = \{u_0, u_1, u_2, \ldots\}$ be a countable dense subset.

Define $v \in \mathbf{Z}_2^{\mathbf{N}}$ as the element whose coordinate at $n = 2^i(4j + 1)$ is equal to the nth coordinate of u_j and whose coordinate at $n = 2^i(4j + 3)$ is different from the nth coordinate of u_j. Apply the homeomorphism $(-) + v$ to move D to a subset $D' = D + v$. D' is still a countable dense subset. It is disjoint from the set $E \subset \mathbf{Z}_2^{\mathbf{N}}$ of all elements which are eventually all 0 or eventually all 1.

We switch now to the view that Cantor space is the lexicographically ordered set $\{0, 1\}^{\mathbf{N}}$ with the open interval topology. When the set E is removed the ordering is a dense ordering without endpoints. Any dense subset must also be densely ordered without endpoints, still with the interval topology. [A1] now finishes the proof.

APPENDIX B

B.1. Syntax

Let Σ be a set of SORTS, Σ^* the set of finite SORT WORDS. We will need an alphabet of VARIABLES $\mathscr{V}$, a SORT ASSIGNMENT $\mathscr{V} \to \Sigma$ for which there are countably infinitely many variables of each sort, a set of PREDICATE SYMBOLS $\mathscr{P}$, and a SORT TYPE (or arity) ASSIGNMENT $\mathscr{P} \to \Sigma^*$. The sort assignment $\mathscr{V} \to \Sigma$ extends naturally to the sort type assignment $\mathscr{V}^* \to \Sigma^*$ on finite words of variables. $\mathscr{P}$ is assumed to include $\top$, $\bot$ (both of the empty sort type) and EQUALITY SYMBOLS E_σ of sort type $\sigma\sigma$ for each sort σ.

We will use CONNECTIVES $\wedge$, $\vee$, $\Rightarrow$, QUANTIFIERS $\exists$, $\forall$, and PUNCTUATORS (,).

B.11. FORMULAE are defined inductively. If $\boldsymbol{x}$ is a finite word of variables of the same sort type as a predicate symbol P, then the word $P\boldsymbol{x}$ is an (atomic) formula. (We will write $x = y$ for $E_\sigma xy$, and $|x|$ for $E_\sigma xx$.) All variables in an atomic formula occur FREELY. If A, B are formulae, $*$ a connective, Q a quantifier, and x a variable, then $(A * B)$ and $(Qx\ A)$ are formulae. Free (occurrences of) variables of $(A * B)$ are the free (occurrences of) variables of A or of B. The variable immediately following a quantifier is an INDEX occurence of the variable. Formula A is the SCOPE of Q in $(Qx\ A)$. All free occurrences of the index variable are changed to BOUND occurrences within the scope of the corresponding quantifier.

ASSERTIONS are expressions $A \rightharpoonup B$, where A, B are formulae. Such an assertion is read A TOLERATES B.

We write $Q_x A$ for $Qx\ A$, $A \Leftrightarrow B$ for $(A \Rightarrow B) \wedge (B \Rightarrow A)$, and $\neg A$ for $A \Rightarrow \bot$. The outer pair of parentheses in a formula is customarily omitted.

B.12. A PRIMITIVE FUNCTIONAL SEMANTICS is given by specifying a set S_σ for each sort σ; a $\{0, 1\}$-valued function $\hat{P}$ for each predicate symbol P of sort type $\sigma_1\sigma_2 \cdots \sigma_n$ so that $\hat{P}(a_1, \ldots, a_n)$ is defined whenever $a_i \in S_{\sigma_i}$, $1 \leq i \leq n$, where $\hat{\top} = 1$, $\hat{\bot} = 0$, $\hat{E}_\sigma(a, b) = 1$ iff

$a = b$. More generally, we specify inductively as follows a $\{0, 1\}$-valued function $\hat{A}$ for each formula A: $(A \wedge B)^\wedge = \min\{\hat{A}, \hat{B}\}$, $(A \vee B)^\wedge = \max\{\hat{A}, \hat{B}\}$, $(A \Rightarrow B)^\wedge = 0$ iff $\hat{A} = 1$ and $\hat{B} = 0$, $(\forall_x A)^\wedge = \min\{\hat{A}: a \in S_\sigma, x \text{ of sort } \sigma\}$, and $(\exists_x A)^\wedge = \max\{\hat{A}: a \in S_\sigma, x \text{ of sort } \sigma\}$, where min and max over the empty set are 1 and 0, respectively.

An assertion $A \rightharpoonup B$ is VALID in a primitive functional semantics (and the semantics is a MODEL for the assertion) iff $\hat{A} \leqslant \hat{B}$ for every instantiation of the variables that occur freely in A or in B. It is understood of course, that instantiations of variables of sort σ range over S_σ.

A THEORY is a set of assertions. A model of a theory **T** is a model of each assertion in **T**. **T** ENTAILS $A \rightharpoonup B$ IN PRIMITIVE FUNCTIONAL SEMANTICS iff every model of **T** is a model of $A \rightharpoonup B$.

Because of the presence of possibly empty domains, this notion of entailment does not satisfy all of the traditional rules of inference. In particular, if A, C have no free variables, and B has at least one free variable, the pair $A \rightharpoonup B$ $B \rightharpoonup C$ does not entail $A \rightharpoonup C$ because $A \rightharpoonup B$ and $B \rightharpoonup C$ are automatically valid if the free variables range over the empty domain. $\forall_x A \rightharpoonup \exists_x A$ also fails over the empty domain.

B.2. Rules of Inference

We wish to replace the notion of entailment given in [B.12] and other more general notions of semantic entailment with a syntactic one.

B.21. Let A, B, C be formulae, and let x, x' be variables. We list the rules of inference of FIRST ORDER LOGIC.

$[\rightharpoonup]$ $\qquad A \rightharpoonup A$.

(Read: It may always be inferred that A tolerates A.)

$[\rightharpoonup\rightharpoonup]$ $\qquad$ *If each variable free in B occurs freely in A or C, then*:

$$\frac{A \rightharpoonup B \qquad B \rightharpoonup C}{A \rightharpoonup C}.$$

(Read: From A tolerates B and B tolerates C infer that A tolerates C.)

$[\top]$ $\qquad A \rightharpoonup \top$.

$[\wedge\rightharpoonup]$ $\qquad A \wedge B \rightharpoonup A, \qquad A \wedge B \rightharpoonup B$.

[$\rightharpoonup\wedge$] $$\frac{A \rightharpoonup B \qquad A \rightharpoonup C}{A \rightharpoonup B \wedge C}\,.$$

[$\bot$] $$\bot \rightharpoonup A\,.$$

[$\rightharpoonup\vee$] $$A \rightharpoonup A \vee B, \qquad B \rightharpoonup A \vee B\,.$$

[$\vee\rightharpoonup$] $$\frac{A \rightharpoonup C \qquad B \rightharpoonup C}{A \vee B \rightharpoonup C}\,.$$

[$\Rightarrow$] $$\frac{A \rightharpoonup (B \Rightarrow C)}{(B \wedge A) \rightharpoonup C}$$

(Inference in either direction.)

If x does not occur freely in B, then:

[$\exists$] $$\frac{\exists_x A \rightharpoonup B}{A \rightharpoonup |x| \wedge B}$$

[$\forall$] $$\frac{B \rightharpoonup \forall_x A}{B \wedge |x| \rightharpoonup A}$$

If A is atomic and A' is the result of replacing an occurrence of x with x', then:

[$\wedge =$] $$A \wedge (x = x') \rightharpoonup A'\,.$$

Thus far, it is possible to think of equality as the graphic equality of variables as symbols. Not so with the last rule:

[$\exists=$] $$|x| \rightharpoonup \exists_{x'}(x' = x) \quad (x, x' \text{ distinct variables}).$$

An assertion $A \rightharpoonup B$ may be INFERRED from a theory **T** (we write $\mathbf{T} \vdash (A \rightharpoonup B)$) iff it may be inferred from the assertions in **T** by finitely many applications of the rules of inference. In that case **T** SYNTACTICALLY ENTAILS $A \rightharpoonup B$ IN FIRST ORDER LOGIC.

B.211. COHERENT LOGIC is obtained by omitting $\Rightarrow$, $\forall$ and their rules, but adding [B.223–4]. REGULAR LOGIC is obtained by further omitting $\bot$, $\vee$ and their rules. HORN LOGIC is obtained from regular logic by omitting $\exists$ and its rules [1.444].

HIGHER ORDER LOGIC is obtained, on the other hand, by specifying an infinite Σ, a distinguished sort $[\sigma]$ and a binary predicate symbol $\in_\sigma$ of sort type $\sigma[\sigma]$ for each sort σ, and legislating

$$[\in\, =] \qquad \forall_x((x \in_\sigma y) \Leftrightarrow (x \in_\sigma y')) \rightharpoonup (y = y')\,,$$

and if y is not free in A

$$[\in \exists] \qquad \top \rightharpoonup \exists_y \forall_x((x \in_\sigma y) \Leftrightarrow A)\,.$$

We write $x \in_\sigma y$ for $\in_\sigma(x, y)$.

Empty sorted theories are often called *propositional theories.*

B.22. Derived rules

$$A \wedge B \rightharpoonup B \wedge A\,, \qquad A \wedge (B \wedge C) \rightharpoonup (A \wedge B) \wedge C\,,$$

$$A \vee B \rightharpoonup B \vee A, \quad \text{and} \quad A \vee (B \vee C) \rightharpoonup (A \vee B) \vee C$$

are easily obtained and will be used tacitly.

B.221. $\top \rightharpoonup |x|$.

BECAUSE: It may be inferred from $\exists_x \top \rightharpoonup \exists_x \top$ by $[\exists]$, $[\wedge\rightharpoonup]$, and $[\rightharpoonup\rightharpoonup]$.

B.222.

$$\frac{\dfrac{A \rightharpoonup |x| \wedge B}{A \wedge |x| \rightharpoonup |x| \wedge B}}{A \wedge |x| \rightharpoonup B}$$

BECAUSE: $[\wedge\rightharpoonup]$, $[\rightharpoonup\wedge]$, and $[\rightharpoonup\rightharpoonup]$ may be applied.

B.223. *If x is not free in B, then* $B \wedge \exists_x A \rightharpoonup \exists_x(B \wedge A)$.

BECAUSE: $\exists_x(B \wedge A) \rightharpoonup \exists_x(B \wedge A)$ yields $A \wedge |x| \rightharpoonup B \Rightarrow \exists_x(B \wedge A)$ by $[\exists]$, [B.222], and $[\Rightarrow]$. Then apply [B.222], $[\exists]$, and $[\Rightarrow]$.

B.224. $A \wedge (B \vee C) \rightharpoonup (A \wedge B) \vee (A \wedge C)$.

BECAUSE: From $A \wedge B \rightharpoonup (A \wedge B) \vee (A \wedge C)$ (an instance of $[\rightharpoonup\vee]$) obtain $B \rightharpoonup A \Rightarrow ((A \wedge B) \vee (A \wedge C))$ by $[\Rightarrow]$.

$C \rightharpoonup A \Rightarrow ((A \wedge B) \vee (A \wedge C))$ may be similarly obtained from $A \wedge C \rightharpoonup (A \wedge B) \vee (A \wedge C)$. Then apply $[\vee \rightharpoonup]$ and $[\Rightarrow]$.

B.225. *From* $A_1 \rightharpoonup A_2$ *infer*

$$B \wedge A_1 \rightharpoonup B \wedge A_2 , \qquad A_1 \wedge B \rightharpoonup A_2 \wedge B ,$$
$$A_2 \Rightarrow B \rightharpoonup A_1 \Rightarrow B , \qquad B \Rightarrow A_1 \rightharpoonup B \Rightarrow A_2 ,$$
$$\exists_x A_1 \rightharpoonup \exists_x A_2 , \quad \textit{and} \quad \forall_x A_1 \rightharpoonup \forall_x A_2 .$$

B.226. We write $A_1 \equiv A_2$ if A_1, A_2 have the same free variables, $A_1 \rightharpoonup A_2$ and $A_2 \rightharpoonup A_1$.

If x is not free in B, then $(B \wedge \exists_x A) \equiv \exists_x (B \wedge A)$.

BECAUSE: $\exists_x A \rightharpoonup \exists_x A$ yields $A \rightharpoonup |x| \wedge \exists_x A$ by $[\exists]$. Infer $B \wedge A \rightharpoonup |x| \wedge B \wedge \exists_x A$ by $[\rightharpoonup \wedge]$, $[\wedge \rightharpoonup]$, and $[\rightharpoonup \rightharpoonup]$. Now use $[\exists]$ to obtain $\exists_x (B \wedge A) \rightharpoonup (B \wedge \exists_x A)$. The other direction was shown in [B.223].

B.227. *If x is not free in B, then*

$$(B \wedge \forall_x A) \rightharpoonup \forall_x (B \wedge A) \quad \textit{and} \quad \exists_x (B \vee A) \rightharpoonup (B \vee \exists_x A) .$$

(Note that $(B \vee \exists_x A) \rightharpoonup \exists_x (B \vee A)$ fails in the primitive functional semantics in the empty model if $B = \top$. Similarly, $\forall_x (B \wedge A) \rightharpoonup (B \wedge \forall_x A)$ fails in the empty model if $B = \bot$.)

B.228. *If x is not free in B, then*

$$(\exists_x A \Rightarrow B) \equiv \forall_x (A \Rightarrow B) \quad \textit{and} \quad (B \Rightarrow \forall_x A) \equiv \forall_x (B \Rightarrow A) .$$

BECAUSE: $(\exists_x A \Rightarrow B) \rightharpoonup \forall_x (A \Rightarrow B)$ may be inferred from $\exists_x (A \wedge (\exists_x A \Rightarrow B)) \rightharpoonup B$ by $[\exists]$, [B.222], $[\Rightarrow]$, and $[\forall]$. This assertion follows from $(\exists_x A \Rightarrow B) \rightharpoonup (\exists_x A \Rightarrow B)$ by $[\Rightarrow]$, [B.226] and $[\rightharpoonup \rightharpoonup]$. $\forall_x (A \Rightarrow B) \rightharpoonup (\exists_x A \Rightarrow B)$ may be inferred from $\forall_x (A \Rightarrow B) \rightharpoonup \forall_x (A \Rightarrow B)$ by $[\forall]$, $[\Rightarrow]$, [B.222], $[\exists]$, [B.226], $[\rightharpoonup \rightharpoonup]$, and $[\Rightarrow]$.

$(B \Rightarrow \forall_x A) \rightharpoonup \forall_x (B \Rightarrow A)$ may be easily obtained from $(B \Rightarrow \forall_x A) \rightharpoonup (B \Rightarrow \forall_x A)$ by $[\Rightarrow]$, $[\forall]$, $[\Rightarrow]$, and $[\forall]$. $\forall_x (B \Rightarrow A) \rightharpoonup (B \Rightarrow \forall_x A)$ follows from $\forall_x (B \Rightarrow A) \rightharpoonup \forall_x (B \Rightarrow A)$ by $[\forall]$, $[\Rightarrow]$, $[\forall]$, and $[\Rightarrow]$.

B.229. $\exists_x \exists_y A \equiv \exists_y \exists_x A$, $\exists_x \exists_x A \equiv \exists_x A$.

If x is not free in A, then

$$\exists_x A \equiv ((\exists_x |x|) \wedge A) \quad \textit{and} \quad \forall_x A \equiv ((\exists_x |x|) \Rightarrow A) .$$

B.22(10). *If A' is the result of replacing some free occurrences of x in A with free occurrences of x' of the same sort as x, then*:

$$A \wedge (x = x') \rightharpoonup A' .$$

BECAUSE: $[\wedge =]$ may be used as a basis for induction on the complexity of A.

B.22(11). *If A' (resp. B') is the result of replacing all free occurrences of x in A (resp. B) with free occurrences of x', then from $A \rightharpoonup B$ one may infer $A' \rightharpoonup B'$.*

BECAUSE: $A' \rightharpoonup A' \wedge \exists_x(x' = x) \rightharpoonup \exists_x(A' \wedge (x' = x)) \rightharpoonup \exists_x(A \wedge (x' = x)) \rightharpoonup \exists_x(B \wedge (x' = x)) \rightharpoonup \exists_x(B' \wedge (x' = x)) \rightharpoonup B' \wedge \exists_x(x' = x) \rightharpoonup B'$.

B.22(12). $[\rightharpoonup\rightharpoonup]$ may now be improved to:

If every sort of a variable that occurs freely in B is a sort of a variable that occurs freely in A or in C, then:

$$\frac{A \rightharpoonup B \qquad B \rightharpoonup C}{A \rightharpoonup C} .$$

BECAUSE: If x is a variable that occurs freely in B but neither in A nor in C, A may be replaced by $A \wedge |x|$ in the assumptions. $A \wedge |x| \rightharpoonup C$ follows by $[\rightharpoonup\rightharpoonup]$. Now replace the free occurrence of x in this assertion by a free occurrence of a free variable y in A of the same sort as x and apply [B.22(11)]. $A \equiv (A \wedge |y|)$, so $A \rightharpoonup C$ follows by $[\rightharpoonup\rightharpoonup]$.

B.22(13). *If $\mathbf{T}$ contains $\top \rightharpoonup \exists_x|x|$ for each sort, then from $\mathbf{T} \vdash (A \rightharpoonup B)$ and $\mathbf{T} \vdash (B \rightharpoonup C)$ one may always infer $\mathbf{T} \vdash (A \rightharpoonup C)$.*

BECAUSE: If A and C have no free variables whose sort is the same as the sort of a free variable in B, and if x is a variable of this sort that occurs in an assertion $\top \rightharpoonup \exists_x|x|$ contained in $\mathbf{T}$, [B.22(12)] allows us to infer $A \wedge |x| \rightharpoonup C$. $\exists_x(A \wedge |x|) \rightharpoonup C$ follows by [B.222] and $[\exists]$. [B.226] yields $A \wedge \exists_x|x| \rightharpoonup C$. Then $A \rightharpoonup C$ follows by the hypothesis, $[\rightharpoonup\wedge]$, and $[\rightharpoonup\rightharpoonup]$.

B.3. Theories as Allegories

Let $\mathscr{W}$ be an alphabet of LETTERS disjoint from the given alphabet $\mathscr{V}$ of variables and let $\mathscr{W} \to \Sigma$ be a sort assignment for which there are countably infinitely many letters of each sort. Let $\mathbf{T}$ be a regular theory (that is, a theory in the syntax of regular logic).

A DERIVED PREDICATE TOKEN is obtained from a formula by, firstly, linearly ordering its free variables, and secondly, replacing all of its variables (free, bound and index) by letters from $\mathscr{W}$ with distinct letters replacing distinct variables, and preserving sort. Note that each derived predicate token has a linear ordering on the letters that have replaced the free variables, and thus that it has a sort type, whereas formulae do not. In referring to derived predicate tokens we will often picture the linear ordering in parentheses: $P(w_1, \ldots, w_n)$.

An INSTANTIATION of a derived predicate token is a formula obtained by replacing all of its letters by variables (distinct variables for distinct letters, and preserving sort). Two derived predicate tokens $R(\mathbf{w})$, $R'(\mathbf{w}')$ of equal sort type name the same DERIVED PREDICATE (of that sort type) iff $\mathbf{T} \vdash R(\boldsymbol{y}) \equiv R'(\boldsymbol{y})$ for an instantiation of R (resp. R') in which $\mathbf{w}$ (resp. $\mathbf{w}'$) is replaced by a word of variables $\boldsymbol{y}$.

B.31. The FREE ALLEGORY $\mathbf{A}_\mathbf{T}$ on a theory $\mathbf{T}$ may be described by [1.2–1.22]: the objects are sort words $\alpha, \beta, \gamma, \ldots$, the proto-morphisms are derived predicates, with source-target predicate given by $\alpha \xrightarrow{P} \beta$ iff the sort type of P is the catenation. Suppose u, v, and w are words of letters of sort type α, β, and γ. If $\alpha \xrightarrow{R(u,v)} \beta$ and $\beta \xrightarrow{S(v,w)} \gamma$, then the composition of proto-morphisms named by $R(u, v)$ and $S(v, w)$ is named by a derived predicate obtained by replacing the variables $x_1, \ldots, x_n$ in the expression $\exists_{x_1} \cdots \exists_{x_n} (R(u, x_1, \ldots, x_n) \wedge S(x_1, \ldots, x_n, w))$ with letters from $\mathscr{W}$. Here the sort type of the word of variables $x_1 \cdots x_n$ is β, the sort type of v. The composition of protomorphisms is not well-defined without the source-target information.

Regarding the allegory structure [2.11], the reciprocal of $\alpha \xrightarrow{R(u,\,v)} \beta$ is named by $\beta \xrightarrow{R(v,\,u)} \alpha$, and the intersection of morphisms from α to β named by $(R \wedge S)\,(u, v)$.

B.311. $\mathbf{A}_T$ *is an allegory.*

BECAUSE: The law of modularity [2.112] follows from $\exists_z(T(x, z) \wedge S(z, y)) \rightharpoonup \exists_z(T(x, z) \wedge S(z, y))$ by [$\exists$], [B.22(11)] (the latter for z, z'), [$\wedge\rightharpoonup$], [$\rightharpoonup\wedge$], and [B.225]. Other conditions are readily verified.

B.312. $\mathbf{A}_T$ *is a unitary pre-tabular allegory.*

BECAUSE: The empty sort word is a unit [2.15]. Regarding pretabularity [2.165, 2.14, 2.153], the maximal morphism from α to β is tabulated by the maps with source $\alpha\beta$ that are named by

$$(x_1 = x_1') \wedge \cdots \wedge (x_n = x_n') \wedge (y_1 = y_1) \wedge \cdots \wedge (y_m = y_m)$$

and by

$$(y_1 = y_1') \wedge \cdots \wedge (y_m = y_m') \wedge (x_1 = x_1) \cdot \cdots \wedge (x_n = x_n),$$

where the words $x_1 \cdots x_n$ and $x_1' \cdots x_n'$ are of sort type α, and the words $y_1 \cdots y_m$ and $y_1' \cdots y_m'$ are of sort type β.

B.313. *If* **T** *is coherent (respectively: first order, higher order) theory, then* $\mathbf{A}_T$ *is a distributive (respectively: division, power) allegory.*

BECAUSE: 0 and $\cup$ are given by $\perp$ and $\vee$ [B.211, 2.21]. Left division and the power allegory structure are given as in the allegory of sets [2.312, 2.411].

B.314. [2.152] allows us to consider any derived predicate token $P(w)$ as a name for a coreflexive morphism of $\mathbf{A}_\mathbf{T}$. Therefore, both objects and morphisms of the unitary tabular allegory $\mathscr{S}\mathit{plit}(\mathscr{C}\mathit{or}\ \mathbf{A}_\mathbf{T})$ [2.14, 2.165] may be taken to be derived predicates.

If **T** *is a regular (respectively: coherent, first order, higher order) theory, then the category* $\mathscr{M}\mathit{ap}(\mathscr{S}\mathit{plit}(\mathscr{C}\mathit{or}\ \mathbf{A}_\mathbf{T}))$ *is regular (respectively: a pre-logos, a logos, a topos).*

BECAUSE: [2.154, 2.212, 2.32, 2.414].

B.315. In the case of the empty theories, that is, for the regular (respectively: coherent, first order, higher order) logic on predicate symbols $\mathscr{P}$, the category $\mathscr{M}\mathit{ap}(\mathscr{S}\mathit{plit}(\mathscr{C}\mathit{or}\ \mathbf{A}))$ is the FREE REGULAR CATEGORY (respectively: FREE PRE-LOGOS, FREE LOGOS, FREE TOPOS) GENERATED BY $\mathscr{P}$. The natural assignment of objects of this category to the predicate symbols is universal:

Let **C** *be a regular category (respectively: a pre-logos, a logos, a topos). Let each predicate symbol of a regular (respectively: coherent, first order, higher order) logic of sort type* $\sigma_1 \cdots \sigma_n$ *be assigned a table in* **C** *whose feet are assigned to sorts* $\sigma_1, \ldots, \sigma_n$. *Then there exists*

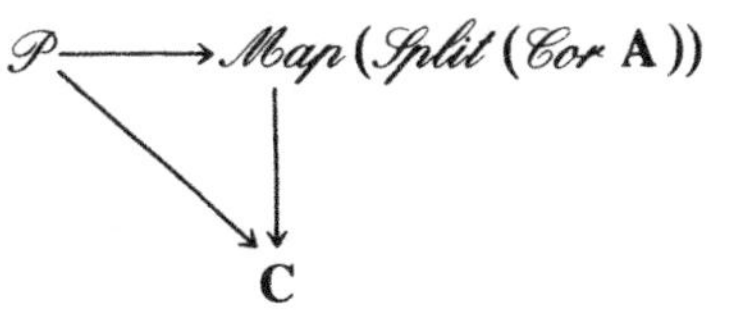

where the unitary representation of regular categories (respectively: pre-logoi, logoi, topoi) is unique up to natural equivalence.

BECAUSE: Such an assignment extends to a representation of allegories $\mathbf{A} \to \mathscr{R}el(\mathbf{C})$, unique up to natural equivalence. Now apply [2.154, 2.164].

B.316. We consider theories of ARITHMETIC, which have a distinguished sort called NUMERICAL SORT, on which there is one unary predicate symbol, **Zero**; one binary predicate symbol, **Succ**; and two ternary predicate symbols, **Add** and **Mult**. A theory of arithmetic must include the assertions

$$T \rightharpoonup \exists_x \mathbf{Zero}(x)\,,$$

$$\mathbf{Zero}(x) \wedge \mathbf{Zero}(y) \rightharpoonup (x = y)\,,$$

$$|x| \rightharpoonup \exists_y \mathbf{Succ}(x, y)\,,$$

$$\mathbf{Succ}(x, y) \wedge \mathbf{Succ}(x, z) \rightharpoonup (y = z)\,,$$

$$|x| \wedge |y| \rightharpoonup \exists_z \mathbf{Add}(x, y, z)\,,$$

$$\mathbf{Add}(x, y, z) \wedge \mathbf{Add}(x, y, u) \rightharpoonup (z = u)\,,$$

$$|x| \wedge |y| \rightharpoonup \exists_z \mathbf{Mult}(x, y, z)\,,$$

$$\mathbf{Mult}(x, y, z) \wedge \mathbf{Mult}(x, y, u) \rightharpoonup (z = u)\,.$$

These assertions allow us to use NUMERICAL CONSTANTS and FUNCTION SYMBOLS in stating the other assertions included in the theory of arithmetic. This is facilitated by referring to terms, that is, those finite words of variables, parentheses, and symbols **0**, **s**, +, and · that may be obtained from **0** and the variables by finitely many applications of the formation rules $\mathbf{s}(p)$, $+(p, q)$, and $\cdot(p, q)$, where p and q are terms. We write $p + q$ for $+(p, q)$ and $p \cdot q$ for $\cdot(p, q)$. If p, q, and r are terms, we write

$$p = \mathbf{0} \quad \text{for } \exists_u((p = u) \wedge \mathbf{Zero}(u))\,,$$

$$q = \mathbf{s}(p) \quad \text{for } \exists_u \exists_v((p = u) \wedge (q = v) \wedge \mathbf{Succ}(u, v))\,,$$

$$r = p + q \quad \text{for } \exists_u \exists_v \exists_{v'}((r = u) \wedge (p = v) \wedge (q = v') \wedge (q = v') \wedge \mathbf{Add}(v, v', u))\,,$$

and

$$r = p \cdot q$$

$$\text{for } \exists_u \exists_v \exists_{v'}((r = u) \wedge (p = v) \wedge (q = v') \wedge \mathbf{Mult}(v, v', u)),$$

where u, v, and v' are mutually distinct variables distinct from any variables in p, q, and r. We also write 1 for $\mathbf{s}(\mathbf{0})$, 2 for $\mathbf{s}(1)$, 3 for $\mathbf{s}(2), \ldots$, and call them numerals.

A theory of arithmetic must further contain assertions abbreviated by:

$$\mathbf{s}(x) = \mathbf{0} \rightharpoonup \bot,$$

$$\mathbf{s}(x) = \mathbf{s}(y) \rightharpoonup x = y,$$

$$\top \rightharpoonup x + \mathbf{0} = x,$$

$$\top \rightharpoonup x + \mathbf{s}(y) = \mathbf{s}(x + y),$$

$$\top \rightharpoonup x \cdot \mathbf{0} = \mathbf{0},$$

$$\top \rightharpoonup x \cdot \mathbf{s}(y) = (x \cdot y) + x,$$

and INDUCTION:

$$A^0 \wedge \forall_x (A \Rightarrow A^+) \rightharpoonup \forall_x A, \quad \text{for each formula } A,$$

where A^0, A^+ are obtained by replacing all free occurrences of x in A by $\mathbf{0}$, $\mathbf{s}(x)$ respectively. The assertions listed in this section are called PEANO AXIOMS.

HIGHER ORDER ARITHMETIC is obtained by letting $\Sigma = \{N, [N], [[N]], \ldots\}$ and requiring (in addition to the assertions of higher-order logic [B.211]) the Peano axioms for sort N, where formulae A in the induction assertions may contain other variables of any sort.

B.317. The reader will recall the notion of a natural numbers object (NNO) in a topos [1.98].

Let $\mathbf{T}$ *be higher order arithmetic. Then* $\mathscr{Map}(\mathscr{Split}(\mathscr{Cor}\ \mathbf{A}_\mathbf{T}))$ *is the free topos with a natural numbers object.*

BECAUSE: $\mathscr{Map}(\mathscr{Split}(\mathscr{Cor}\ \mathbf{A}_\mathbf{T}))$ is a topos [B.314]. We first outline the argument that the sort N with equality on N, $\mathbf{0}: 1 \to N$ and $\mathbf{s}: N \to N$ given in [B.317] constitutes a NNO in $\mathscr{Map}(\mathscr{Split}(\mathscr{Cor}\ \mathbf{A}_\mathbf{T}))$. Use induction in higher order arithmetic to show that N is a disjoint union of the images of **0** and **s**, that the equivalence closure of **s** is the entire $N \times N$, and that N has the Peano property [1.987]. Now apply [1.98(10)] and [1.954].

For the universality, suppose **B** is a topos with a NNO $1 \xrightarrow{e} M \xrightarrow{t} M$. By [B.314], it suffices to find a representation of power allegories $\mathbf{A}_{\mathrm{T}} \to \mathscr{R}el(\mathbf{B})$ that sends N to M. Let **L** be the empty higher order theory on the predicate symbols **Zero, Succ, Add,** and **Mult**. Assign e and t to **Zero** and **Succ**, respectively. Use [1.983] to define addition and multiplication on M; assign them to **Add** and **Mult**, respectively. Let F be the induced representation of power allegories $\mathbf{A}_{\mathrm{L}} \to \mathscr{R}el(\mathbf{B})$ [B.315]. Clearly, for each assertion $B \rightharpoonup C$ listed in [B.317] except Induction, we have $F(B) \subset F(C)$. We have to establish this fact in regard to Induction as well.

Because we are considering higher order logic [B.211], we may assume that in Induction, formula A is of the form $x \in_N u$. Let B be the formula $0 \in u \wedge \forall_y (y \in u \Rightarrow s(y) \in u) \Rightarrow x \in u$. We have to show that $F(B)$ is the entire $M \times [M]$ in **B**. Clearly $F(B^0) = F(\forall_x(B \Rightarrow B^+)) = [M]$. Now consider the topos **B**. Writing e for $[M] \to 1 \xrightarrow{e} M$, we have

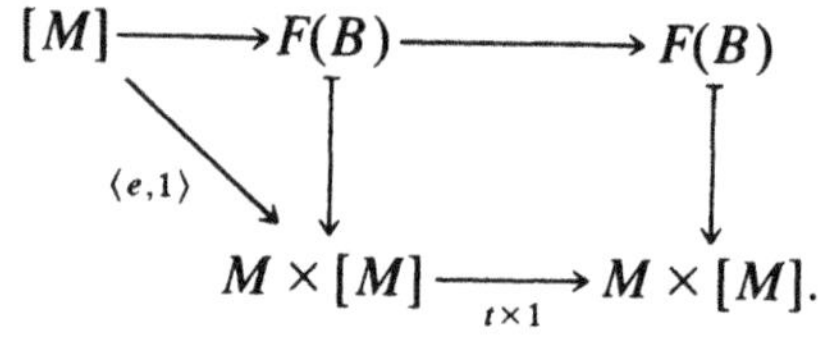

Using [1.981] we obtain

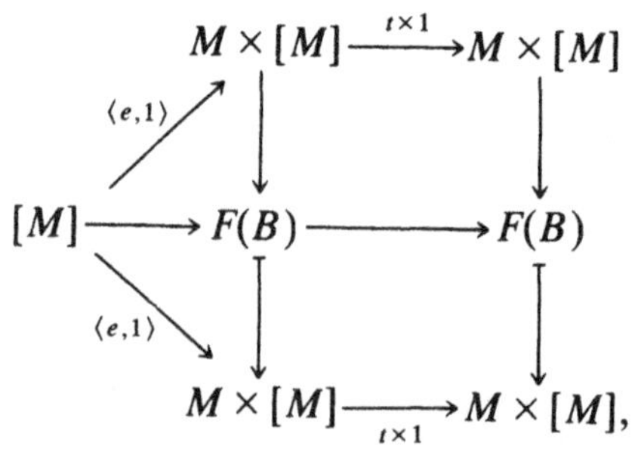

and the composition of vertical maps must be the identity by the uniqueness condition on a NNO, so $F(B)$ is entire.

Therefore F induces a representation of power allegories $\mathbf{A}_{\mathrm{T}} \to \mathscr{R}el(\mathbf{B})$ and we may apply [2.154] and [2.164] as in [B.315].

B.318. The symbols of the syntax of arithmetic may be replaced in a one-to-one fashion by natural numbers. The prime factorization theorem may be used to code nonempty finite sequences of natural numbers by natural numbers. Thus formulae, assertions, and, indeed, inferences may be uniquely represented by natural numbers, and the properties of syntax such as inference and inconsistency (that is, the condition that $\top \rightharpoonup \bot$ may be inferred) become number-theoretic properties. In fact, as Gödel showed in 1931, inference and inconsistency are thus coded as recursively enumerable number-theoretic properties. See [2.438].

B.32. The sconing argument [1.(10)] shows that in various free categories, 1 is a coprime projective. By [B.315, B.316–7] this may be restated in terms of syntax. If A' is the result of replacing all free occurrences of x in a formula A with free occurrences of x', we write $\top \rightharpoonup \exists!_x A$ for $\top \rightharpoonup \exists_x A$ and $A \wedge A' \rightharpoonup (x = x')$. We also abbreviate an assertion $\top \rightharpoonup B$ by B.

Let **T** *be regular (respectively: coherent, first order, higher order) logic, first order arithmetic or higher order arithmetic.*

Suppose there are no free variables in formulae A, B. If $\mathbf{T} \vdash (A \vee B)$, *then* $\mathbf{T} \vdash A$ *or* $\mathbf{T} \vdash B$; *the* DISJUNCTION PROPERTY.

Suppose A has exactly x free. If $\mathbf{T} \vdash \exists_x A$, *then there exists a formula C with exactly x free such that* $\mathbf{T} \vdash \exists!_x(C \wedge A)$; *the* EXISTENCE PROPERTY. *If, in addition, x is of the numerical sort, then C may be chosen as $x = n$ for some numeral n; the* NUMERICAL EXISTENCE PROPERTY.

In the case of first order arithmetic, a complete argument would require a precise discussion of arithmetic in logoi.

The representations of the free allegories in their scones correspond to various proof-theoretic slash interpretations introduced by Kleene, Friedman, and Aczel.

B.4. The Completeness Theorems

The representation theorems in chapters 1 and 2 imply the completeness theorems for regular, coherent and first-order logic.

B.41. A THEORY **T** SEMANTICALLY ENTAILS AN ASSERTION $A \rightharpoonup B$ IN A UNITARY ALLEGORY **E**, $\mathbf{T} \models_{\mathbf{E}} (A \rightharpoonup B)$, iff $M(A) \subset M(B)$ for each unitary representation $M\colon \mathbf{A}_{\mathbf{T}} \to \mathbf{E}$. (Because of [B.222] we may assume that A and B have the same free variables.)

For an assertion P, we write $\mathbf{T} \models_C P$ instead of $\mathbf{T} \models_{\mathcal{R}el(C)} P$.

B.411. Specifying $\mathbf{E} = \mathscr{S}$ yields primitive functional semantics; it is readily verified that $\mathbf{T} \models_{\mathscr{S}} P$ iff $\mathbf{T}$ entails P in primitive functional semantics [B.12]. The traditional *tarskian semantics* is a further specialization that deals only with those coherent theories that, firstly, contain assertions $T \rightharpoonup \exists_x |x|$ for each sort, and secondly, have the property that for each formula A there exists a unique formula A^- with the same free variables such that $\mathbf{T}$ contains assertions $T \rightharpoonup (A \vee A^-)$ and $(A \wedge A^-) \rightharpoonup \bot$. In any such BOOLEAN THEORY $A \Rightarrow B$ may be defined as $A^- \vee B$, and $\forall_x A$ as $(\exists_x A^-)^-$. Then A^- may in turn be defined as $A \Rightarrow \bot$, that is, $\neg A$.

B.42. *Let* $\mathbf{T}$ *be a regular (respectively, coherent) theory. Then* $\mathbf{T} \models_{\mathscr{S}} P$ *iff* $\mathbf{T} \vdash P$.

BECAUSE: Unitary (respectively, distributive unitary) representations of the allegory $\mathbf{A_T}$ in the allegory of sets are collectively faithful [2.167, 2.154, 1.55, 2.218, 1.635].

B.421. GÖDEL'S COMPLETENESS THEOREM for boolean theories is a special case of [B.42] because a representation of pre-logoi automatically preserves the boolean structure [1.641]:

Let $\mathbf{T}$ *be a boolean first order theory. Then* $\mathbf{T} \models_{\mathscr{S}} P$ *iff* $\mathbf{T} \vdash P$.

B.43. *Let* $\mathbf{T}$ *be a first order theory, boolean or not, on countably many predicate symbols. Then* $\mathbf{T} \models_{\mathscr{Sh}(\mathbb{R})} P$ *iff* $\mathbf{T} \vdash P$.

Because: [1.74, 2.33].

More generally, [2.331] yields:

Let X be a metrizable space without isolated points, and let $\mathbf{T}$ *be a first order theory on countably many predicate symbols. Then* $\mathbf{T} \models_{\mathscr{Sh}(X)} P$ *iff* $\mathbf{T} \vdash P$.

No space suffices for all first order theories. Given a space X, let α be an ordinal larger than the cardinality of the set of all open subsets of X. Let $\mathbf{T}$ be a first order empty-sorted theory whose predicate symbols are the ordinals $< \alpha$, and which consists of the assertions $\gamma \rightharpoonup \delta$, all $\gamma \leqslant \delta < \alpha$; $(\gamma \Rightarrow \delta) \rightharpoonup \delta$, all $\delta < \gamma < \alpha$; and $(\gamma \Rightarrow \bot) \rightharpoonup \bot$, all $\gamma < \alpha$. Then there are ordinals $\delta < \alpha$ such that $\mathbf{T} \models_{\mathscr{H}(X)} (\top \rightharpoonup \delta)$ although it is not the case that $\mathbf{T} \vdash (\top \rightharpoonup \delta)$.

On the other hand, [1.75] implies:

For every first order theory **T** *there is a compact, totally disconnected space X without isolated points such that* $\mathbf{T} \models_{\mathscr{H}(X)} P$ *iff* $\mathbf{T} \vdash P$.

B.5. Zermelo–Fraenkel Set Theory

ZERMELO–FRAENKEL SET THEORY (ZF) is a boolean, one-sorted, first order theory on binary predicate symbols $\in$ and $=$, and nullary predicate symbols $\top$ and $\bot$. We write $x \in y$ for $\in(x, y)$ (pronounced 'x is in y'). The theory consists of the assertions $\top \rightharpoonup B$, where B is one of the formulae:

Extent $\forall_x(x \in u \Leftrightarrow x \in v) \Rightarrow (u = v)$,

Pairing $\exists_x(u \in x \wedge v \in x)$,

Union $\exists_x\forall_y\forall_z(z \in u \wedge y \in z \Rightarrow y \in x)$,

Comprehension $\exists_x \forall_y(y \in x \Leftrightarrow y \in u \wedge A)$, where A is a formula, and x is not free in A,

Infinity $\exists_x(\exists_u u \in x \wedge \forall_y(y \in x \Rightarrow \exists_z(z \in x \wedge y \in z)$,

Power Set $\exists_x\forall_y(\forall_z(z \in y \Rightarrow z \in u) \Rightarrow y \in x)$,

Foundation $\forall_x(\forall_y(y \in x \Rightarrow A') \Rightarrow A) \Rightarrow \forall_x A$, where A is a formula with no free occurrences of y, and A' is obtained from A by replacing all free occurences of x with free occurrences of y,

Collection $\forall_x(x \in u \Rightarrow \exists_y A) \Rightarrow \exists_v\forall_x(x \in u \Rightarrow \exists_y(y \in v \wedge A)$, where A is a formula, and v is not free in A.

B.51. In the next section we shall describe an interpretation of ZF in the boolean quotient [2.521] of $\mathscr{Rel}(\mathscr{S}^{\mathbf{C}^{\circ}})$, for any small **C**, thus in any boolean Grothendieck topos [1.84]. This interpretation is closely related to a special case of the FOURMAN–HAYASHI INTERPRETATION in a Grothendieck topos. When **C** is a poset, our interpretation is reduced to a SCOTT–SOLOVAY BOOLEAN-VALUED MODEL.

The Fourman-Hayashi interpretation assigns values of a Grothendieck topos to formulae in the syntax of ZF with no free variables, such that a terminator is assigned to all formulae A with no free variables for which the assertion $\top \rightharpoonup A$ may be inferred in ZF without referring to booleanness. This is accomplished by considering the well-founded part of the Grothendieck topos, that is, the minimal full subcategory closed under all the operations of a complete topos. Its objects are the subobjects of V_α, α an ordinal, where V_α is the colimit of $[V_\lambda]$, $\lambda < \alpha$, along the standard embeddings. The connectives and the restricted quantifiers

$\exists_x(x \in V_\alpha \wedge A)$ and $\forall_x(x \in V_\alpha \Rightarrow A)$ are interpreted as in any power allegory, by considering V_α as a sort. Arbitrary quantification may be reduced to these by taking suprema and, respectively, infima over the ordinals.

B.511. We shall define a cumulative hierarchy of left **C**-sets [1.273]. (We shall neglect the distinction between the objects of **C** and the left **C**-sets corresponding to the contravariant representable functors.) For each ordinal α let

$$V_\alpha^{\mathbf{C}} = \bigcup_{\lambda<\alpha} \{\langle c, y\rangle : c \in |\mathbf{C}|,\ y \textit{ a subobject of } c \times V_\lambda\}\,,$$

with the obvious **C**-action. Let $V^{\mathbf{C}} = \bigcup_\alpha V_\alpha^{\mathbf{C}}$. In the next definition, u, v, w are elements of V; c, c', c'' are objects of **C**, and f, g are morphisms of **C**. Furthermore, it is understood that the **C**-action is defined whenever referred to. Let

$c \Vdash T$, never $c \Vdash \perp$, for all c,

$c \Vdash (v \in u)$ iff for each $f: c' \to c$ there exist $g: c'' \to c'$ and w such that $\langle c'', w\rangle \in gfu$ and $c'' \Vdash (gfv = w)$,

$c \models (u = v)$ iff for each $f: c' \to c$ and each w, $\langle c', fw\rangle \in fu$ implies $c' \Vdash (fw \in fv)$, and $\langle c', fw\rangle \in fv$ implies $c' \Vdash (fw \in fu)$,

$c \Vdash (A \wedge B)$ iff $c \Vdash A$ and $c \Vdash B$,

$c \Vdash (A(u) \vee B(u))$ iff for each $f: c' \to c$ there exists $g: c'' \to c'$ such that $c'' \Vdash A(gfu)$ or $c'' \Vdash B(gfu)$,

$c \Vdash (A(u) \Rightarrow B(u))$ iff for each $f: c' \to c$, $c' \Vdash A(fu)$ implies $c' \Vdash B(fu)$,

$c \Vdash \exists_x A(x, u)$ iff for each $f: c' \to c$ there exist $g: c'' \to c'$ and w such that $c'' \Vdash A(w, gfu)$,

$c \Vdash \forall_x A(x, u)$ iff for each $f: c' \to c$ and each w, $c' \Vdash A(w, fu)$,

$c \Vdash (A \rightharpoonup B)$ iff $c \Vdash (A \Rightarrow B)$.

We say that $A \rightharpoonup B$ *holds in the Fourman-Hayashi interpretation if* $c \Vdash (A \Rightarrow B)$, for all c.

Every assertion that may be inferred from ZF *holds in the Fourman-Hayashi interpretation.*

BECAUSE: It is helpful to consider two stages in interpreting an assertion. Consider an auxiliary non-boolean theory **T** on the predicate symbols of ZF consisting of Pairing, Union, Comprehension, Infinity, Foundation, and the following substitutes for Power Set and Collection:

$$\forall_s \exists_p \forall_x \exists_y (y \in p \wedge \forall_z (z \in y \Leftrightarrow z \in x \wedge z \in s))\,,$$

$$\forall_s \exists_t \forall_x (x \in s \wedge \exists_y A \Rightarrow \exists_y (y \in t \wedge A))\,.$$

Let $\neg A$ stand for $A \Rightarrow \bot$. ZF may be translated into **T**:

$$(x \in y)^{\sim} \quad \text{as}$$

$$\neg\neg\exists_z(z \in y \wedge \forall_t(t \in x \Rightarrow (t \in z)^{\sim}) \wedge \forall_t(t \in z \Rightarrow (t \in x)^{\sim}))\,,$$

$$\top^{\sim} \quad \text{as}\ \top\,, \qquad \bot^{\sim} \quad \text{as}\ \bot\,,$$

$$(A \wedge B)^{\sim} \quad \text{as}\ A^{\sim} \wedge B^{\sim}\,,$$

$$(A \vee B)^{\sim} \quad \text{as}\ \neg\neg(A^{\sim} \vee B^{\sim})\,,$$

$$(A \Rightarrow B)^{\sim} \quad \text{as}\ A^{\sim} \Rightarrow B^{\sim}\,,$$

$$(\exists_x A)^{\sim} \quad \text{as}\ \neg\neg\exists_x(A^{\sim})\,,$$

$$(\forall_x A)^{\sim} \quad \text{as}\ \forall_x(A^{\sim}), \quad \text{and}$$

$$(A \rightharpoonup B)^{\sim} \quad \text{as}\ A^{\sim} \rightharpoonup B^{\sim}\,.$$

The reader may first verify that $\mathbf{T} \vdash P^{\sim}$ whenever $\mathrm{ZF} \vdash P$. Next, **T** may be interpreted in $\mathcal{R}el(\mathcal{S}^{\mathbf{C}^{\circ}})$ by a simplification of the Fourman-Hayashi interpretation:

$$c \Vdash' (v \in u) \quad \text{iff} \quad \langle c, v\rangle \in u\,,$$

$$c \Vdash' (u = v) \quad \text{iff} \quad u = v\,,$$

$$c \Vdash' (A(u) \vee B(u)) \quad \text{iff} \quad c \Vdash A(u) \text{ or } c \Vdash' B(u)\,,$$

$$c \Vdash' \exists_x A(x, u) \quad \text{iff} \quad \text{there exists } v \text{ such that } c \Vdash' A(v, u)\,,$$

and the other cases as before. The reader may verify that for any assertion P, if $\mathbf{T} \Vdash P$, then $c \Vdash' P$, for all c. Observe, however, that P holds in the Fourman interpretation iff $c \Vdash' P^{\sim}$, for all c.

B.52. *The Continuum Hypothesis may not be inferred in* ZF.

BECAUSE: If **C** is the poset obtained by deleting the bottom from the free boolean algebra given in [2.453], then the Continuum Hypothesis does not hold in the interpretation [B.511].

B.53. *The Axiom of Choice may not be inferred in* ZF.

BECAUSE: If **C** is the category described in [2.46], then IAC fails in the well-founded part [B.5] of the boolean topos discussed in [2.56], and therefore the Axiom of Choice does not hold in the interpretation [B.511].

(The interpretation given in [B.511] may be formulated and inferred in ZF for any finite subtheory of ZF. Therefore, we have used exactly the consistency of ZF as an assumption in [B.52–3].)

SUBJECT INDEX

Alternate words are marked by an asterisk.

Printed and bound by CPI Group (UK) Ltd, Croydon, CR0 4YY
03/10/2024
01040430-0014